Practical Bayesia

T0258241

A Primer for Physica. ....

Science is fundamentally about learning from data, and doing so in the presence of uncertainty. This volume is an introduction to the major concepts of probability and statistics, and the computational tools for analysing and interpreting data. It describes the Bayesian approach, and explains how this can be used to fit and compare models in a range of problems. Topics covered include regression, parameter estimation, model assessment, and Monte Carlo methods, as well as widely used classical methods such as regularization and hypothesis testing. The emphasis throughout is on the principles, the unifying probabilistic approach, and showing how the methods can be implemented in practice. R code (with explanations) is included and is available online, so readers can reproduce the plots and results for themselves. Aimed primarily at undergraduate and graduate students, these techniques can be applied to a wide range of data analysis problems beyond the scope of this work.

Coryn A.L. Bailer-Jones was educated at Oxford and Cambridge universities. He has worked on modelling the processing of metals and has done research into the properties of low mass stars and brown dwarfs. He is a senior staff member at the Max Planck Institute for Astronomy in Heidelberg, where he leads a group working on the analysis of data from the Gaia survey mission. He also teaches at Heidelberg University in statistics and physics. His main scientific interests are statistical inference, stars and our Galaxy, and the impact of astronomical phenomena on the Earth.

Practical Bayesian Inference

A Primer for Physical Scientists

CORYN A.L. BAILER-JONES

Max Planck Institute for Astronomy, Heidelberg

CAMBRIDGE
UNIVERSITY PRESS

CAMBRIDGE
UNIVERSITY PRESS

University Printing House, Cambridge CB2 8BS, United Kingdom

One Liberty Plaza, 20th Floor, New York, NY 10006, USA

477 Williamstown Road, Port Melbourne, VIC 3207, Australia

4843/24, 2nd Floor, Ansari Road, Daryaganj, Delhi – 110002, India

79 Anson Road, #06–04/06, Singapore 079906

Cambridge University Press is part of the University of Cambridge.

It furthers the University's mission by disseminating knowledge in the pursuit of
education, learning, and research at the highest international levels of excellence.

www.cambridge.org
Information on this title: www.cambridge.org/9781107192119

First published 2017

Printed in the United Kingdom by TJ International Ltd. Padstow Cornwall

A catalogue record for this publication is available from the British Library.

Library of Congress Cataloguing in Publication Data
Names: Bailer-Jones, Coryn A. L., author.
Title: Practical Bayesian inference : a primer for physical scientists /
Coryn A.L. Bailer-Jones, Max-Planck-Institut für Astronomie, Heidelberg.
Description: Cambridge, United Kingdom ; New York, NY : Cambridge University
Press, 2017. | Includes bibliographical references and index.
Identifiers: LCCN 2016059505 | ISBN 9781107192119 (hbk. ; alk. paper) | ISBN 1107192110
(hbk. ; alk. paper) | ISBN 9781316642214 (pbk.) | ISBN 1316642216 (pbk.)
Subjects: LCSH: Bayesian statistical decision theory. | Mathematical physics.
Classification: LCC QC20.7.B38 B35 2017 | DDC 519.5/42–dc23
LC record available at https://lccn.loc.gov/2016059505

ISBN 978-1-107-19211-9 Hardback
ISBN 978-1-316-64221-4 Paperback

Additional resources for this publication at www.cambridge.org/9781107192119

Contents

Preface

Science is fundamentally about learning from data, and doing so in the presence of uncertainty. Uncertainty arises inevitably and avoidably in many guises. It comes from noise in our measurements: we cannot measure exactly. It comes from sampling effects: we cannot measure everything. It comes from complexity: data may be numerous, high dimensional, and correlated, making it difficult to see structures.

This book is an introduction to statistical methods for analysing data. It presents the major concepts of probability and statistics as well as the computational tools we need to extract meaning from data in the presence of uncertainty.

Just as science is about learning from data, so learning from data is nearly synonymous with data modelling. This is because once we have a set of data, we normally want to identify its underlying structure, something we invariably represent with a model. Fitting and comparing models is therefore one of the cornerstones of statistical data analysis. This process of obtaining meaning from data and reasoning is what is meant by *inference*.

Alas, statistics is all too often taught as a set of seemingly unconnected, ad hoc recipes. Having identified what appears to be a relevant statistical test from a menu – according to the properties of your data and your assumptions – you then apply a procedure that delivers some numerical measure of significance. This kind of approach does little to promote your confidence in the result, and it will leave you lost when your assumptions aren't on the menu. My goal in this book is to show that the process of analysing and interpreting data can be done within a simple probabilistic framework. Probability is central to interpreting data because it quantifies both the uncertainty in the data and our confidence in models derived from them. I will show that there are basic principles for tackling problems that are built on a solid probabilistic foundation. Armed with this know-how you will be able to apply these principles and methods to a wide range of data analysis problems beyond the scope of this book.

This book is aimed primarily at undergraduate and graduate science students. Knowledge of calculus is assumed, but no specific experience with probability or statistics is required. My emphasis is on the concepts and on illustrating them with examples, using both analytical and numerical methods. This is not a maths book, and while I hope I am reasonably rigorous, I omit some more complex derivations and formal proofs. I also hope this book will be useful for more experienced practitioners – in particular those with limited or no exposure to Bayesian methods – by providing an overview of the main concepts and techniques.

Many real-world problems in data analysis must be solved with a computer. One of my goals is to show how theoretical ideas can be converted into practical solutions. To support this I include code in the R language (https://www.r-project.org). R is easy to learn and use, has a convenient environment for plotting, and is widely used within and outside academia. I encourage you to use and modify the code provided, for which some basic knowledge of R will be necessary. This is not a guide to doing statistics in R, however, and you can follow everything in the book without actually using the programs.

Before delving in, let me give you a roadmap. The first chapter covers the basic ideas of probability and distributions. Chapter 2 looks at the concepts of estimation, measurement, and uncertainty. Together, these two chapters provide essential background material for what follows. Chapter 3 is an introduction to probabilistic inference, for which I use simple problems to show how the measured data and prior information can be, and must be, combined. This chapter introduces many of the concepts in inference that are expanded upon later in the book. Chapter 4 looks at least squares and maximum likelihood in the context of curve fitting. This is a straightfoward problem, but illustrates many of the concepts needed to tackle more complex problems. Chapters 5 and 6 look at parameter estimation for one and two parameter problems respectively, by calculating the posterior probability density over the parameters. Some problems have analytic solutions based on so-called conjugate priors or the use of standard integrals. In these chapters I will also discuss the issue of assigning priors and how to summarize distributions. In chapter 7 I describe a method for approximating the posterior, and I will introduce the method of density estimation, which we use to describe distributions we have sampled from. Whereas low dimensional posteriors can often be found by brute force evaluation on a grid, higher dimensional problems are better solved by sampling from the distribution. Chapter 8 introduces Monte Carlo methods for doing sampling; we look in particular at the Metropolis method. This can also be used for integration, so here I also summarize the uses of integration in inference. Chapter 9 applies the Metropolis method to find the posterior probability density function for higher dimensional problems, again taking curve fitting problems as the examples. Here I will further investigate what to do (remaining in a probabilistic framework) when we have outliers or errors on both axes. In chapter 10 I look at frequentist hypothesis testing and discuss its uses and weaknesses. This will be contrasted in chapter 11 with the Bayesian approach to model comparison using the marginal likelihood, where I will also mention some alternative metrics of model suitability. In the final chapter I look at some of the techniques available for tackling more complicated problems. These include cross-validation, regularization, basis functions, and bootstrapping.

Readers who are familiar with basic statistical concepts and who only want an introduction to probabilistic (Bayesian) methods can read chapters 3, 5, 6, 9, and 11, as well as chapter 8 if you are not familiar with Monte Carlo methods.

Most of the notation I use should be clear from the context, but here is a quick summary of my main conventions. Where relevant to define a base I use log for base 10 and ln for the natural logarithm. (Annoyingly in R, these are `log10` and `log` respectively.) Lower case letters can indicate scalars or vectors. Where necessary to make a distinction, as in matrix algebra, I use bold face for vectors. Matrices are written with upper case letters. An $N \times J$ matrix has N rows and J columns. A vector with J elements is by default a column vector, i.e. $J \times 1$. The transpose is indicated with ^{T} as in X^{T}. Curly brackets are used to indicate a sample, so $\{x\}$ is a set of values of x. $P(x)$ usually indicates a probability density function (PDF) over x. In the few places where we deal with discrete distributions, $P(x)$ will refer to an actual probability; a probability mass function (PMF). It will be clear from the context which is meant. $P(x < a)$ is the probability that x is less than a. An asterisk highlights that a PDF or PMF in unnormalized: P^*. Integrals and sums are implicitly over the full range of the variable unless indicated otherwise, although to avoid ambiguity I indicate the range on sums in some cases (such as when forming numerical averages). Where necessary to distinguish between a quantity x and an estimate thereof, I use a hat – $\hat{x}$ – for the latter.

R code and its screen output are included using the `monospace font`. Shorter pieces of code are provided in the text as part of the discussion. Longer scripts are separated out and are also available online from `http://www.mpia.de/~calj/pbi.html`. A few longer scripts are only available online. You are encouraged to run the examples and to explore by modifying the model parameters and data. The code is provided freely (you can use it in your own work) but without warranty. Most of the plots in this book were produced using the code provided.

No public work can be – or should be – written in isolation. For their useful comments and constructive suggestions I would like to thank Rene Andrae, Brendon Brewer, Morgan Fouesneau, Iskren Georgiev, Nikos Gianniotis, David Hogg, Mike Irwin, Željko Ivezić, Ronald Läsker, Klaus Meisenheimer, Sara Rezaei, Hans-Walter Rix, Jan Rybizki, Luis Sarro, Björn Malte Schäfer, Gregor Seidel, and Ted von Hippel. My special thanks go to my partner Sabina Pauen for supporting (and enduring) me during the writing of this book, and to my parents, Anne Jones and Barrie Jones, from whom I learned so much, and to whom I dedicate this book.

Error and uncertainty are as unavoidable in life as in science. My best efforts notwithstanding, this book inevitably contains errors. Notification of these is appreciated, and a list of corrections will be maintained at `http://www.mpia.de/~calj/pbi.html`.

Probability basics

This chapter reviews the basic ideas of probability and statistics that we will use in the rest of this book. I will outline some fundamental concepts, introduce some of the most common discrete and continuous probability distributions, and work through a few examples. But we'll start with an example that illustrates the role of information in solving problems.

1.1 The three doors problem

You are a contestant in a game show, presented with three closed doors. Behind one of the doors – chosen at random without your knowledge – is a prize car. Behind each of the other doors is a goat. Your objective is to reveal and thus win the car (the assumption is you're less keen on goats). You first select a door at random, which you do not open. The game show host, who knows where the car is, then opens one of the *other* doors to reveal a goat (she would never show you the car). She then gives you the opportunity to change your choice of door to the other closed one. Do you change?

This is a classic inference problem, also known as the Monty Hall problem. If you've not encountered it before – and even if you have – do think about it before reading on.

Your initial probability of winning (before any door is opened) is $1/3$. The question is whether this probability is changed by the actions of the game show host. It appears that her actions don't change anything. After all, she will always show you a goat. In that case you may think that your probability of winning remains at $1/3$ whether or not you change doors. Or you may think that once the host has revealed a goat, then because there are only two doors left, your chance of winning has changed to $1/2$ whether you change or not. In either of these cases changing doors does not improve your chances of winning.

In fact, if you change to the other closed door your chance of winning increases to $2/3$. This may seem counter-intuitive, but it is explained by how the host behaves. If you initially selected the door with the car behind it, then her choice of which door to open is random. But if you initially selected a door with a goat behind it, then her choice is not random. She is in fact forced to open the door with the other goat behind it. Let's call the door you initially select "door 1". The three possible arrangements are as follows.

door 1	door 2	door 3	door opened by host	result if staying	result if changing
car	goat	goat	2 or 3	car	goat
goat	car	goat	3	goat	car
goat	goat	car	2	goat	car

In two of the three initial possible choices you would select a goat, so if you change you will win the car in both cases. In the third case you select the car, so will lose if you change doors. Thus your chance of success is $2/3$. If you don't change then you will only win the car if your initial choice was indeed the car, a probability of $1/3$.

The outcome of this inference depends on how the game show host behaves: she always opens one of the other doors to reveal a goat, and does so at random if she can.

1.1.1 Earthquake variation

The fundamental point of the three doors problem is that your understanding of the problem, and therefore your decision of what to do, depends crucially on your model for the door opening (i.e. the game show host's behaviour). A variant on the problem should make this clear.

Just after making your initial choice, and before the game show host has had a chance to respond, there is an earthquake that causes one of the doors you did not select to open, revealing a goat. After recovering from the shock and seeing the goat behind the door, the game show host decides to continue, and offers you the opportunity to change. Do you?

If we assume that the earthquake opened a door at random, then it was just chance that it opened a door with a goat behind it. (Had it revealed the car, presumably the host would have declared the game invalid.) The difference now is that the choice of door to open could never have been forced by your initial selection, as it was in the original game. So the probability that the car lies behind one of the two remaining doors is $1/2$, and your chance of winning is $1/2$ whether you change or not.

If you are not convinced that you should change doors in the original problem, consider the case of one hundred doors, again with only one prize. After making your initial choice (let's define that as door 1 again), the game show host will now open 98 other doors (to reveal a lot of goats), leaving closed doors 1 and, let's say, 76. Would you stay with door 1 or switch to door 76? I shall present a formal analysis at the end of the chapter (section 1.10).

The take-home message of this problem is that not just data, but also background information, are important for drawing a conclusion. The background information here is our understanding of how the game show host operates. This information is in fact vital because the data – being shown a goat – will always be the same. Changing the background information will generally change how we interpret data and therefore what we conclude.

1.2 Probability

1.2.1 Fundamental properties

What is probability?

Probability is a way of quantifying our state of knowledge of a proposition. Examples of propositions are "this is the spectrum of a star", "it will rain tomorrow", "model M is correct", and "the period of oscillation lies between 1 and 1.1 seconds". Our state of knowledge is inseparable from our belief: the probability reflects what *we* think about the proposition, based on information, data, etc., that *we* have. Other people with other information may well assign a different probability.

Let A and B be two propositions which have probabilities $P(A)$ and $P(B)$ respectively. Without being too formal, the following are the essential axioms of probability.[1] Probabilities are real numbers that satisfy

$$0 \leq P(A) \leq 1. \tag{1.1}$$

Denote as A' the complement of A (i.e. A doesn't happen or isn't true). Then

$$P(A') = 1 - P(A). \tag{1.2}$$

If two propositions A and B are *mutually exclusive* then

$$P(A \text{ or } B) = P(A) + P(B) \quad \text{(exclusive propositions)} . \tag{1.3}$$

If $\{E_i\}$ is the set of *mutually exclusive* and *exhaustive* propositions (no others are possible), then

$$\sum_i P(E_i) = 1. \tag{1.4}$$

The *conditional probability* of A occurring given that B is true (e.g. A = "rain", B = "cloudy") is written $P(A|B)$. The *joint probability* that both are true is written $P(A, B)$. It follows logically that we can factorize this and write it as

$$P(A \text{ and } B) \equiv P(A, B) = P(A|B)P(B) = P(B|A)P(A). \tag{1.5}$$

If A and B are mutually exclusive, $P(A, B) = 0$. If A and B are independent, $P(B|A) = P(B)$ and $P(A|B) = P(A)$, in which case

$$P(A, B) = P(A)P(B) \quad \text{(independent propositions)}. \tag{1.6}$$

Equation 1.5 is simple, yet it is one of the most important for inference. It can be rearranged and written as

$$P(A|B) = \frac{P(A, B)}{P(B)}. \tag{1.7}$$

Note that $P(A|B) \neq P(B|A)$. If you're not convinced, take an example: think of

---

[1] Much has been written on the foundations of probability theory. Two important contributions are by Kolmogorov (1933) and Cox (1946).

A = "being female", B = "being pregnant". Suppose that the average human female gives birth to two children in her life (gestation period $9/12$ of a year) and the life expectancy is 80 years, then for a female chosen at random, $P(B|A) \simeq 2 \times (9/12)/80 \simeq 0.02$. Different assumptions will give a slightly different result, but unless you work in a maternity ward it will be small. Yet I think all will agree that $P(A|B) = 1$, for all common definitions of "female" and "pregnant".

In general it must hold that

$$P(A) = P(A|B)P(B) + P(A|B')P(B') \qquad (1.8)$$

because B is either true or is not true, so all possibilities are covered. More generally, if $\{B_i\}$ is the set of all possible propositions, i.e. $\sum_i P(B_i) = 1$, then it follows that

$$P(A) = \sum_i P(A, B_i)$$

$$= \sum_i P(A|B_i)P(B_i). \qquad (1.9)$$

This is called the *marginal probability* of A, where "marginal" just means "average", here over the $\{B_i\}$.

1.2.2 Some problems

Drawing cards

We draw two cards from a well-shuffled deck of 52 playing cards. What is the probability they are both aces given that (a) we replace the first card, (b) we don't?

$$\text{(a)} \ \frac{4}{52}\frac{4}{52} = \frac{1}{169} \simeq 0.0059. \qquad (1.10)$$

This is also the probability of drawing an ace then a king, or a three then a nine, etc. It's only we who attach meaning to it being an ace both times.

$$\text{(b)} \ \frac{4}{52}\frac{3}{51} = \frac{1}{221} \simeq 0.0045. \qquad (1.11)$$

Firing missiles

A missile has a probability $p = 0.1$ of destroying its target. What is the probability that the target is destroyed when we fire two missiles?

It is tempting to say 0.2. But this is obviously wrong, because if we instead had $p = 0.7$ we would not say the probability is 1.4. Logically this is because we cannot destroy the target twice. Mathematically it's because we do not add probabilities for non-exclusive events. The question we are really asking is "what is the probability that the target is destroyed by either missile or by both missiles?". In the general case of n missiles it would take a while to work out all combinations of some missiles hitting the target and others not. In such problems it is easier to work out the opposite probability, namely the probability that the target is not destroyed at all. This is $(1 - p)^n$. The probability that *at least one* missile hits the target is then $1 - (1 - p)^n$. For $n = 2$ this is 0.19 with $p = 0.1$, and 0.91 with $p = 0.7$.

When writing the probability that proposition A or B is true, we need to be careful whether the "or" is exclusive or not. Non-exclusive means A or B or both could be true, so it follows that

$$P(A \text{ or } B) = 1 - P(A', B') \qquad \text{(non-exclusive propositions)}. \qquad (1.12)$$

This is the missile example, in which the events are also independent so $P(A', B') = P(A')P(B')$ and $P(A') = P(B') = 1 - p$. If the events are exclusive, equation 1.3 applies instead.

1.2.3 Frequency

Probabilities are sometimes equated with frequencies. For example, with a deck of 52 cards we could say that the probability of drawing the ace of spades is $1/52$, perhaps because we imagine repeating this process many times. This frequency interpretation is often useful, but it doesn't always work, as we can see with unique events. For example, if we ask what is the probability that a helicopter will make an emergency landing outside my house next 21 May, we cannot resort to frequencies of actual past or even imagined events. We must instead resort to similar events. But what counts as similar is a matter of personal judgement and experience, so has no unique answer. Likewise, when it comes to inferring something from data, such as the probability that the mass of a particle lies in a particular range, then the frequency interpretation is of little help. It is often more useful to think of probability as a degree of belief.

1.2.4 Bayes' theorem and the principle of inference

If we rearrange the equation for joint probabilities (equation 1.5), we get

$$P(B|A) = \frac{P(A|B)P(B)}{P(A)}. \qquad (1.13)$$

This is called *Bayes' theorem*. It was once referred to as *inverse probability*, because it relates the probability of B given A to the probability of A given B. It is fundamental in inference for the following reason. Taking M as a model (or hypothesis) and D as some data that we have obtained to test the model, then we can write Bayes' theorem as

$$P(M|D) = \frac{P(D|M)P(M)}{P(D)}. \qquad (1.14)$$

When we set up an experiment we can normally determine $P(D|M)$, the probability of observing the data (or perhaps a quantity derived from the data) under the assumption that the model is true. For example, we might be looking for the presence of spectral line ($D = $ line is present) under the assumption that it is produced by a particular trace gas in an exoplanet's atmosphere. We then set up a model M for the atmosphere and our measurements, and compute the probability $P(D|M)$ that this would allow us to observe the line. Note that this is not the probability that the model is true, given the data. That quantity is $P(M|D)$ (recall the example above about pregnancy and females). Observing the line does not necessarily indicate the presence of the gas, because there may be other origins of the line (i.e. other models).

To infer $P(M|D)$ we use Bayes' theorem. Equation 1.14 tells us that we need two other quantities in addition to $P(D|M)$. The first is $P(M)$, the unconditional probability that the model M is true. This is called the *prior* probability, prior here meaning "prior to using the data". The other quantity we need is $P(D)$, the unconditional probability of the data, meaning its probability independent of any particular model. Let $\{M_i\}$ be the complete set of all models (mutually exclusive and exhaustive). Using equation 1.9 we can write

$$P(D) = \sum_i P(D|M_i)P(M_i). \tag{1.15}$$

We now have all terms needed to compute $P(M|D)$, the *posterior probability* that M is true. This differs from the prior probability in that it is conditional on the data. Bayes' theorem underpins much of inference, so I will have a lot more to say about what it is and how we use it. We will return to it again in chapter 3.

1.2.5 Discrete and continuous probability distributions

So far we have considered discrete variables, whereby $P(B_i)$ gives the actual probability of proposition B_i being true or of event B_i occurring. We sometimes refer to the set of probabilities over all events as the *probability mass function* (PMF).

For a continuous variable x we instead deal with the *probability density function* (PDF) $P(x)$. This is a density – probability per unit x – not a probability. $P(x)dx$ is the infinitesimal probability of x in the range x to $x + dx$, so a finite probability is obtained by integrating over a region. Specifically, the probability of x lying between x_1 and x_2 is

$$\int_{x_1}^{x_2} P(x)\,dx. \tag{1.16}$$

Probability is dimensionless: it has no units. Therefore the density $P(x)$ has units $1/x$. Note that $P(x|y)$ is a PDF in x, not in y. The thing after the bar just gives conditional information. The variables x and y need not have the same units, which is another reason why $P(x|y) \neq P(y|x)$ in general. I will use the same symbol – upper case P – to refer to both PMFs and PDFs. It will be clear from the context which I mean.

The range of the variable x over which the PDF (or PMF) is non-zero is called the *support* of the distribution. If it is non-zero over all real values, the distribution is said to have *infinite support*. If it has a bound, e.g. it is non-zero only for positive values, it is referred to as having *semi-infinite support*.

1.2.6 Normalization

A bona fide PDF is always normalized, i.e.

$$\int_{-\infty}^{+\infty} P(x)\,dx = 1 \tag{1.17}$$

and likewise

$$\sum_i P(x_i) = 1 \tag{1.18}$$

for discrete distributions. Distributions that cannot be normalized are called *improper*. An example is the uniform distribution with infinite support (section 1.4.4). It is not normalizable because its integral is infinite. We will nonetheless see that improper distributions can be useful in certain contexts.

We will sometimes work with unnormalized distributions, which I will indicate using an asterisk P^*. For an unnormalized PDF $P^*(x)$ the normalization constant is $\int P^*(x)\,dx$, which is independent of x (because we've integrated over it). The normalized PDF is therefore

$$P(x) = \frac{P^*(x)}{\int P^*(x)\,dx} \tag{1.19}$$

and the normalized discrete distribution (probability mass function) is

$$P(x_i) = \frac{P^*(x_i)}{\sum_i P^*(x_i)}. \tag{1.20}$$

1.3 Expectation, variance, and moments

We often want to summarize probability distributions with just a few numbers. Two summary statistics of particular interest are the mean, which indicates the location of the distribution, and the standard deviation, which is a measure of its width. Many of the most commonly occurring distributions are fully defined by these two parameters.

The *expectation value* or *mean* of a continuous variable x with a PDF $P(x)$ is defined as

$$E[x] = \int x\,P(x)\,dx. \tag{1.21}$$

For a discrete variable with PMF $P(x_i)$ it is defined as

$$E[x] = \sum_i x_i\,P(x_i). \tag{1.22}$$

The expectation value is sometimes written as μ. The expectation value of a function $f(x)$ is defined as

$$E[f(x)] = \int f(x)\,P(x)\,dx. \tag{1.23}$$

If we have a set of N data points drawn from a continuous distribution, then we can use equation 1.22 as an approximation for equation 1.21. Each data point occurs once, so implicitly $P(x_i) = 1/N$, and

$$E[x] \simeq \frac{1}{N} \sum_{i=1}^{N} x_i = \bar{x}, \tag{1.24}$$

which is called the *sample mean*. The *law of large numbers* says that in the limit $N \to \infty$ the sample mean tends towards the (true) mean.

Some useful properties of the expectation operator follow from its definition. Let x and y be random variables and a a fixed constant.

$$E[a] = a \tag{1.25}$$
$$E[ax] = aE[x] \tag{1.26}$$
$$E[x + y] = E[x] + E[y] \tag{1.27}$$
$$E[E[x]] = E[x] \tag{1.28}$$
$$E[xE[y]] = E[x]E[y] \tag{1.29}$$

but in general $E[xy] \neq E[x]E[y]$.

The *variance* of a random variable x with expectation value $E[x]$ is defined as

$$\begin{aligned}
\mathrm{Var}(x) &= \int (x - E[x])^2 P(x)\, dx \\
&= E[(x - E[x])^2] \\
&= E[x^2 - 2xE[x] + E[x]^2] \\
&= E[x^2] - 2E[x]E[x] + E[x]^2 \\
&= E[x^2] - E[x]^2. \tag{1.30}
\end{aligned}$$

We can remember this as "the variance is the expectation of the square minus the square of the expectation" (or just "mean square minus square mean"). If we know the true mean μ, then for a set of N data points $\{x_i\}$, the variance can be approximated by

$$\mathrm{Var}(x) \simeq \frac{1}{N} \sum_{i=1}^{N} (x_i - \mu)^2. \tag{1.31}$$

This is an approximation because of the finite sample size. If we don't know μ then we might want to estimate the variance as

$$V_x = \frac{1}{N} \sum_{i=1}^{N} (x_i - \bar{x})^2 = \overline{x^2} - \bar{x}^2. \tag{1.32}$$

However, if we calculate the expected value of this we find

$$E[V_x] = \frac{N - 1}{N} \mathrm{Var}(x) \tag{1.33}$$

which is not equal to the true variance for finite N. It is therefore a biased estimate; specifically it is an underestimate (see section 2.1 for a discussion of estimator bias). The origin of this bias is that V_x uses the estimated (sample) mean of the data (equation 1.24), whereas the definition of variance uses the true mean. As we have computed this bias we can correct for it, to give the *sample variance*, which is defined as[2]

$$\widehat{\text{Var}}(x) = \frac{1}{N-1} \sum_{i=1}^{N} (x_i - \bar{x})^2 \tag{1.34}$$

which is an unbiased estimate of the true variance.

The *standard deviation* is defined as

$$\sigma = \sqrt{\text{Var}(x)} \tag{1.35}$$

which for a set of data of known mean is

$$\sigma \simeq \sqrt{\frac{1}{N} \sum_{i=1}^{N} (x_i - \mu)^2} \tag{1.36}$$

where the approximation symbol has again been used because we have a finite-sized sample. The *sample standard deviation* is the square root of the sample variance (equation 1.34)

$$\hat{\sigma} = \sqrt{\frac{1}{N-1} \sum_{i=1}^{N} (x_i - \bar{x})^2}. \tag{1.37}$$

The variance and standard deviation of a set of data are computed in R by the functions var and sd respectively. They both use the $N-1$ term.

As variance is a squared quantity it tells you nothing about the asymmetry of a distribution. This we can measure with the *skew*

$$\gamma = \frac{1}{N\sigma^3} \sum_{i=1}^{N} (x_i - \mu)^3 \tag{1.38}$$

which is a dimensionless number. A positive value for the skew means the data have an asymmetry about the mean with a tail to positive values. Here μ and σ are the true mean and true standard deviation respectively. In practice we would estimate these from the data, so we should use a slightly different expression for the skew that involves a small correction analogous to the one applied for the sample variance. The *kurtosis* is the next higher power and measures how centrally concentrated a distribution is

$$\kappa = \frac{1}{N\sigma^4} \sum_{i=1}^{N} (x_i - \mu)^4 - 3. \tag{1.39}$$

It is also dimensionless. The -3 is in the definition so that a Gaussian distribution has zero kurtosis. More negative values are more centrally concentrated.

---

[2] Here I use the hat symbol to distinguish the sample variance from the variance. I also do this in equations 1.37 and 1.66 to indicate sample standard deviation and sample covariance, respectively. More generally the hat symbol is often used to indicate an estimate.

Mean, variance, skew, and kurtosis are related to the first four *moments* of a distribution. The kth moment of $P(x)$ is defined as

$$\int x^k P(x)\, dx \tag{1.40}$$

for integer $k \geq 0$ (the zeroth moment is the normalization). This is sometimes called the kth raw moment to distinguish it from the kth *central moment*, which is defined as

$$\int (x - \mu)^k P(x)\, dx. \tag{1.41}$$

The kth *sample moment* of a set of data $\{x_i\}$ drawn from the distribution is

$$\frac{1}{N} \sum_{i=1}^{N} x_i^k. \tag{1.42}$$

There is no $N - 1$ correction term in the definition of these (non-central) sample moments. Although we can always calculate the raw moments of a set of data, not all distributions have defined moments. We shall see an example of this in section 2.3.2.

1.4 Univariate probability distributions

We turn now to some of the most common univariate probability distributions. The first two of these, the binomial and Poisson distributions, describe discrete variables. The others describe continuous variables. There are of course many other distributions, and several more will be defined in later chapters as we encounter them (see the index under "distributions" for a full list). The derivation of some of these distributions relies on combinations and permutations, so you may want to read section 1.7 before proceeding.

1.4.1 Binomial

This probability distribution describes processes in which an event can have only one of two possible outcomes. Examples include tossing a coin, detecting something at a security check, or winning the lottery. Let p be the probability of one event, call it "success"; $1 - p$ is the probability of the other event ("failure"). If the trial is repeated independently n times, then we are interested in the probability of getting exactly r successes, which we can label $P(r\,|\,p, n)$. Suppose that the first r trials are successes and the remainder, $n - r$, are all failures. As the trials are independent, the probability of this sequence is just the product of the probabilities, $p^r (1 - p)^{n-r}$. This is the probability of just one particular sequence of trials. The number of unique sequences with this probability is the number of ways of selecting r from n (without replacement), which is $_nC_r$ (see section 1.7). As these sequences are mutually exclusive we just sum the probability $_nC_r$ times. This gives us the *binomial distribution*

$$P(r|p,n) = {}_nC_r\, p^r (1-p)^{n-r} = \binom{n}{r} p^r (1-p)^{n-r} = \frac{n!}{r!(n-r)!} p^r (1-p)^{n-r}.$$

(1.43)

This is the probability of getting r successes from n trials if the probability of one success is p, where of course $0 \le r \le n$.

Here is an example: $P(r|p,n)$ is the probability of drawing r red balls blindly from a set of n red and white balls, in which the fraction of red balls is p. The draws are done one by one with replacement (so p remains constant).[3]

Using equations 1.22 and 1.30 we can show that the expected number of successes is

$$E[r] = np$$

(1.45)

(which agrees with our intuition) and the variance is

$$\mathrm{Var}(r) = np(1-p).$$

(1.46)

Examples of the binomial distribution are shown in the left panel of figure 1.1. The right panel shows instead how the probability density varies as a function of p for a given r; this is not a density function in p. The plots were produced using the following R script, which uses the function dbinom to compute the densities.

R file: binomial_distribution.R

```
##### Plot the binomial distribution

# Plot P vs. r for fixed n for a range of p
n <- 10
r <- 0:n
pseq <- c(0.1, 0.2, 0.5, 0.8, 0.9)
pdf("dbinom1.pdf", 4, 4)
par(mfrow=c(1,1), mgp=c(2.0,0.8,0), mar=c(3.5,3.5,1,1), oma=0.1*c(1,1,1,1))
plot(r, r, type="n", xlim=c(0,max(r)), ylim=c(0,0.4), xlab="r",
     ylab="P(r | p,n)")
for (p in pseq) {
  points(r, dbinom(x=r, size=n, prob=p), pch=20)
  lines(r,  dbinom(x=r, size=n, prob=p), lty=2)
```

---

[3] If we did the selection without replacement then the probability that there are r red balls in a sample of n balls, which we draw without replacement from an original set of N balls of which m were red, is

$$P(r|n,m,N) = \frac{\binom{m}{r}\binom{N-m}{n-r}}{\binom{N}{n}}$$

(1.44)

which is called the *hypergeometric distribution*. Here $\binom{m}{r}$ is the number of ways of selecting r red balls from the total m red balls, $\binom{N-m}{n-r}$ is the number of ways in which the remaining $n-r$ white balls can be chosen from the $N-m$ white balls, and $\binom{N}{n}$ is the number of ways in which we could select any sample of n balls (of any colour). All of these selections are without replacement.

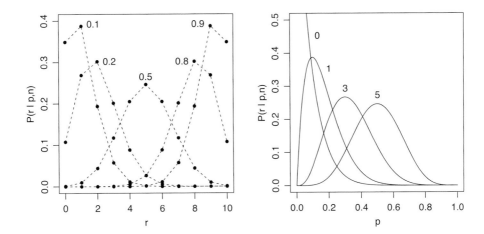

Fig. 1.1 The binomial distribution for $n = 10$. Left: $P(r\,|\,p, n)$ vs r for the five values of p indicated. The distribution is discrete: points are joined with dashed lines just to help identification of the points with common p. Right: $P(r\,|\,p, n)$ vs p for the four values of r indicated. Note that this is still a probability distribution over r, not p.

```
}
text(c(1,2,5,8,9), c(0.39,0.30,0.285,0.30,0.39), pseq, pos=c(4,4,1,2,2))
dev.off()

# Plot P vs. p for fixed n for a range of r
p    <- seq(from=0, to=1, by=0.001)
rseq <- c(0,1,3,5)
n    <- 10
pdf("dbinom2.pdf", 4, 4)
par(mfrow=c(1,1), mgp=c(2.0,0.8,0), mar=c(3.5,3.5,1,1), oma=0.1*c(1,1,1,1))
plot(p, p, type="n", xlim=range(p), ylim=c(0,0.5), xlab="p",
     ylab="P(r | p,n)")
for (r in rseq) {
  lines(p,  dbinom(x=r, size=n, prob=p))
}
text(c(0.08,0.15,0.25,0.45), c(0.45,0.35,0.29,0.27), rseq, pos=4)
dev.off()
```

The binomial distribution is a discrete distribution, so $P(r\,|\,p, n)$ (i.e. dbinom) is a probability mass function; it gives actual probabilities.

Example: coin tossing

What is the probability of getting exactly two heads in six tosses of a fair coin?

$$P(r = 2\,|\,p = 1/2, n = 6) = \binom{6}{2}\left(\frac{1}{2}\right)^2\left(1 - \frac{1}{2}\right)^4 = \frac{15}{64} \simeq 0.23. \tag{1.47}$$

1.4.2 Poisson

The binomial distribution describes events in which there is a definite event taking place that has a two-way result: it's either a "success" or a "failure"; something happens or it doesn't. Yet a lot of natural processes are only one-way, by which I mean they are only identifiable by having happened. Examples are lightning strikes and α particle emission from a radioactive source. We cannot count non-events because we cannot identify a sequence of events where something is supposed to happen or not. Suppose that on average these events occur at a rate λ, so that λ is the expected number of events in some unit time interval. We would like to find the probability that we get r events in this interval.

We can describe this as the limit of a binomial process. If we divide the time series into n divisions, then in the limit as the divisions become very small such that n gets large, we can write $p = \lambda/n$. The binomial distribution is

$$P(r\,|\,\lambda/n, n) = \frac{n!}{r!(n-r)!}\left(\frac{\lambda}{n}\right)^r\left(1-\frac{\lambda}{n}\right)^{n-r}. \tag{1.48}$$

We now take n to infinity to produce a continuum of events. As $n \to \infty$ with r finite,

$$\frac{n!}{(n-r)!} = n(n-1)(n-2)\ldots(n-r+1) \to n^r \tag{1.49}$$

(because each term tends towards n) and

$$\left(1-\frac{\lambda}{n}\right)^{n-r} \to \left(1-\frac{\lambda}{n}\right)^n \to e^{-\lambda} \tag{1.50}$$

which is a definition of e (Euler's constant). Inserting these two terms into equation 1.48, n^r cancels and we end up with the *Poisson distribution*

$$P(r\,|\,\lambda) = \frac{e^{-\lambda}\lambda^r}{r!} \quad \text{where} \quad \lambda > 0,\ r \geq 0. \tag{1.51}$$

This is the probability of getting r events if the mean expected number is λ, i.e. $E[r] = \lambda$. Note that although r is an integer, λ does not have to be. An important property of this distribution is that its variance is $\text{Var}(r) = \lambda$, i.e. equal to the mean. It is a discrete distribution so $P(r\,|\,\lambda)$ is a probability mass function.

The following R code plots the distribution (see figure 1.2). The larger λ the closer the Poisson distribution becomes to the Gaussian distribution (notwithstanding the fact that the Poisson distribution is only defined for integer r).

```
pdf("dpois1.pdf", 4, 4)
par(mfrow=c(1,1), mgp=c(2.0,0.8,0), mar=c(3.5,3.5,1,1), oma=0.1*c(1,1,1,1))
r <- 0:20
plot(r, r, type="n", xlim=c(0,max(r)), ylim=c(0,0.4),
  xlab="r", ylab=expression(paste(, "P(r | ", lambda, ")")))
for (lambda in c(1,2,3.5,10)) {
  points(r, dpois(x=r, lambda=lambda))
  lines(r,  dpois(x=r, lambda=lambda), lty=2)
}
dev.off()
```

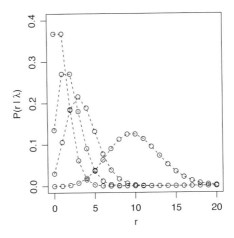

The Poisson distribution for $\lambda = 1, 2, 3.5, 10$ (maxima from left to right). The distribution is discrete: points are joined with dashed lines just to help identify the points with common λ.

A useful property of the Poisson distribution is

$$
P(r+1\,|\,\lambda) = \frac{e^{-\lambda}\lambda^{r+1}}{(r+1)!}
$$

$$
= \frac{e^{-\lambda}\lambda^{r}}{r!}\,\frac{\lambda}{r+1}
$$

$$
= P(r\,|\,\lambda)\frac{\lambda}{r+1} \tag{1.52}
$$

and

$$
P(0;\lambda) = e^{-\lambda}. \tag{1.53}
$$

It can also be proven that if two Poisson processes have means λ_a and λ_b, then the probability of getting a total of r events from the two processes (without distinguishing which came from where – consider counting radioactive decays from two isotopes) is described by a Poisson distribution with mean $\lambda_a + \lambda_b$.

Example: radioactive decay

Consider a radioactive source with half-life $t_{1/2}$. If n_0 is the initial number of radioactive atoms, then the number of radioactive atoms left after time t is

$$
n = n_0\,e^{-t/\tau} = n_0\,e^{-\lambda t} \tag{1.54}
$$

where $\tau = t_{1/2}/\ln 2$. The mean (expected) number of decays per unit time is $\lambda = 1/\tau$. The theoretical distribution of the number of decays per time is a Poisson distribution. To see this from a set of data, we record the times at which decays occur, then divide up the observed time span into equal-sized time intervals. We then count how many of these

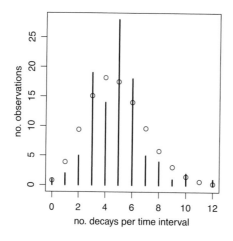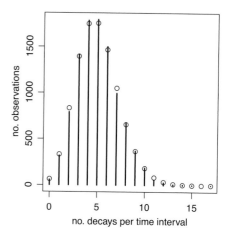

Fig. 1.3
The radioactive decay example showing the frequency (scaled probability mass function) of decays per unit time interval for the simulated data (vertical bars), and for the Poisson distribution (open circles) with λ determined from the data. The left panel is the result for 100 time intervals; the right panel is to a much longer sequence of data with 10 000 time intervals. In both cases the true value of λ (used in the simulation) is 5.

intervals have 0, 1, 2, 3, etc., decays. When normalized, this is approximately a Poisson distribution with mean $1/\tau$.

The R code below demonstrates this using a simulation of radioactive decay. Let's assume a source has a decay rate of $\lambda = 5$ per unit time interval. The number of decays in a unit time interval can be simulated by drawing once from a Poisson distribution with this mean. I repeat this $\texttt{nint} = 100$ times to simulate a sequence of observation intervals. Based just on these data I then count how many of these intervals have 0, 1, 2, 3, etc., decays. I do this using the function $\texttt{table}$. The resulting frequency distribution is shown in the left panel of figure 1.3. I then overplot a Poisson density distribution with its mean derived from the data (which is $\texttt{mean(ndecay)} = 4.81$), and which I have scaled to give the expected number of counts per time interval. If I increase the amount of data – the number of time intervals – the data follow the Poisson distribution more closely. This is shown in the right panel of the figure, which uses $\texttt{nint} = 10\,000$ (the empirical mean is 5.01). Note that doubling the length of the time interval would double both λ and the typical number of events per time interval, so the Poisson distribution would look more like a Gaussian.

R file: $\texttt{poisson.R}$

```
##### Compare data drawn from a Poisson with its theoretical distribution

pdf("dpois2.pdf", 4, 4)
par(mfrow=c(1,1), mgp=c(2.0,0.8,0), mar=c(3.5,3.5,1,1), oma=0.1*c(1,1,1,1))
truelambda <- 5 # = 1/tau
nint <- 100    # number of time intervals
```

```
set.seed(200)
ndecay <- rpois(n=nint, lambda=truelambda) # no.decays in each time interval
nobs   <- table(ndecay) # frequency distribution of ndecay
x <- 0:max(ndecay)
# multiply Poisson density by nint to get expected counts
plot(x, dpois(lambda=mean(ndecay), x=x)*nint, xlim=range(x),
     ylim=c(0,max(nobs)), xlab="no. decays per time interval",
     ylab="no. observations")
points(nobs)
dev.off()
```

Example: why do we build large telescopes?

The main reason for building telescopes with large mirrors is to be able to detect faint sources. The larger the telescope mirror the more photons we gather from the source (for a given exposure time). But why do we need more photons? The reason is that the emission of photons is governed by the Poisson distribution: the number of photons emitted will differ from the expected number due to the finite variance of the distribution. This is noise. How reliably we can detect something is determined by the signal-to-noise ratio (SNR), which is proportional to the signal r divided by its standard deviation σ_r. For the Poisson process $\sigma_r = \sqrt{r}$, so the SNR is proportional to $\sqrt{r}$, i.e. it increases with the number of photons collected.[4] By building a large telescope, we get a larger SNR in a given exposure time.

1.4.3 Beta

A useful PDF for a quantity p bound to lie between 0 and 1 is the beta distribution. This is described by two shape parameters, α and β. Its PDF is

$$P(p) = \frac{1}{B(\alpha, \beta)} p^{\alpha-1}(1-p)^{\beta-1} \quad \text{where} \quad \alpha > 0, \ \beta > 0, \ 0 \le p \le 1 \quad (1.55)$$

where

$$B(\alpha, \beta) = \int_0^1 p^{\alpha-1}(1-p)^{\beta-1} dp, \quad (1.56)$$

the *beta function*, is the normalization constant. Figure 1.4 shows some examples of the distribution plotted using the dbeta function in R (shape1 is α and shape2 is β). The mean, mode (maximum), and variance of the beta distribution are

$$\text{mean} = \frac{\alpha}{\alpha + \beta} \quad (1.57)$$

$$\text{mode} = \frac{\alpha - 1}{\alpha + \beta - 2} \quad \text{for} \quad \alpha > 1, \ \beta > 1 \quad (1.58)$$

$$\text{variance} = \frac{\alpha\beta}{(\alpha + \beta)^2(\alpha + \beta + 1)}. \quad (1.59)$$

---

[4] The standard deviation is actually equally to the square root of the *expected* number of photons, but we don't know that, so we use r as our best estimate thereof.

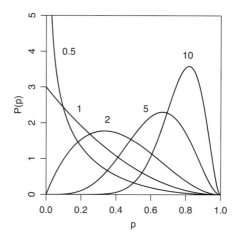

Fig. 1.4   The beta distribution for $\alpha = 0.5, 1, 2, 5, 10$ with $\beta = 3$.

Notice how the dependence on p is similar in form to that of the binomial distribution. They are in fact the same when $\alpha = r + 1$ and $\beta = n - r + 1$. We shall see the relevance of this in section 5.1.

1.4.4 Uniform

A uniform distribution $\mathcal{U}(a, b)$ is constant between a and b and zero outside. In order for it to be normalized the value of the distribution must be $1/(b - a)$. Its mean and median are obviously $(a + b)/2$, but its mode is not uniquely defined (it could have any value between a and b). Using equation 1.30 we find the variance to be $(b - a)^2/12$.

Sometimes we make use of a uniform distribution which has infinite support (no bounds) or which has semi-infinite support (e.g. a lower bound at zero). Such a distribution is constant over an infinite range. It cannot be normalized and so is an improper distribution (section 1.2.6).

1.4.5 Gaussian

The Gaussian or normal distribution is probably the best known and most commonly used distribution in the physical sciences. Its domain is the set of all real numbers and its density function is

$$P(x) = \frac{1}{\sigma\sqrt{2\pi}} \exp\left[-\frac{(x - \mu)^2}{2\sigma^2}\right] \quad \text{where} \quad \sigma > 0. \tag{1.60}$$

This is sometimes abbreviated with the notation $\mathcal{N}(\mu, \sigma)$. The notation $\epsilon \sim \mathcal{N}(\mu, \sigma)$ means that ϵ is a random number drawn from this distribution. The mean μ and standard deviation σ fully characterize the Gaussian. This PDF is symmetric and the mean is equal to the mode and the median. Approximately 68% of the probability mass lies between $\mu - \sigma$ and $\mu + \sigma$, the "1σ range". The full-width at half-maximum (FWHM) is $2\sqrt{2\ln 2}\,\sigma \simeq 2.35\sigma$.

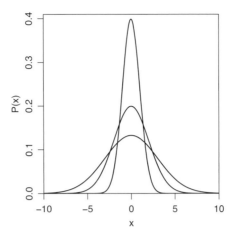

Fig. 1.5 The Gaussian (normal) distribution with zero mean and standard deviation of 1, 2, and 3 (larger is wider/lower).

Examples of the distribution are shown in figure 1.5. As the distribution is normalized, a larger σ means a lower maximum. The *standardized Gaussian* is the distribution $\mathcal{N}(0, 1)$.

There are at least three reasons why the Gaussian is ubiquitous: (1) it has some convenient properties, making it easy to use; (2) the central limit theorem (to be discussed in section 2.3); (3) the principle of maximum entropy says that if we have a continuous variable for which we know only its mean and variance, then the Gaussian is the distribution that makes the fewest assumptions; it is the most conservative choice.

1.4.6 Gamma

The gamma distribution has semi-infinite support: it is non-zero only for $x \geq 0$. It is characterized by two parameters, the shape k and the scale θ. Its PDF is

$$P(x) = \frac{1}{\Gamma(k)\theta^k} x^{k-1} e^{-x/\theta} \quad \text{where} \quad k > 0, \ \theta > 0, \ x \geq 0 \tag{1.61}$$

and $\Gamma(k)$ is the *gamma function*, defined as

$$\Gamma(k) = \int_0^\infty x^{k-1} e^{-x} \, dx \tag{1.62}$$

which can be seen as a generalization of the factorial function for non-integer k, because $\Gamma(k) = (k-1)!$ for integer k. (The gamma function is gamma in R.) The mean and variance of the gamma distribution are $k\theta$ and $k\theta^2$ respectively. It has a mode only when $k \geq 1$, and then at $(k - 1)\theta$. When $k < 1$ the function tends to infinity as x goes to zero, but the area under the curve of course remains equal to one. Examples of the distribution are shown in figure 1.6. The gamma distribution may be convenient for defining distributions over quantities that cannot be negative.

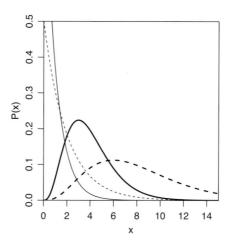

Fig. 1.6   The gamma distribution for two values of its shape k (thin = 1, thick = 4) and scale θ (solid = 1, dashed = 2) parameters.

1.4.7 Cauchy

The Cauchy distribution (also called the Lorentz distribution) is characterized by a location parameter a and a width (or scale) parameter b. Its PDF is

$$P(x) = \frac{1}{\pi b[1 + (\frac{x-a}{b})^2]} = \frac{b}{\pi[b^2 + (x - a)^2]} \quad \text{where} \quad b > 0. \qquad (1.63)$$

If you plot it (figure 1.7), it looks quite harmless. It has a mode (and median) at $x = a$ and is symmetric about this. But it is the canonical example of a pathological distribution, because it has no mean and no standard deviation. Why? The distribution has a finite zeroth moment (equation 1.40) so it is normalizable. But it turns out that all its higher moments are either undefined or infinite (we'll investigate this further in section 2.3.2). We can nonetheless characterize the width of the distribution using the FWHM, for example. It turns out this is $2b$, so the parameter b is the half-width at half-maximum (HWHM) (and it is also the interquartile range, defined in section 1.5).

Figure 1.7 plots two different Cauchy distributions. Owing to the normalization, a larger width b means a lower maximum. The plot compares the Cauchy distribution with Gaussians with the same FWHM. For values nearer to the centre the Gaussian is broader, but as we move into the lower probability density regions the Cauchy is much wider. The Cauchy has "heavier tails" than the Gaussian. Other than being nicely pathological, the Cauchy distribution is also the distribution of the variable which is the ratio of two independent Gaussian variables. It is sometimes used to describe the broadening of spectral lines.

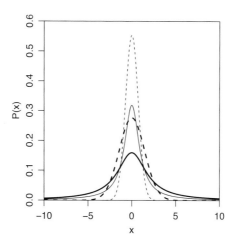

Fig. 1.7
The Cauchy distribution (solid lines) with zero mode and b equal to 1 (thin) and 2 (thick). The dashed lines shows a Gaussian with zero mode and the same FWHM as the two Cauchy distributions.

1.5 Cumulative distribution function and quantile function

Given a continuous, one-dimensional PDF $P(x)$, the *cumulative distribution function* (CDF) is the probability that x is less than some value q. For a PDF with infinite support this is

$$C(q) = P(x < q) = \int_{-\infty}^{q} P(x)\,dx. \tag{1.64}$$

If $P(x)$ is only defined over a limited range (x_{min}, x_{max}), the lower limit of the above integral is x_{min}. $C(q)$ is a monotonically non-decreasing[5] function of q, and as $P(x)$ is normalized, $C(q)$ lies between 0 and 1. The derivative of the CDF is the PDF. The CDF is only defined for univariate probability distributions.

The inverse function of the CDF is the *quantile function* $Q(p)$. It gives the value of x below which the integral of $P(x)$ is equal to p. These two functions are shown in figure 1.8 for Gaussian and Cauchy distributions. Of particular interest as a measure of the location of a distribution is the 50% quantile ($p = 0.5$), the *median*: this is the value of x for which half the integrated probability lies below that value and half above. Together with the 25% and 75% quantiles these are the *quartiles*. The difference between the 75% and 25% quantiles, $Q(p = 0.75) - Q(p = 0.25)$, is the *interquartile range* (IQR), and is sometimes used as a measure of the width of a distribution.

Given a set of data, $Q(p)$ can be estimated by sorting the data and finding the point

---

[5] A function is *monotonic* if its gradient is always either non-negative ("monotonically non-decreasing") or non-positive ("monotonically non-increasing"), both of which allow the gradient to be zero. A function is *strictly monotonic* if its gradient is either always positive or always negative, i.e. it also cannot be zero.

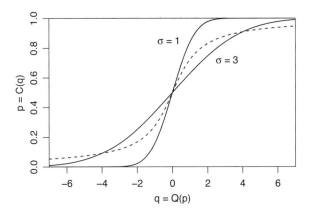

Fig. 1.8 The cumulative distribution function $C(q)$ and the quantile function $Q(p)$ for two zero-mean Gaussian distributions (solid lines) with $\sigma = 1$ (steeper function) and $\sigma = 3$, and for a Cauchy distribution (dashed line) with mode zero and the same FWHM as the Gaussian with $\sigma = 1$.

which puts a fraction p of the data below that point. For a finite set of data, there will not usually be a value of x that corresponds to the required p. We must therefore smooth or interpolate the data in some way. There are various different ways of doing this, so not all algorithms will return the same results.

For many standard distributions R has a set of built-in functions for calculating quantities from the distribution. For example, for the normal distribution they are

- dnorm(x) $= P(x)$, the probability density function (PDF)
- pnorm(q) $= p = \int_{-\infty}^{q} P(x)\, dx$, the cumulative distribution function (CDF)
- qnorm(p) the quantile function, which is the inverse function of pnorm(q)
- rnorm draws a random number from the distribution.

The CDF of a discrete set of data can be calculated in R using cumsum and the quantiles can be found using quantile. The parameter called type in the latter determines how the quantile is defined for a discrete data set.

1.6 Multiple variables

1.6.1 Covariance

Often we have problems involving multiple variables, and want to measure how closely two variables vary with one another. The *covariance* between two variables x and y is defined as

$$
\begin{aligned}
\mathrm{Cov}(x,y) &= \iint P(x,y)(x - E[x])(y - E[y])\, dx\, dy \\
&= E[(x - E[x])(y - E[y])] \\
&= E[xy - yE[x] - xE[y]] + E[x]E[y] \\
&= E[xy] - E[x]E[y].
\end{aligned}
\tag{1.65}
$$

We can remember this as "expectation of the product minus product of expectations". It generalizes the definition of variance in equation 1.30. In analogy to the sample variance, the *sample covariance* is what we should compute from a set of data when we estimate the means from the data. This is

$$
\widehat{\mathrm{Cov}}(x,y) = \frac{1}{N-1} \sum_{i=1}^{N} (x_i - \overline{x})(y_i - \overline{y}).
\tag{1.66}
$$

It is often helpful to normalize the covariance according to the standard deviation in each variable, σ_x and σ_y, which leads to the definition of the *correlation coefficient*

$$
\rho(x,y) = \frac{\mathrm{Cov}(x,y)}{\sigma_x \sigma_y}
\tag{1.67}
$$

$$
= \frac{\overline{xy} - \overline{x}\,\overline{y}}{\sigma_x \sigma_y}.
\tag{1.68}
$$

The *sample correlation coefficient* is computed using the sample covariance and the sample standard deviations (equation 1.37). The correlation coefficient lies in the range $-1 \leq \rho \leq +1$. A value of $\rho = +1$ corresponds to perfect correlation: the two variables are equal apart from some scale factor (which cancels in the ratio) and an offset (which is removed by the mean). Similarly $\rho = -1$ corresponds to perfect anticorrelation. If $\rho = 0$ the variables are uncorrelated.

The R functions that calculate the covariance and correlation between two vectors of variables are cov and cor respectively.

If we have two or more variables then we can form a *covariance matrix*, in which element c_{ij} is the covariance between variable i and variable j. Thus the diagonal elements of the matrix are the variances and the off-diagonal elements are the covariances (we will see an example in the next section). This is a symmetric matrix.

1.6.2 Multivariate probability distributions

When we have multiple variables we are often interested in their *joint probability distribution*. Let $\mathbf{x}$ be a J-dimensional vector of the variables, the mean of which is $\boldsymbol{\mu}$ (also a J-dimensional vector). The J-dimensional Gaussian PDF is defined as

$$
P(x_1, \ldots, x_J) = \frac{1}{(2\pi)^{J/2}|\Sigma|^{1/2}} \exp\left(-\frac{1}{2}(\mathbf{x} - \boldsymbol{\mu})^{\mathsf{T}} \Sigma^{-1}(\mathbf{x} - \boldsymbol{\mu})\right)
\tag{1.69}
$$

where Σ is the $J \times J$ covariance matrix of the data, $|\Sigma|$ is its determinant, and $(\mathbf{x} - \boldsymbol{\mu})^{\mathsf{T}}$ denotes the transpose of the $J \times 1$ column vector $(\mathbf{x} - \boldsymbol{\mu})$. (My vectors are column vectors by default.) The argument of the exponential is of course a scalar. The dimension of this probability density is $(\prod_i x_i)^{-1}$, as can also be seen by the fact that the determinant of the covariance enters as the square root in the denominator. If the variables are independent of one another then the covariance matrix is diagonal. In two dimensions, writing $\mathbf{x} = (x, y)$, the covariance matrix is

$$\Sigma = \begin{bmatrix} \sigma_x^2 & \rho\sigma_x\sigma_y \\ \rho\sigma_x\sigma_y & \sigma_y^2 \end{bmatrix} \tag{1.70}$$

and the distribution is

$$P(x, y) = \frac{1}{2\pi|\Sigma|^{1/2}} \exp\left(-\frac{1}{2}(\mathbf{x} - \boldsymbol{\mu})^{\mathsf{T}}\Sigma^{-1}(\mathbf{x} - \boldsymbol{\mu})\right) \tag{1.71}$$

where $|\Sigma| = \sigma_x^2\sigma_y^2(1 - \rho^2)$. The locus of points with constant probability density is given when the argument of the exponential is constant. Consider for convenience the zero mean Gaussian with zero covariance, in which case

$$\Sigma^{-1} = \begin{bmatrix} \frac{1}{\sigma_x^2} & 0 \\ 0 & \frac{1}{\sigma_y^2} \end{bmatrix}. \tag{1.72}$$

Performing the matrix multiplications we see that the locus of points of constant probability is

$$\frac{x^2}{\sigma_x^2} + \frac{y^2}{\sigma_y^2} = \text{constant}. \tag{1.73}$$

This is the equation for an ellipse. We can show that if the covariance is non-zero then this just tilts the ellipse. So we can always transform a two-dimensional Gaussian distribution with non-zero covariance into a two-dimensional Gaussian with zero covariance by rotating the axes (and likewise for higher dimensional Gaussians). The following code plots a bivariate Gaussian with non-zero covariance both as a three-dimensional perspective mesh and as a contour plot. This is shown in figure 1.9.

R file: 2D_gaussian.R

```
##### Plot bivariate Gaussian as a 3D mesh plot
##### and as contours of constant probability density

library(mvtnorm) # for dmvnorm
sigma.x <- 1
sigma.y <- 1
rho <- 0.5 # correlation coefficient
Cov <- matrix(data=c(sigma.x^2, rho*sigma.x*sigma.y, rho*sigma.x*sigma.y,
                     sigma.y^2), nrow=2, ncol=2)
Nsig  <- 3.5
Nsamp <- 100
x <- seq(from=-Nsig*sigma.x, to=Nsig*sigma.x, length.out=Nsamp)
y <- seq(from=-Nsig*sigma.y, to=Nsig*sigma.y, length.out=Nsamp)
z <- matrix(dmvnorm(x=expand.grid(x,y), mean=c(0,0), sigma=Cov),
            nrow=length(x), ncol=length(y))
z <- z/max(z)

pdf("2D_gaussian_3Dmesh.pdf", 4, 4)
```

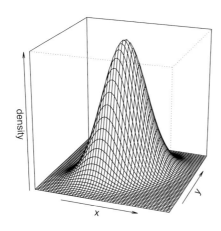
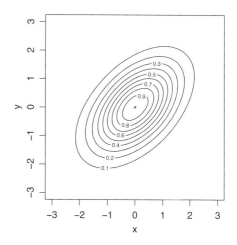

Fig. 1.9 A bivariate Gaussian with $\rho = 0.5$ and unit standard deviation in x and y, shown as a three-dimensional perspective mesh (left) and as contours of constant probability density (right), labelled with the density relative to the peak.

```
par(mfrow=c(1,1), mar=c(1,1,1,1), oma=c(0,0,0,0), mgp=c(2.2,0.8,0), cex=1.0)
persp(x=x, y=y, z=z, phi=20, theta=20, d=5, zlab="density")
dev.off()

pdf("2D_gaussian_contours.pdf", 4, 4)
par(mfrow=c(1,1), mgp=c(2.0,0.8,0), mar=c(3,3,1,1), oma=0.1*c(1,1,1,1))
contour(x, y, z, asp=1, xlim=c(-3,3), ylim=c(-3,3), xlab="x", ylab="y")
dev.off()
```

Given a two-dimensional PDF we might want to know $P(x\,|\,y\!=\!y_0)$. This is the *conditional* PDF on x given $y = y_0$ and is a one-dimensional distribution. Equation 1.7 gives this for discrete variables. For continuous variables it is

$$P(x\,|\,y\!=\!y_0) \;=\; \frac{P(x, y\!=\!y_0)}{P(y_0)}. \tag{1.74}$$

This is a slice through the two-dimensional PDF at constant y, with normalization constant $P(y_0)$. If necessary, we can work out this normalization constant numerically (see chapter 5).

We may also want to know $P(x)$ regardless of the value of y. This is the *marginal* PDF of x over y. We can think of it as viewing the two-dimensional PDF along the y-axis. From the laws of probability this is

$$P(x) \;=\; \int_{-\infty}^{+\infty} P(x, y)\, dy$$

$$ \;=\; \int_{-\infty}^{+\infty} P(x\,|\,y)P(y)\, dy \tag{1.75}$$

which is known as marginalizing over y. The corresponding marginalization for discrete variables was given by equation 1.9.

1.7 Combinations and permutations

Expressions for the number of ways of combining and arranging objects come up in many places, including – as we have seen – in the derivation of some probability distributions.

Unique pairings of objects

You have n objects. How many ways are there of selecting unique pairs (without caring about the ordering)? Think of an $n \times n$ array. Every point in the array is a pairing, except for the leading diagonal, where an object is paired with itself. The two halves on each side of the diagonal are identical (A paired with B is indistinguishable from B paired with A). So the number of ways must be $(n^2 - n)/2 = n(n - 1)/2$.

Unique orderings

How many ways can you order r objects (e.g. the three letters abc)? There are r options for the first object, $r - 1$ for the second, $r - 2$ for the third, etc. Therefore it is $r!$.

Selection with replacement, order relevant

Imagine a word written in an alphabet with n different letters. How many unique words can you make with r characters? The first character has n different possibilities, so does the second, third, etc. Therefore it is n^r. This is the number of ways of selecting r objects from a set of n *with replacement*, where the order is relevant.

Selection without replacement, order relevant

You select r unique objects from a set of n $(r \le n)$ and place them in a line. How many unique arrangements are there of this line? There are n ways of selecting the first, $n - 1$ for the second, ..., and $n - r + 1$ for selecting the rth. This is *selection without replacement*. So the total number of ways of selecting is $n(n - 1)(n - 2) \ldots (n - r + 1)$. The number of arrangements is called the *permutations* $_nP_r$, and can be written

$$_nP_r = \frac{n!}{(n - r)!} \quad \text{where} \quad n \ge 0, r \le n. \tag{1.76}$$

Note that $_nP_n = n!/0! = n!$ (as $0! = 1$ by definition). We are concerned here with the order of the objects, so an arrangement abc is distinct from bca.

Selection without replacement, order irrelevant

You now select the r objects in the previous case but you are no longer concerned about the order. The number of different unique samples you can select must be smaller than $_nP_r$ by a factor of $r!$, because this is the number of unique orderings of r objects. The number

```
                                        1
                                  1           1
                            1           2           1
                      1           3           3           1
                1           4           6           4           1
          1           5           10          10          5           1
    1           6           15          20          15          6           1
```

Fig. 1.10   The first few rows of Pascal's triangle. The nth row (counting the top one as $n = 0$) gives the values of "n choose r" for all r between 0 and n inclusive (equation 1.77). The nth diagonal (counting the first one as $n = 1$) gives the values of "n multichoose r" for all $r \geq 0$ (equation 1.78).

of ways of selecting r objects from a set of n without regard to the order of selection is called the number of *combinations* $_nC_r$, and is

$$_nC_r = \frac{_nP_r}{r!} = \binom{n}{r} = \frac{n!}{r!(n-r)!} \quad \text{where} \quad n \geq 0, \, r \leq n \qquad (1.77)$$

which we often call "n choose r". It is also called the binomial coefficient, because it occurred in the derivation of the binomial distribution (see equation 1.43). This is of course also selection without replacement. The set of values for all $r \leq n$ for a given n are the values in the rows of Pascal's triangle (figure 1.10).

Selection with replacement, order irrelevant

The number of ways of choosing r objects from n with replacement, in which we don't care about the order, is

$$\binom{n+r-1}{r} = \frac{(n+r-1)!}{r!(n-1)!} \quad \text{where} \quad n > 0, \, r \geq 0 \qquad (1.78)$$

which is sometimes called "n multichoose r". For example, the number of ways of choosing two letters from the set abc is six: aa, ab, ac, bb, bc, cc. The order doesn't matter: ba is identical to ab so is not counted separately, for example. If we choose three letters, there are ten ways. The set of values for all r for a given n are given in the diagonals of Pascal's triangle (figure 1.10), and are also known as the figurate numbers.[6]

---

[6] There are various ways to derive this. One is to think of an n-dimensional space with integer steps along each axis. The number sought is the number of points on the flat (hyper)surface that have an L^1 distance of r from the origin (counting just the positive quadrant). The L^1 distance between two points is the sum of projected distances along each axis, $\sum_i |\Delta x_i|$. That is, we can take any number of (positive) steps along each axis we like (we can select from each n as often as we like) subject to the constraint that the total number of steps (choices) is r. We don't care about the path (the order of selection). For $n = 2$ the "surface" is a straight line, and with $r = 2$ there are just three points (choices) on this line: $(2, 0)$, $(1, 1)$, $(0, 2)$. For $n = 3$ the surface is a triangle, and the number of points with L^1 distance from the origin of $r = 2$ is six. For $n = 4$ the surface is the surface of a four-dimensional triangle, etc.

The birthday problem

In a large room full of people, how many people do you have to ask before there is a 50% chance that any two or more of them share a common birthday?

As with the missiles problem it is easier to work out the probability of the complementary event. Assume that $n = 365$ birthdays are equally probable (not actually true, but it's fine for this problem). You ask r people. Let

$$A = n(n - 1)\ldots(n - r + 1) = \frac{n!}{(n - r)!} = \text{number of ways for } r \text{ people} \quad (1.79)$$

$$\text{to share no common birthday}$$

and

$$B = n^r = \text{number of ways of assigning } n \text{ birthdays to } r \text{ people.} \quad (1.80)$$

The probability of there being no common birthday among the r people is A/B, so the probability of at least one common birthday is $1 - A/B$. Setting this equal to 0.5 and solving for r is a bit complicated. But as $1 - A/B$ increases monotonically with r, and r is an integer, it's easier instead to write a program to compute the probability for a range of values of r. We find that the probability of no common birthday for $r = 22$ is 0.476 and for $r = 23$ it is 0.507. So we need to ask 23 people. This might seem low, but that is because we are interested in *any* pair of birthdays.

How many people do you instead have to ask before there is a 50% chance that at least one of them shares *your* birthday?

This will be larger because we are interested in only one specific pair of birthdays. Intuition may suggest $n/2 \simeq 183$. But this is wrong because when we ask people we sample birthdays with replacement: the tenth person we ask might have the same birthday as the third person we asked. Again we work out the complementary probability. The probability that each person we ask doesn't share your birthday is $(n - 1)/n$, so the probability that none of the r people asked share your birthday is $[(n - 1)/n]^r$. Thus the probability that at least one of these people shares your birthday is $1 - [(n - 1)/n]^r$. We can solve by taking logs

$$\log 0.5 = r \log\left(\frac{n - 1}{n}\right) \quad (1.81)$$

the solution to which is $r = 253$ (rounded up to the nearest integer). This is more than the naive $n/2$ "without replacement" estimate.

1.7.1 Using R

The factorial is provided by `factorial`. It even works on non-integers because it is implemented in R using the gamma function (equation 1.62), as $n! = \Gamma(n+1)$. In some analytic calculations you might want to use Stirling's approximation to $n!$ for large n

$$n! \simeq \sqrt{2\pi n}\, n^n e^{-n}. \quad (1.82)$$

The number of combinations $_nC_r$ is given by the function `choose(n,r)`. You can use `choose(n,r)*factorial(r)` to compute $_nP_r$. If you want to list the actual permutations and combinations, use `combinations` and `permutations` in the `gtools` package.

1.8 Random number generation

We shall see in later chapters that the practice of inference often requires us to draw random numbers from a distribution. We can do this with computer algorithms despite the fact that we use computers as deterministic machines. These algorithms create extremely long sequences of numbers which are apparently random, but which are in fact determined by a random number seed: use the same seed and you get the same sequence of numbers. Although not random, the sequence has many properties of randomness, such as having no shortscale correlations, having a distribution and a Fourier spectrum consistent with random, etc. Such sequences of numbers are *pseudo random*, meaning that they are random enough for our purposes (but often not for other applications such as cryptography).

A simple algorithm for generating pseudo random numbers is the *linear congruential generator*, which generates a sequence of integers $\{x_i\}$. Given large positive integers a, b, and m, the sequence is defined by the recurrence relation

$$x_{i+1} = \text{Mod}(ax_i + b, m) \tag{1.83}$$

where Mod is the modulus operator. This gives the remainder after integer division of the first argument by m, i.e.

$$\text{Mod}(x, m) = x - \text{IntegerPart}(x/m) \times m. \tag{1.84}$$

For example $\text{Mod}(17, 5) = 2$. This can be achieved on a digital computer by truncating the number of bits used in the storage. The algorithm creates a sequence of integers between 0 and $m - 1$ which will repeat after no more than m steps. If the values of a, b, and m are chosen sensibly, then this maximum can be achieved and each integer will appear once. The seed just determines the starting point. Of course, in any practical application, m should be much larger than the number of samples required (a typical value is 2^{64}).

How do we use such a sequence of numbers to draw from (sample) a distribution? For a uniform distribution we simply need to scale the sequence range to the range of our distribution. Provided m is large enough this will produce a set of real numbers with high enough numerical precision such that they are practically indistinguishable from numbers which really have been drawn from the distribution. When we scale to the distribution $\mathcal{U}(0, 1)$, for example, the smallest difference between any two pseudo random numbers obtained in this way is $1/m$.

For other univariate distributions we can in principle use the cumulative distribution function (CDF), which was defined in section 1.5, and ranges from 0 to 1. Suppose we want to draw from the PDF $P(x)$. We first draw a random number from $\mathcal{U}(0, 1)$. Transforming this via the inverse of the CDF gives a sample which has been drawn from $P(x)$. This can be understood in reference to figure 1.8: a uniform distribution along the vertical axis (the

CDF) will be transformed to a non-uniform one along the horizontal axis. There will be relatively few values of q far from zero – the tails of these distributions – because they are spread out over a larger range by the transformation. This method is convenient if the inverse of the CDF (the quantile function) is fast to compute, which is the case for some standard distributions. The R functions like `rnorm`, `rpois`, and `rbinom` select random numbers from standard distributions for you. You can test them by using a histogram to plot the distribution of the sample, for example as follows.

```
x <- rnorm(1e5)
hist(x, breaks=100, prob=TRUE)
xp <- seq(from=-4, to=+4, by=0.01)
lines(xp, dnorm(xp))
```

To draw random variables from a multivariate Gaussian distribution you can use `mvrnorm` in the `MASS` package, for which you need to specify the mean vector and the covariance matrix.

In practice, more sophisticated algorithms than the linear congruential generator are used in order to produce much longer sequences of pseudo random numbers. The default algorithm in R is the Mersenne-Twister algorithm, with a repeatability length of about 10^{6000}.

To sample a given set of numbers you can use the function `sample`, which will draw samples at random from a defined set, either with or without replacement. The function `set.seed` is used to define the seed. For a given seed you will always get the same sequence of numbers.

```
sample(10) # samples ten times from 1:10 without replacement
sample(x=10, size=3) # draws three samples
sample(x=10, size=3) # gives a different sequence
set.seed(100)
sample(x=10, size=3)
set.seed(100)
sample(x=10, size=3) # now gives the same sequence
sample(x=10, size=15, replace=TRUE) # samples with replacement
sample(c(-7,0,4,56,-76,128,17), size=3)
```

I advise that you set the seed whenever your data analysis procedure involves random number generation. You should modify the seed to ensure that there is no relevant sensitivity to this randomness, but recording the seed ensures repeatability and thus easier bug detection. Pay particular attention to how seeds and random number sequences are dealt with if you are doing parallel processing.

Drawing random numbers from an arbitrary PDF is difficult, because the CDF may not be easy to find or the PDF may not be easy to normalize. Drawing from multivariate PDFs is generally non-trivial. For this we can use Monte Carlo methods, which we will start to explore in chapter 8.

1.9 Change of variables

1.9.1 One-dimensional

Suppose we have a univariate PDF $P(x)$ over x, and want to express it as a PDF over y, where $y = f(x)$. Consider a small interval δx around the point x that corresponds to the small interval δy around the point y. In making the transformation we need to conserve probability,

$$P(x)\,\delta x = P(y)\,\delta y. \tag{1.85}$$

In the limit of infinitesimally small intervals this becomes

$$P(x) = P(y)\left|\frac{dy}{dx}\right|. \tag{1.86}$$

The term dy/dx is called the *Jacobian*. We take the modulus because we are interested only in the ratio of the lengths, not their signs: probabilities and probability densities are always positive (or zero).

Example

If $y = \ln x^a$ then $dy/dx = a/x$. Thus a distribution uniform in $\ln x$ has $P(x) \propto 1/x$. Indeed, for any $a \neq 0$ we see that

$$P(\ln x^a) = \text{constant} \quad \Leftrightarrow \quad P(x) \propto \frac{1}{x}. \tag{1.87}$$

Using a different base logarithm only changes the constant of proportionality.

We also see that if $P(x) \propto x^b$, then $P(\ln x) \propto x^{b+1}$. Setting $b = -1$ we see that the only power law distribution for $P(x)$ that produces a distribution uniform in $P(\ln x)$ is $P(x) \propto x^{-1}$.

1.9.2 Multi-dimensional

The change of variables generalizes to multiple dimensions (here J). We have $P(\mathbf{x})$ where $\mathbf{x} = (x_1, x_2, \ldots, x_J)^\mathsf{T}$ and want to transform this to $P(\mathbf{y})$ where $\mathbf{y} = (y_1, y_2, \ldots, y_J)^\mathsf{T}$. The J-dimensional volume element in the $\mathbf{x}$ space is $\delta x_1 \delta x_1 \ldots \delta x_J$ which I write as $\delta^J V_{\mathbf{x}}$, and the corresponding volume element in the $\mathbf{y}$ space is $\delta^J V_{\mathbf{y}}$. To conserve probability we have

$$P(\mathbf{x})\delta^J V_{\mathbf{x}} = P(\mathbf{y})\delta^J V_{\mathbf{y}} \tag{1.88}$$

and it can be shown that

$$\delta^J V_{\mathbf{y}} = \left|\frac{\partial(y_1, y_2, \ldots, y_J)}{\partial(x_1, x_2, \ldots, x_J)}\right| \delta^J V_{\mathbf{x}} \tag{1.89}$$

which involves the determinant of the Jacobian matrix, which is the $J \times J$ matrix of first partial derivatives with elements (row i, column j) $\partial y_i / \partial x_j$. Thus

$$P(\mathbf{x}) = P(\mathbf{y}) \left| \frac{\partial(y_1, y_2, \ldots, y_J)}{\partial(x_1, x_2, \ldots, x_J)} \right|. \tag{1.90}$$

One can only take the determinant of a square matrix, so the above only shows how we can transform from one J-dimensional space to another J-dimensional space. Sometimes we will want to reduce the dimensionality in a transformation (we cannot increase it). In that case we can marginalize over the unwanted variables, as is shown in the following example.

Example

Consider the transformation from two-dimensional Cartesian coordinates (x, y) to two-dimensional radial coordinates (r, θ),

$$\begin{aligned} x &= r \cos \theta \\ y &= r \sin \theta. \end{aligned} \tag{1.91}$$

We get

$$\left| \frac{\partial(x, y)}{\partial(r, \theta)} \right| = \left| \begin{array}{cc} \cos \theta & -r \sin \theta \\ \sin \theta & r \cos \theta \end{array} \right|$$

$$= r(\cos^2 \theta + \sin^2 \theta) = r. \tag{1.92}$$

We could have taken instead a geometric approach to show that $\delta x \delta y = r \delta r \delta \theta$, in which case we see from equation 1.89 that the Jacobian determinant is r. Suppose that $P(x, y)$ is a bivariate, zero mean, isotropic Gaussian. This is

$$P(x, y) = \frac{1}{2\pi\sigma^2} \exp\left[-\frac{(x^2 + y^2)}{2\sigma^2} \right] \tag{1.93}$$

because it is just the product of two independent Gaussians. Using equation 1.90 we can write the PDF in the radial coordinates as

$$P(r, \theta) = \frac{r}{2\pi\sigma^2} \exp\left[-\frac{r^2}{2\sigma^2} \right]. \tag{1.94}$$

To get the marginal PDF of r we integrate over θ

$$P(r) = \int_0^{2\pi} P(r, \theta) \, d\theta = \frac{r}{\sigma^2} \exp\left[-\frac{r^2}{2\sigma^2} \right]. \tag{1.95}$$

The marginal PDF of θ is of course just $1/2\pi$ (the distribution is isotropic so must be uniform in θ).

In this example the maximum of $P(x, y)$ is at $(x, y) = (0, 0)$, which corresponds to $r = 0$. But the maximum of $P(r, \theta)$ is at $r = \sigma$ for all θ, as you can verify by differentiation or just by plotting. This difference arises because we are looking at *density* functions, which here are probability per unit area. The density function in Cartesian coordinates (equation 1.93) varies as $\exp(-r^2)$. When we express this in radial coordinates the Jacobian stretches the area elements in proportion to r. The combination of these two dependencies is to give a maximum of the density function in radial coordinates at $r > 0$.

It follows from equation 1.86 that probability density functions are generally not invariant under transformations of their variables. This is because the Jacobian is usually a function of the variable too. Thus statistics of the density function, such as the mean or maximum, are in general also non-invariant. An exception arises when the transformation is strictly monotonic, which ensures that the transformation is invertible (one-to-one). It then follows from the definition of the cumulative density function that the quantiles (of univariate density functions), such as the median, are invariant.

1.10 The three doors problem revisited

I opened this chapter with the classic three doors problem. While a more formal analysis is not necessary for understanding the problem, writing down formulae for the relevant probabilities helps us to solve more general problems, such as one with multiple doors, multiple cars, or a different behaviour of the game show host.

The problem can be formulated probabilistically in a number of ways (and some are less useful than others). I define door 1 as the one we choose, and door 2 as the one which is opened by the game show host (to always show a goat). Let W be the proposition *the car is behind door 1*, and C be the proposition *we select the car when changing doors (from 1 to 3)*. From the rules of probability we can write

$$P(C) = P(C|W)P(W) + P(C|W')P(W') \tag{1.96}$$

where W' is the complement of W, i.e. the car is not behind door 1. For the original problem as described, $P(W) = 1/3$, $P(W') = 2/3$, $P(C|W) = 0$, $P(C|W') = 1$. Thus $P(C) = 2/3$.

This approach generalizes easily to the case of n doors and the game show host opening k doors to reveal as many goats ($0 \le k \le n - 2$). There is still one car. What is the probability of winning if we switch to another closed door at random? C is now the proposition *we select the car when changing doors (from 1 to one specific closed door)*. In equation 1.96 we now have $P(W) = 1/n$, $P(W') = (n - 1)/n$, $P(C|W) = 0$, $P(C|W') = 1/(n - k - 1)$. Thus

$$P(C) = \frac{1}{n - k - 1} \frac{n - 1}{n}. \tag{1.97}$$

If $k = n - 2$ (all doors but no. 1 and another are opened), $P(C) = (n - 1)/n$. This gives $2/3$ for the original problem ($n = 3$). With $k \ge 1$ then $(n - 1)/(n - k - 1) > 1$, so $P(C) > 1/n$. The probability of wining is increased over the initial probability (which is $1/n$) whenever one or more doors are opened. Hence we should always switch doors.

## 2

Estimation and uncertainty

In this chapter we look at estimators and how to use them to characterize a distribution. Relevant to this are the concepts of estimator bias, consistency, and efficiency. I shall discuss measurement models and measurement uncertainty and note the difference between distribution properties and estimates thereof. We will learn about the central limit theorem, how and when we can reduce errors through repeated measurements, and we will see how we can propagate uncertainties.

2.1 Estimators

An estimator is something that characterizes a set of data. This is often done in order to characterize the *parent distribution*, the distribution the data were drawn from. We often distinguish between a point estimator, which is a single value such as the mean or mode, and an interval estimator, which characterizes a range, such as the standard deviation or interquartile range. What we use as an estimator depends on what we want to characterise and what we know about the parent distribution.

Suppose we want to learn about the distribution of the heights of a particular species of tree, given a sample $\{x\}$ of the heights of N such trees in a forest. Specifically, we would like to estimate the mean of the parent distribution, which here means the set of all trees in the forest at this time (as this is finite it is sometimes called the *parent population*). How useful any estimate is depends on how the sample of trees was selected: were they taken from across the entire forest rather than in a particularly shady region? Were the tallest trees missed because they were harder to measure (or have already been cut down)? But let us assume that we have a sample that is representative of the forest in some useful sense. There are potentially many ways we could use this sample to estimate the mean of the parent distribution. We could

(1) sum all x and divide by N (arithmetic mean)
(2) sum the first K measurements ($K < N$) and divide by K
(3) sum all x and divide by $N - 1$
(4) average the smallest and largest values of x (midrange)
(5) sort all x and take the middle value; average the two middle ones if N is even (median)
(6) bin the data and take the most frequent (mode)
(7) disregard the data and report the value as 8 m (justified by a model for tree growth, perhaps).

All of these are estimators. But not all of them are good estimators. Ideally we want estimators that are *unbiased*, *consistent*, and *efficient*.

An estimator is said to be *unbiased* if the expectation value of the estimator is equal to the true value. As the name suggests, the expectation value is what we expect the result to be if we had the entire parent population and no uncertainty in the measurements (but see the formal definition in section 1.3). We don't demand that the estimator give the true value; that is generally unachievable. But an unbiased estimate is equal to the true value in some average sense: it is neither systematically too high nor too low. Sometimes our samples are not drawn from a discrete population, as they are in the forest example, but are measurements of a continuous distribution, such as when we measure the temperature in different parts of a lake. Here the parent population is effectively infinite in size, and the expectation value is given by an integral (equation 1.21).

An estimator is *consistent* if its value is expected to converge on the true value in the limit of a large amount of data. This is desirable because it means that as we gather more data we get an estimate closer to the truth.

Efficiency measures the variance of an estimator and is a relative term. The smaller the variance the more efficient it is.

Whether a particular estimator is unbiased and consistent, and how efficient it is, depends on the parent distribution. If the parent distribution is Gaussian, then of the estimators listed above, numbers 1, 4, and 5 are unbiased. Number 2 will be unbiased if the ordering is random. Number 3 is biased because dividing by $N - 1$ overestimates the mean. Number 7 is biased, unless it happens to be the true value. Whether number 6 is biased or not depends on how we binned the data. Estimator 1 is consistent. So is 3, because there is no difference between N and $N - 1$ in the limit $N \to \infty$. As the mean, median, and mode are all equal for a Gaussian, estimators 5 and 6 are consistent (provided we used ever narrower bins for the mode calculation), as is number 4. Estimators 2 and 7, on the other hand, are inconsistent because they do not change as we gather more data. Note that an estimator could be unbiased but inconsistent (number 2) or it could be biased but consistent (number 3). Estimator 1 has a smaller variance than estimator 2 because it is using more of the data. Number 4 likewise has a large variance for a Gaussian distribution. Note that despite its otherwise dubious properties, estimator 7 has the smallest variance: zero.

Sometimes the most efficient estimator is biased, or an unbiased estimator may not be very efficient. Which is the "best" estimator depends on your goals and on the parent distribution. As we shall see in section 4.8 (where we shall also define bias precisely) the error of an estimator can be written in terms of its bias and its variance.

2.2 Noise, measurement errors, and measurement models

Uncertainty arises in a number of ways when dealing with data. First, measurements cannot be made with arbitrary precision. When measuring the length of something with a ruler, the precision is limited to some fraction of the difference between two neighbouring tick marks. It is further limited by how well these tick marks were machined in the first place. Second, there are invariably sources of interference – noise – when measuring. If we are trying to weigh a very light object by hanging it from a spring, air currents and mechanical vibrations will perturb the spring, preventing us from measuring a constant value. Third, many real-world processes are stochastic. For example, the number of photons received from a source in a given time interval is not deterministic, but is described by a Poisson distribution (section 1.4.2). In two different equal-length time intervals we will not necessarily receive the same number of photons. This is an intrinsic uncertainty which is unrelated to the measurement process.

All of this means that what we measure is not the true value of something, but rather an estimate of it. A useful way to think of this is to consider that the measurement process has a probability distribution, which we call the *measurement model*. This gives the probability of measuring some particular value of the data, given some parameters of the process and the measurement procedure. With a Poisson process, for example, if the expected number of photons received in some time interval is λ, then the measurement model is the Poisson distribution $P(r \,|\, \lambda)$. This gives the probability of measuring r when the expected (but unknown) value is λ. The measurement model is also called a *noise model*.

Measurement models are fundamental to inference and are usually described with the likelihood, a concept I will introduce in the next chapter.

A common measurement model is the Gaussian distribution (section 1.4.5). It is described by two parameters, the mean and the standard deviation. If we are given a measurement a and told it has a Gaussian uncertainty $\pm b$, we interpret this to mean that the measurement has a Gaussian distribution with mean a and standard deviation b. How should we understand this, given that we only have one measurement? From a frequency perspective it means that if we could somehow make an infinite number of (noisy) measurements, then we expect these measurements to show this distribution, which would have (for example) 68% of the measurements lying in the range $(a - b, a + b)$. But as we only made the measurement once, what we really mean is that if the true mean is a then there is a probability of 0.68 that a measurement will lie in the range $(a - b, a + b)$.

Note that we have done something subtle but important in this interpretation. We have taken the measurement a as an estimator of the mean of the signal and then used this to make a statement about the measurement process. That is, we have assumed that the measurements are described by a Gaussian, the standard deviation of which is known, and have then estimated its mean from this single measurement. You would not do this if you had good reason to believe that a is not a good estimator, e.g. if you think it is an outlier. In general, finding the mean – or indeed any other parameter of a distribution (such as the standard deviation) – is a process of inference: we use a set of data to make estimates of the values of the parameters. We will look at how to do this in some detail in later chapters, for example in section 3.5, as well as in section 4.4.5 for a Poisson distribution and in section 6.2 for a Gaussian distribution. In fact, the idea that we cannot measure parameters directly but can only infer them from data will appear frequently in later chapters.

The important point for now is that a measurement plus error bar can be thought of as summarizing a probability distribution that describes the measurement process. This works because we often adopt standard probability distributions as measurement models, and many of these standard distributions, like the Gaussian, can be summarized by one point estimate and one width parameter.

2.3 The central limit theorem and $\sqrt{N}$ reduction from repeated measurements

Suppose we have N independent random variables $\{x_i\}$ for $i = 1, 2, \ldots N$, each drawn from a distribution with finite mean μ_i and finite variance V_i. Let

$$y = \sum_i x_i. \tag{2.1}$$

It follows from the definition of expectation (section 1.3) that

$$E[y] = \sum_i \mu_i \tag{2.2}$$

$$\mathrm{Var}(y) = \sum_i V_i \tag{2.3}$$

i.e. the expected mean is the sum of the means and the (expected) variance is the sum of the variances. It can further be proven that in the limit of very large N the distribution of y becomes a Gaussian. This is the *central limit theorem* (CLT). It essentially says that if we add together lots of independent variables, their distribution becomes a Gaussian with mean and variance as specified above.[1] If the variables were not independent then the expression for $\mathrm{Var}(y)$ would include additional terms to account for the covariance (section 1.6.1).

The central limit theorem is of great practical value because it means that if a measured quantity is affected by many independent noise sources, i.e. it is the sum of lots of independent random variables, then the measurement can be described with a Gaussian distribution. This simplifies many analyses and explains why the Gaussian distribution is so widespread. Note that there is no requirement that the original distributions be Gaussian or even identical. They just have to be independent and to have finite means and variances.

If we now define z as the mean of the N independent variables ($z = y/N$), it follows directly from the above that

$$E[z] = \frac{1}{N} \sum_{i=1}^{N} \mu_i \tag{2.4}$$

$$\mathrm{Var}(z) = \frac{1}{N^2} \sum_{i=1}^{N} V_i \quad \text{or} \quad \sigma_z = \frac{1}{N} \sqrt{\sum_{i=1}^{N} \sigma_i^2} \tag{2.5}$$

where $\sigma_z = \sqrt{\mathrm{Var}(z)}$ is the standard deviation in z and $\sigma_i = \sqrt{V_i}$. Why do we have

---

[1] One could argue that in the limit $N \to \infty$ the mean and variance of y would become infinite and so it makes little sense to talk of a distribution at all. But we need not be so pedantic when we realise that in practice the central limit theorem is a very good approximation even for relatively small N, as we will see in section 2.3.1.

a $1/N^2$ factor rather than a $1/N$ factor in the variance of z? It follows directly from the definition of variance:

$$\begin{aligned}
\mathrm{Var}(z) &= E[(z - E[z])^2] \\
&= E\left[\left(\frac{y}{N} - E\left[\frac{y}{N}\right]\right)^2\right] \\
&= \frac{1}{N^2} E[(y - E[y])^2] \\
&= \frac{1}{N^2} \mathrm{Var}(y) \\
&= \frac{1}{N^2} \sum_{i=1}^{N} V_i.
\end{aligned}$$
(2.6)

If all of the $\{x_i\}$ come from a common distribution with mean μ and variance σ^2, then setting $\mu_i = \mu$ and $\sigma_i = \sigma$ for all i, it follows that

$$E[z] = \frac{1}{N} N\mu = \mu$$
(2.7)

$$\mathrm{Var}(z) = \frac{1}{N^2} N\sigma^2 = \frac{\sigma^2}{N} \quad \text{or} \quad \sigma_z = \frac{\sigma}{\sqrt{N}}.$$
(2.8)

This result is of great practical relevance. It says that if we make N independent measurements of a quantity, each of which has the same uncertainty (standard deviation) of σ, then if we estimate the quantity by the mean of our measurements $\bar{x}$, this has an uncertainty (standard deviation) of $\sigma/\sqrt{N}$. In other words, repeated measurements reduce uncertainty. This fact underpins much of the scientific methodology for improving the precision of measurements.

2.3.1 Demonstration of the central limit theorem

We can demonstrate the central limit theorem using real numbers drawn from a uniform distribution $\mathcal{U}(0, 10)$ (this is very non-Gaussian!). We draw 10^4 numbers and plot their distribution using a histogram with the following.

```
hist(runif(1e4)*10)
```

This distribution has an expected mean of 5 and expected standard deviation of $(10 - 0)/\sqrt{12} = 2.89$ (see section 1.4.4). If the central limit theorem applies, then the distribution of the average of N numbers drawn from this should tend towards a Gaussian as N gets large, with mean 5 and standard deviation $2.89/\sqrt{N}$.

To discover the distribution empirically we draw N numbers from this distribution at random and calculate their mean. We repeat this many times and plot their distribution as a histogram. This is done by the following R code for $N = 1, 2, 5$ with the results shown in figure 2.1. As N gets larger, the distribution looks more like a Gaussian, the mean gets nearer to 5, and the standard deviation gets smaller. We see from the plot that the approximation is quite good even for small N. The computed standard deviations of the three distributions are 2.859, 2.050, 1.299 for $N = 1, 2, 5$ respectively. This compares

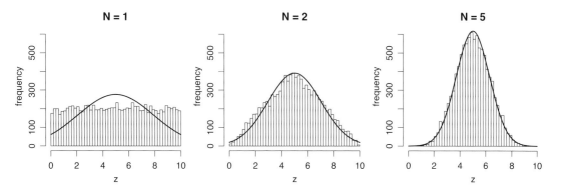

Fig. 2.1 Each histogram is the distribution of the mean of N numbers drawn from the uniform distribution $\mathcal{U}(0, 10)$. The smooth line is the Gaussian limit for the distribution expected from the central limit theorem.

to the standard deviation from the central limit theorem of 2.887, 2.041, 1.291 respectively. Note that because dnorm returns a normalized Gaussian (the area under the curve is unity), but I have plotted the data as a frequency, I have to scale the former to have the same total area as the histogram. This area is equal to the number of data points (Nsamp) multiplied by the histogram bin width (binwidth), which is 1000 in the example above. Alternatively I could have used used the option freq=FALSE in hist to plot the histogram as a probability density.

R file: CLT_uniform.R

```
##### Apply central limit theorem to draws from uniform distribution

pdf("CLT_uniform.pdf", 12, 4)
par(mfrow=c(1,3), mgp=c(2,0.8,0), mar=c(3.5,3.5,1,0), oma=0.1*c(1,1,5,5),
    cex=1.2)
set.seed(200)
for(N in c(1,2,5)) {
  Nsamp <- 10000
  z <- numeric(Nsamp)
  for (i in 1:Nsamp){
    z[i] <- mean(runif(N)*10)
  }
  binwidth <- 0.2
  hist(z, xlim=c(0,10), ylim=c(0,640), breaks=seq(0,10,binwidth),
       ylab="frequency", main=paste("N =",N))
  cat("mean, sd = ", mean(z), sd(z), "\n")
  # overplot Gaussian with mean and sd from CLT
  gsd <- (10/sqrt(12))*1/sqrt(N)
  x <- seq(0,10,0.01)
  y <- dnorm(x, mean=5, sd=gsd)*Nsamp*binwidth
  lines(x, y, lwd=2)
}
dev.off()
```

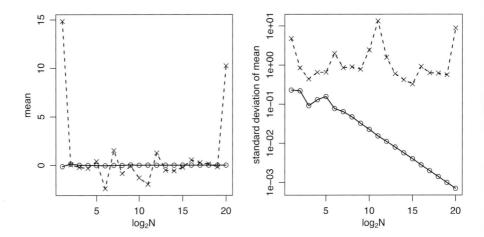

Fig. 2.2 The variation of the mean (left) and standard deviation of the mean (right) for N variables drawn from a Cauchy distribution (crosses and dashed line) and a Gaussian distribution (circles and solid line). Note the log scales.

2.3.2 A counter-example: estimating the mean of the Cauchy distribution

The central limit theorem is valid only if its assumptions are met. These are that the variables are independent and drawn from a distribution (or distributions) with finite mean and variance. Here is an example where not all of these assumptions are met.

We would like to estimate the location parameter a of the Cauchy distribution (section 1.4.7). The PDF is

$$P(x) = \frac{b}{\pi[b^2 + (x-a)^2]}. \tag{2.9}$$

The following R script draws N numbers at random, and independently, from a Cauchy distribution with $a = 0$ and unit half-width at half-maximum (HWHM), $b = 1$. As the distribution is symmetric we might expect the mean of these samples to be a good estimator of a. We investigate this by plotting this mean, as well as the standard deviation of this mean, as a function of N. These are shown in the left and right panels, respectively, of figure 2.2 using dashed lines and crosses. I calculate the standard deviation of the mean according to the central limit theorem, which tells us it is $\sigma/\sqrt{N}$ (equation 2.8), where σ is the sample standard deviation (computed using sd in R). For comparison I compute and plot the same quantities for samples drawn from a Gaussian distribution with the same mode and HWHM (plotted as solid lines and circles).

We see that the Gaussian distribution does what we expect from the central limit theorem. The mean converges (very quickly) towards its true value, and the standard deviation of the mean (also called the standard error in the mean) decreases[2] as $1/\sqrt{N}$. For the

---

[2] The right panel plots the log base 10 of the standard deviation of the mean σ_{mean} against $\log_2 N$. According

Cauchy distribution the mean is also close to zero, but it oscillates a lot. This is because the sample standard deviation does not decrease with increasing N, as can be seen in the right panel.

The reason for this behaviour of the Cauchy-drawn samples is that the central limit theorem does not apply to the Cauchy distribution. This is because the central limit theorem assumes finite values of the mean and variance of the distribution from which the data are drawn. Recall that the definition of the variance of a distribution $P(x)$ is (equation 1.30)

$$\mathrm{Var}(x) = E[x^2] - E[x]^2$$
$$= \int x^2 P(x)\,dx - \left(\int x\,P(x)\,dx\right)^2. \tag{2.10}$$

Thus if $P(x)$ does not drop off faster than $1/x^2$ with increasing $|x|$, the first interval will not converge. The Cauchy distribution drops off exactly as $1/x^2$ for large $|x|$, so its variance is infinite. In fact the mean – the second integral above – is not defined (and neither are the higher central moments) so the variance is not even defined. This is why we are careful to refer to a as the location parameter of the Cauchy distribution, and not as the mean.

Given a set of data drawn from a Cauchy distribution, we can better estimate the location parameter a and the HWHM b using the median and inter-quartile range, respectively.

R file: `cauchy_mean_estimation.R`

```
##### Is the mean a convergent estimator for the location of a Cauchy?

# Calculate mean and CLT prediction of its standard deviation, of a sample
# of numbers drawn from (i) Cauchy, (ii) Gaussian, both with mode=0, HWHM=1,
# with 2^lognsamp[i] no. of samples. Plot these against lognsamp[i]

set.seed(100)
lognsamp <- 1:20
GaussMean  <- vector(length=length(lognsamp))
CauchyMean <- vector(length=length(lognsamp))
GaussSD    <- vector(length=length(lognsamp))
CauchySD   <- vector(length=length(lognsamp))
for(i in 1:length(lognsamp)) {
  s <- rnorm(2^lognsamp[i], mean=0, sd=1/(2*log(2))) # sd=HWHM/2ln2=0.721
  GaussMean[i]  <- mean(s)
  GaussSD[i]    <- sd(s)/sqrt(length(s)) # standard deviation in mean
  s <- rcauchy(2^lognsamp[i], location=0, scale=1) # scale=HWHM
  CauchyMean[i] <- mean(s)
  CauchySD[i]   <- sd(s)/sqrt(length(s)) # standard deviation in mean
}

pdf("cauchy_mean_estimation.pdf", 8, 4)
par(mfrow=c(1,2), mgp=c(2.0,0.8,0), mar=c(3.5,3.5,1,1), oma=0.1*c(1,1,1,1))
plot( lognsamp, GaussMean, ylim=range(c(GaussMean, CauchyMean)), type="n",
  xlab=expression(paste(log[2], N)), ylab="mean")
lines( lognsamp, GaussMean,  lwd=1.5)
```

to the central limit theorem $\sigma_{\mathrm{mean}} \propto 1/\sqrt{N}$. Therefore $\log \sigma_{\mathrm{mean}} \propto -(\log_2 N)/(2\log_2 10)$, i.e. the variation should be linear with a gradient of -0.151. Ignoring the few wiggles at low N due to small number statistics, you can confirm that this is indeed the gradient of the solid line.

```
points(lognsamp, GaussMean)
lines( lognsamp, CauchyMean, lwd=1.5, lty=2)
points(lognsamp, CauchyMean, pch=4)
plot(  lognsamp, GaussMean, ylim=range(c(GaussSD, CauchySD)), log="y",
          type="n", xlab=expression(paste(log[2], N)),
          ylab="standard deviation of mean")
lines( lognsamp, GaussSD,  lwd=1.5)
points(lognsamp, GaussSD)
lines( lognsamp, CauchySD, lwd=1.5, lty=2)
points(lognsamp, CauchySD, pch=4)
dev.off()
cbind(lognsamp, GaussSD, CauchySD)
```

2.4 Population properties vs sample estimates

It is important to understand the distinction between the variance of a distribution and the variance of an estimate of a property of that distribution, such as the mean. Assume that we have a set of N quantities $\{x_i\}$ drawn from a distribution with unknown mean and standard deviation. As already introduced in section 1.3, we can *estimate* these from the data by computing

$$\overline{x} = \frac{1}{N} \sum_{i=1}^{N} x_i \tag{2.11}$$

$$\hat{\sigma} = \sqrt{\frac{1}{N-1} \sum_{i=1}^{N} (x_i - \overline{x})^2}. \tag{2.12}$$

These are the *sample mean* and *sample standard deviation* respectively.[3] Note that the order of magnitude of the sample mean and standard deviation do not change as N increases: in equation 2.12 $(x_i - \overline{x})$ will have some typical size, call it δx, which is independent of N, so $\hat{\sigma} \sim \delta x \sqrt{N/(N-1)} \rightarrow \delta x$ in the limit $N \rightarrow \infty$. This is of course what we want, namely a *consistent* estimator.

According to the central limit theorem, if we draw N numbers from a distribution which has a population (true) standard deviation σ, and average these to get the sample mean $\overline{x}$, then this estimate of the mean has a standard deviation $\sigma/\sqrt{N}$ (equation 2.8).[4] As we don't usually know σ in advance, we estimate it from the data with $\hat{\sigma}$ (equation 2.12). In that case our estimate of the uncertainty in $\overline{x}$ is $\hat{\sigma}/\sqrt{N}$, which is the *standard deviation in the mean*

$$\text{SEM} = \frac{\hat{\sigma}}{\sqrt{N}} = \sqrt{\frac{1}{N(N-1)} \sum_{i=1}^{N} (x_i - \overline{x})^2}. \tag{2.13}$$

---

[3] As noted previously, the presence of $N - 1$ rather than N in the second equation can be understood as correcting for the sample bias.

[4] You can equally well replace the statement "draw N numbers from a distribution" with "make N measurements".

This is sometimes called the *standard error in the mean* instead. It is the precision of our estimate $\bar{x}$, and it clearly decreases as N gets larger. That is, our *inference* of the value of $\bar{x}$ gets more precise as we take more data. We will see in section 6.2 that this result is a direct consequence of the inference process when the data have been drawn from a Gaussian.

Naturally our estimate of the standard deviation (equation 2.12) also gets more precise as we take more data, and it can be shown that the variance in $\hat{\sigma}$ is approximately $\hat{\sigma}^2/[2(N-1)]$, i.e. its standard deviation is approximately $\hat{\sigma}/\sqrt{2(N-1)}$. The idea of a standard deviation in a standard deviation may seem strange at first, but it's perfectly okay to ask what the standard deviation ("uncertainty") is in any quantity estimated from data.

Some books, publications, and computer programs do not distinguish carefully between "sample standard deviation" and "standard deviation". Some just define the latter using $1/N$ and then either apply $N/(N-1)$ corrections to reach the former or just say N is large so it doesn't matter. You have been warned! The R function sd(x) gives the *sample* standard deviation of the vector of values x (equation 2.12), which of course must estimate the mean from the data given.

2.5 The mean is not necessarily the most efficient estimator

I stated in section 2.1 that a good estimator is one that is unbiased, consistent, and efficient. Achieving all of these simultaneously can be impossible in practice, and there is no universal estimator – such as the mean, median, or mode for characterizing the central value of a distribution – which one can single out as always being "optimal". What makes a good estimator depends on the probability density function the data were drawn from, and often this is itself unknown and must be approximated from the data.

The central limit theorem tells us that by averaging N independent measurements of a quantity we can reduce its standard deviation by a factor of $\sqrt{N}$ compared to a single measurement (provided its assumptions are met). However, the central limit theorem does not claim that the average is the most efficient estimator, and sometimes it is not, as we will now see.

Consider the uniform distribution $\mathcal{U}(-1,+1)$. This has an expectation value $E[x]$ of zero, and a standard deviation σ of $1/\sqrt{3}$ (see section 1.4.4). We draw a sample $\{x\}$ of size N from this distribution and wish to use these to estimate $E[x]$. An obvious estimator is the mean of $\{x\}$. An alternative is the midrange, which is the average of the highest and lowest values in $\{x\}$. In the limit of infinite data we expect them both to be zero, because they are both consistent estimators. To compare their efficiencies we examine how their standard deviations vary with N. We proceed as follows.

Draw a sample of size N from the distribution and compute the two estimators, the mean and midrange. Repeat this K times (I use $K = 1000$) and compute the mean and standard

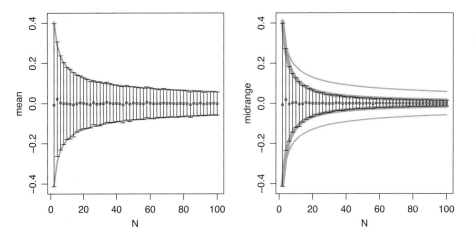

Fig. 2.3 The left panel shows the mean (circles) and standard deviation of the mean (error bars) for the mean estimator, $(1/N)\sum x_i$, of a sample of size N drawn from the uniform distribution $\mathcal{U}(-1, +1)$. The right panel shows the same but for the midrange estimator, $(x_{\max} + x_{\min})/2$. The thin grey lines (identical in both panels) show the expected standard deviation of the mean estimator, $\sigma/\sqrt{N}$ (where $\sigma = 1/\sqrt{3}$ is the standard deviation of the uniform distribution). The thick grey lines in the right panel – which more or less overlap the ends of the error bars – show the expected standard deviation of the midrange estimator, $\sigma\sqrt{6/(N+2)(N+1)}$.

deviation of the two estimators over these K sets.[5] We then plot this as a point and an error bar for each estimator at this value of N, as is done in figure 2.3. We then repeat this for a range of different values of N. R code to do this is given below.

We see in figure 2.3 how the standard deviation of the midrange estimator (right panel) drops more rapidly with N than the standard deviation of the mean estimator (left panel). It can be shown that the standard deviation of the midrange estimator (for the uniform distribution) varies as $\sigma\sqrt{6/(N+2)(N+1)}$, which for large N is approximately $\sqrt{6}\,\sigma/N$. This drops with increasing N considerably faster than the standard deviation of the mean, which is $\sigma/\sqrt{N}$.

The point of this example is not to show that the central limit theorem is wrong. The central limit theorem applies to the uniform distribution, but it does not apply to the midrange estimator because it is not an average of N estimators. The point of this example is instead to show that there is nothing very special about the mean, and that estimators can have an efficiency that improves faster than $1/\sqrt{N}$. The properties that estimators have depend on the underlying distribution.

I use the function `Vectorize` in the following R code to generate an in-place version of `runif` which can accept a vector argument. I use this to generate nsamp ($= K$) vectors of

---

[5] If we used only one set for each N we would hardly see the effect we want to demonstrate, because the estimators are very noisy for small N.

size n (= N) in each cycle of the loop, which is stored in the matrix samp. The function apply then computes the mean and midrange (over n) for each of the nsamp vectors.

R file: midrange_estimator.R

```
##### Show that midrange estimator of the uniform distribution has a
##### standard error which drops as ~1/N. Cf. mean which drops as 1/sqrt(N).

library(gplots) # for plotCI
midrange <- function(x) {0.5*sum(range(x))}
set.seed(150)

nmax  <- 1e2
nsamp <- 1e3
sigma <- 2/sqrt(12) # standard deviation of U(-1,+1)
nVec  <- seq(from=2, to=nmax, by=2)
est <- matrix(NA, nrow=nmax, ncol=4) # mean(mu), mean(mr), sd(mean), sd(mr)
for(n in nVec) { # samp is a n*nsamp matrix
  samp <- Vectorize(runif, "n")(n=rep(n, nsamp), min=-1, max=1)
  mu <- apply(samp, 2, mean)      # vector size nsamp, mean estimator
  mr <- apply(samp, 2, midrange) # vector size nsamp, midrange estimator
  est[n,1] <- mean(mu)
  est[n,2] <- sd(mu)
  est[n,3] <- mean(mr)
  est[n,4] <- sd(mr)
}

pdf("midrange_estimator.pdf", 8, 4)
par(mfrow=c(1,2), mgp=c(2.0,0.8,0), mar=c(3.5,3.5,1,1), oma=0.1*c(1,1,1,1))
plot(nVec, est[nVec,1], type="n", ylim=c(-0.45, 0.45), yaxs="i",
     xlab="N", ylab="mean")
lines(nVec,  sigma*1/sqrt(nVec), col="grey60", lw=2.5)
lines(nVec, -sigma*1/sqrt(nVec), col="grey60", lw=2.5)
plotCI(nVec, est[nVec,1], uiw=est[nVec,2], gap=0, cex=0.5, add=TRUE)
plot(nVec, est[nVec,3], type="n", ylim=c(-0.45, 0.45), yaxs="i",
     xlab="N", ylab="midrange")
lines(nVec,  sigma*1/sqrt(nVec), col="grey60", lw=2)
lines(nVec, -sigma*1/sqrt(nVec), col="grey60", lw=2)
lines(nVec,  sigma*sqrt(6/((nVec+2)*(nVec+1))), col="grey60", lw=6)
lines(nVec, -sigma*sqrt(6/((nVec+2)*(nVec+1))), col="grey60", lw=6)
plotCI(nVec, est[nVec,3], uiw=est[nVec,4], gap=0, cex=0.5, add=TRUE)
dev.off()
```

2.6 Outliers and robust statistics

Real data are often affected by outliers. These are measurements that do not follow an assumed noise model. Consider taking an image of an empty region of the night sky with a digital camera. Our simple model might be that the sky is uniform across the image, with small variations about a constant value due to photon statistics (Poisson noise) and various sources of detector and electronics noise (Gaussian). We could bundle all of these into a Gaussian noise model, perhaps justified by the central limit theorem on the grounds that combining lots of independent random effects will be approximately Gaussian. One could calculate a mean and standard deviation, then plot a Gaussian over a suitable histogram of the pixel measurements. The agreement may be quite good overall, but chances are that there are some pixels in this image that have very deviant values: extremely high, zero, or even negative. These *outliers* can come about for a large number of reasons, and depend on the physical set up and the data processing that has been done. In this example, origins of outliers could be: stars in the field; scattered light in the camera; cosmic rays hitting the detector; broken pixels; processing errors. The values involved could be so large that they would distort the calculation of the mean, standard deviation, or other statistics. The reason that they are outliers – and so need special treatment – is that they do not originate from the specified noise model.

Broadly speaking there are two things we can do with outliers: remove them or model them. Physical knowledge can be used to remove some outliers, for example if we know that flux values cannot be negative. But one must be careful about this, as some processing of the data may mean that such values are in fact acceptable. For example, if a so-called dark image has been subtracted from the sky image (done because the detector gives a signal even when not exposed to light), then because the dark image itself will contain noise we will now have negative values in the dark-subtracted sky image.

The effect of outliers can can be ameliorated by adopting a more robust statistic. For example, rather than using the mean, which weights all of the data equally, the median can be used to estimate the mean for distributions where they are theoretically equal. One could consider using a weighted mean (section 2.8.7), but this requires reliable uncertainty estimates, whereas the very nature of outliers normally means that these are absent. Instead of using the standard deviation, which is a squared quantity and thus strongly dependent on deviations from the mean, one could calculate the mean absolute deviation, the full-width at half-maximum, or the interquartile range, and then scale this to what the standard deviation would be for a theoretical distribution. The literature is full of suggestions of such robust estimators.

Another approach is to adopt an iterative approach to clip outliers. For example, to estimate the mean and standard deviation of a set of data that we believe is Gaussian but has outliers we can

(1) calculate the median and scaled interquartile range as initial estimates of the mean $\bar{x}$ and standard deviation $\hat{\sigma}$, respectively;
(2) using all of the data, remove ("clip") those values that lie more than $n\hat{\sigma}$ from $\bar{x}$, where n depends on how much of the data you expect to be outliers, but is typically 3–5;
(3) calculate the mean $\bar{x}$ and standard deviation $\hat{\sigma}$ from the remaining data;
(4) iterate steps 2 and 3 until convergence is reached, i.e. no more points are clipped.

We use the median instead of the mean in the first iteration because this provides a more robust estimation when the distribution is skewed by the outliers. It corresponds to an extreme clipping, as the median clips all points except the central one.

There are many ad hoc approaches to robust statistics and dealing with outliers, many of which need to be tailored to specific applications and which take into account what we know about potential outliers. In section 9.3 we will see how we can use a mixture model to simultaneously identify and model outliers when fitting a model to data without having to clip them.

2.7 Errors, accuracy, and precision

Even within statistics the term *error* does not have a rigorous definition, as its broad usage in different books and publications makes clear.

In some contexts "error" might mean the difference between the measured value (or the inferred value) and the true value. This is a useful concept, although as the true value is usually unknown (unless we are doing simulations) this error cannot usually be computed.

A closely related term is *residual*, which usually refers to the difference between a measured value and a model prediction of this value (neither of which is usually equal to the true value). However, the word "error" is often use synonymously with "residual".

The term "error" is often used in a more colloquial sense to mean the uncertainty in a measurement; it is frequently used to indicate the standard deviation of the measurement model, as discussed in section 2.2. One often distinguishes between random errors and systematic errors.

Random errors refer to differences between measurements or potential measurements of the same thing. Such variations arise from the fact that the measurement process does not always deliver exactly the same value (the probability distribution describing the measurements has a non-zero width), for reasons discussed in section 2.2. We normally think of random errors as having zero mean, in which case repeated, independent measurements often allow us to reduce the uncertainty in the thing we are measuring in accordance with the central limit theorem (section 2.3).

Systematic errors, in contrast, mean that measurements are consistently offset (biased) with respect to the true value. The expectation value of the difference between the measured and true value is then not zero: no matter how many measurements we average we will not beat down systematic errors. If a weighing scale reads 5 mg when there is nothing on it, then all measurements will be systematically wrong by 5 mg. We can sometimes detect and remove systematic errors in the experimental set up by measuring objects for which we know the true values and calibrating accordingly (here by adjusting the zero point of the weighing scales).

If x' is the true, unknown value for something, then an example of a measurement model which includes both random and systematic errors is $\hat{x} = x' + \mathcal{N}(\Delta x, \sigma_x)$, where $\hat{x}$ indicates our measurement, or an estimate obtained through some analysis. We can think of σ_x as the standard deviation of the random errors and Δx as the systematic error. If we know this measurement model we can estimate σ_x from a set of measurements of $\hat{x}$ (e.g. by computing their sample standard deviation). But we cannot infer Δx, because it is degenerate with the unknown value x'. Of course, if we were told Δx then we could remove it by subtracting it from all the measurements. But we could never remove the actual random error from each measurement, because it is stochastic and remains unknown for each measurement.

Many scientists make the following distinction: a *precise* measurement is one which has a small random error and so has a small variance. An *accurate* measurement is one which has a small systematic error and so a small bias.

2.8 Propagation and combination of uncertainties

We are often given, or have determined, the uncertainty in one variable, but would like to know what the corresponding uncertainty is in a function of that variable. Here we look at a few simple yet common situations.

2.8.1 Linear function of one variable

Suppose we have a variable x with variance $\mathrm{Var}(x)$, or equivalently standard deviation $\sigma_x = \sqrt{\mathrm{Var}(x)}$. We take a linear function of x

$$f = a + bx \tag{2.14}$$

where a and b are fixed constants. What is the standard deviation of f?

We can derive this exactly from the definition of variance and the properties of the expectation operator (section 1.3)

$$
\begin{aligned}
\mathrm{Var}(f) &= E[(f - E[f])^2] \\
&= E[(a + bx - E[a + bx])^2] \\
&= E[(bx - bE[x])^2] \\
&= b^2 E[(x - E[x])^2] \\
&= b^2 \mathrm{Var}(x).
\end{aligned}
\tag{2.15}
$$

Taking the square root to get an equation for the standard deviations gives

$$\sigma_f = |b|\sigma_x. \tag{2.16}$$

This is logical: from a dimensional point of view there must be a linear scale factor b in the conversion. a shouldn't play any role, as it is just a constant offset on all data points. So apart from a potential numerical factor, the above solution should have been obvious.

2.8.2 Arbitrary function of one variable

We can apply the same approach to propagate the uncertainties to a more general, nonlinear function $f(x)$. We do this by making a Taylor expansion of the function about its mean

$$f(x) \simeq f(x_0) + (x - x_0)\frac{df}{dx}\bigg|_{x=x_0}. \tag{2.17}$$

As $E[f(x)] = f(x_0)$ and with $\delta x = x - x_0$, we have[6]

$$\text{Var}(f) = E[(f - E[f])^2]$$

$$\simeq E\left[\left(\delta x \frac{df}{dx}\right)^2\right] = \left(\frac{df}{dx}\right)^2 \text{Var}(x) \quad \text{so}$$

$$\sigma_f \simeq \left|\frac{df}{dx}\right| \sigma_x. \tag{2.19}$$

This approximation only holds for small uncertainties, such that the derivative does not change much with x. If we had a larger uncertainty, $x - x_0$ would be large and this first-order Taylor expansion would break down (we would need higher order terms). But we've not made any assumption about a Gaussian distribution of the uncertainties.

Example: angular errors

If the (small) uncertainty in an angle α is δ_α radians, what is the corresponding uncertainty in $\sin \alpha$?

From equation 2.19 it is $\delta_\alpha |\cos \alpha|$. Note that this is maximum when the angle is 0 or π radians, but zero when the angle is $\pi/2$. This is because when $\alpha = \pi/2$ the sine function has its maximum and is flat, so small changes in the angle have negligible effect on the sine.

2.8.3 Linear function of two variables

Now consider

$$f = a + bx + cy \tag{2.20}$$

for a, b and c fixed constants and x and y random variables. Proceeding as before

$$\begin{aligned}
\text{Var}(f) &= E[(f - E[f])^2] \\
&= E[(a + bx + cy - E[a + bx + cy])^2] \\
&= E[(bx + cy - bE[x] - cE[y])^2] \\
&= b^2(E[x^2] - E[x]^2) + c^2(E[y^2] - E[y]^2) + 2bc\,(E[xy] - E[x]E[y]) \\
&= b^2\text{Var}(x) + c^2\text{Var}(y) + 2bc\,\text{Cov}(x, y) \tag{2.21}
\end{aligned}$$

in which some tedious algebra has been omitted between lines 3 and 4. If x and y are uncorrelated then this reduces to

$$\sigma_f^2 = b^2\sigma_x^2 + c^2\sigma_y^2. \tag{2.22}$$

---

[6] Alternatively write

$$f(x) \simeq f(x_0) - x_0 \frac{df}{dx}\Big|_{x=x_0} + x \frac{df}{dx}\Big|_{x=x_0} \tag{2.18}$$

which is just $f \simeq a + bx$ where the first two terms are a (independent of x) and $b = df/dx$. So from equation 2.15 we have $\sigma_f \simeq |b|\sigma_x$

If $b = c = 1$, i.e. if f is an unweighted linear sum of x and y, then

$$\sigma_f^2 = \sigma_x^2 + \sigma_y^2. \tag{2.23}$$

This is often summarized by saying "errors sum in quadrature".

2.8.4 Arbitrary function of two variables

Expanding a general two-parameter function $f(x, y)$ as a Taylor series up to first order gives

$$f(x, y) \simeq f(x_0, y_0) + (x - x_0)\frac{\partial f}{\partial x} + (y - y_0)\frac{\partial f}{\partial y} \tag{2.24}$$

where the partial derivatives are computed at (x_0, y_0). This has the form of equation 2.20 (b and c are the derivatives). Equation 2.21 therefore tells us that

$$\text{Var}(f) \simeq \left(\frac{\partial f}{\partial x}\right)^2 \text{Var}(x) + \left(\frac{\partial f}{\partial y}\right)^2 \text{Var}(y) + 2\frac{\partial f}{\partial x}\frac{\partial f}{\partial y}\text{Cov}(x, y). \tag{2.25}$$

2.8.5 Arbitrary function of many variables

We can generalize the result in the previous section to a function of J variables $f(x_1, \ldots, x_j, \ldots, x_J)$ using the J-dimensional first-order Taylor expansion, in which case it's best to write down the result using matrix algebra in terms of the covariance matrix (see section 1.6.1). We get

$$\text{Var}(f) \simeq (\nabla f)^\mathsf{T} C_x \nabla f \tag{2.26}$$

where C_x is the $J \times J$ covariance matrix of the data with terms $\sigma_{x_j}^2$ on the leading diagonal, and ∇f is a J-dimensional (column) vector of the gradients $\partial f/\partial x_j$. $\text{Var}(f)$ is the resulting variance in f (a scalar). If the variables are uncorrelated this reduces to

$$\sigma_f^2 \simeq \sum_j \left(\frac{\partial f}{\partial x_j}\right)^2 \sigma_{x_j}^2. \tag{2.27}$$

In the case of having K functions $f_1, \ldots, f_k, \ldots, f_K$ of the J variables, then in addition to a variance in each f_k there will also, in general, be covariances between them. We can write these as the $K \times K$ covariance matrix C_f (the leading diagonal of which is the variances). Applying again the first-order Taylor expansion we get

$$C_f \simeq A C_x A^\mathsf{T} \tag{2.28}$$

where A is the $K \times J$ matrix of partial first derivatives – the *Jacobian matrix* – in which the element in the kth row and jth column is $\partial f_k/\partial x_j$. This approximation of the covariance becomes exact if the functions are linear in the $\{x_j\}$.

2.8.6 Fractional errors

If $f = xy$ and x and y are independent, then from equation 2.25 we have

$$\text{Var}(f) = y^2 \text{Var}(x) + x^2 \text{Var}(y) \tag{2.29}$$

which is exact because the second and higher derivatives are zero. Dividing by f^2 gives

$$\left(\frac{\sigma_f}{f}\right)^2 = \left(\frac{\sigma_x}{x}\right)^2 + \left(\frac{\sigma_y}{y}\right)^2 \tag{2.30}$$

which we can remember by saying "fractional errors add in quadrature". You can check that we get the same result for the ratio of two variables x/y (although then the result is just approximate, because $\partial^2 f/\partial y^2 \neq 0$).

2.8.7 Weighted mean

If we have positive weights $\{w_i\}$ for independent measurements $\{x_i\}$ then their weighted mean is

$$\hat{\mu}_w = \frac{1}{\sum w_i} \sum_i w_i x_i. \tag{2.31}$$

We can use the result in section 2.8.5 to show that the variance in $\hat{\mu}_w$ is

$$\text{Var}(\hat{\mu}_w) = \frac{1}{(\sum_i w_i)^2} \sum_i w_i^2 \sigma_i^2 \tag{2.32}$$

where $\{\sigma_i^2\}$ are the variances of the individual measurements. The square root of $\text{Var}(\hat{\mu}_w)$ is the standard deviation (or standard error) in the weighted mean.[7]

Often one takes N measurements of some quantity x, each of which has its own uncertainty estimate. How should we combine these into an estimate of x which takes the uncertainties into account, and what is the resulting uncertainty? Let's suppose we know the standard deviations $\{\sigma_i\}$ of the measurements. We can weight the measurements according to the inverse variance, $w_i = 1/\sigma_i^2$. From the above we then have that

$$\hat{\mu}_w = \frac{1}{\sum_i 1/\sigma_i^2} \sum_i \frac{x_i}{\sigma_i^2} \tag{2.33}$$

and the variance in this weighted mean is

$$\text{Var}(\hat{\mu}_w) = \frac{1}{\sum_i 1/\sigma_i^2}. \tag{2.34}$$

As we will see later (section 4.4.1), this weighting is "best" in terms of being the maximum likelihood solution for Gaussian data. Be aware that the variances $\{\sigma_i^2\}$ are the true variances, not those estimated from the data. Variances estimated from the data themselves

---

[7] This is a biased estimator of the standard deviation, because we have estimated $\hat{\mu}_w$ from the data. One can derive an unbiased estimator, but in the interests of simplicity we will just live with this bias in this section. It will be of little practical significance once N is larger than about 10 or so. Let's not forget that we are anyway dealing with *estimates* from noisy data.

have variances (uncertainties), and if we used these we would derive a different expression for the mean (and its variance). This is the case no matter what distribution the data are drawn from. Note the particular case of the Poisson distribution, for which the true variance equals the true mean. Having calculated an estimate of that mean, it is tempting to use this as the weight in the above.

It is also important to realise that $\text{Var}(\hat{\mu}_w)$ is the variance of (i.e. a measure of our uncertainty in) the weighted mean. It is *not* the variance of the weighted sample. If we want a measure for the "spread" in the weighted sample (whatever that means exactly), we could generalize the definition of variance to include a weight, and quote

$$\frac{1}{\sum_i w_i} \sum_i w_i (x_i - \hat{\mu}_w)^2.\tag{2.35}$$

With $w_i = 1/\sigma_i^2$, this takes into account both the spread in the measurements and their individual uncertainties.

Note that $\text{Var}(\hat{\mu}_w)$ is independent of the values $\{x_i\}$: moving the data points doesn't change how precisely we can determine the weighted mean from them. This is consistent with what we get when we don't use weights. Equation 2.34 then gives the variance in the mean as σ^2/N, where σ^2 is the assumed common variance of the data points. This is just what we got from the central limit theorem (section 2.3).

3   Statistical models and inference

In this chapter I will introduce the principles of probabilistic inference. We will see how to set up models and will learn about the prior, likelihood, posterior, and evidence. I will show how inference works in practice using two simple examples: model comparison in the context of medical testing, and parameter estimation in astronomy. All of the issues covered in this chapter form the basis for deeper exploration in later chapters.

3.1 Introduction to data modelling

We perform experiments or make observations in order to learn about a phenomenon. We may describe the resulting data by calculating statistics and making plots. Such data explorations and summaries are useful – even essential – to get a feel for the data, but they are just a first step. To interpret the data we usually have to model them.

Typically we can only observe a phenomenon in part, and the data we obtain on it are noisy. Inference is the process of making general statements about a phenomenon, via a model, using noisy and incomplete data. The model represents the data in a form that gives us scientific meaning.

To do inference we must describe both the phenomenon itself and the measurement process. Consider modelling the orbit of a planet around its host star. We first define a relevant model M. This might describe the orbit as an ellipse (Keplerian orbit), as opposed to an oval or rosette. The model will have some parameters θ that describe the specific properties of the model. In the elliptical orbit case this would include the size (semi-major axis) and shape (eccentricity) of the ellipse, as well as its orientation in space. But when we observe the motion of a planet about a star, we do not observe directly the shape of the orbit or any of the other parameters. We instead see the planet at different positions (and with different velocities) at different times. The *generative model* (also called the *forward model*) is the theoretical entity that generates (or simulates) the observable data from the model parameters. Normally this is a mathematical equation. In this example it would be a deterministic equation derived from the physical laws of mechanics and gravity: given the model parameters and time of the observation, the position and velocity of the planet can be simulated exactly. Other generative models may be stochastic and only predict a distribution. An example is the kinematic theory of gases, which predicts the distribution of the speeds of gas molecules, but not the speeds of individual molecules.

The generative model predicts the data we would measure (if our model represents reality) in the absence of noise. Yet measurements are always noisy: we won't observe the planet at exactly the position predicted by the model due, for example, to the finite reso-lution of our instruments or the blurring caused the Earth's atmosphere. We need to know how the measurement process affects our data. For this purpose we must also define a *mea-surement model*, sometimes also called a *noise model* (discussed already in section 2.2). This describes a probability distribution over possible observations given the ideal (noise-free) data and generally has some parameters too. A widely used example is a Gaussian in the variable x (equation 1.60) in which the mean is the ideal data (predicted by the gen-erative model), x is the data we measure, and the standard deviation is the typical size of the uncertainty in the measurement. This probability distribution is in fact the likelihood, which we shall discuss properly in section 3.3 below.

The key to data modelling is to use the data together with the generative model and measurement model to make consistent, probabilistic inferences. Broadly speaking, there are three different things we will want to do given some data D.

(1) **Parameter estimation**. For a specified model M with parameters θ, infer the values of the model parameters, or more specifically, infer the PDF $P(\theta | D, M)$. This quantity is known as the *parameter posterior PDF*. In terms of our planet orbit example, this means infer the (multidimensional) PDF over the orbit parameters, and perhaps the parameters of the measurement model too.

(2) **Model comparison**. Given a set of different models $\{M_i\}$, find out which one is best supported by the data. Ideally this means finding $P(M_i | D)$, the *model posterior prob-ability*. But we shall see that ratios of $P(D | M_i)$ for the various M_i are often adequate. In our example, this corresponds to finding the absolute or relative probabilities of different models for the shape of the orbit (ellipse, oval, rosette, etc.).

(3) **Prediction**. Given a model M (which may have been identified/fitted from the data), predict the data at some new location. Having determined that the planet does have an elliptical orbit, for example, find the PDF over the position of the planet at some future time.

Much of the rest of this book is concerned with showing how we put these ideas into practice. I start in the next section by looking at model comparison for the case of models having no parameters (or rather, we ignore the details of their parametrization). I will then introduce the concept of parameter estimation in section 3.3, and will apply this to a simple one-parameter model in section 3.5. More complicated problems and more sophisticated inference methods will be introduced in the following chapters.

3.2 Bayesian model comparison

3.2.1 Theory

Medical tests are done to find out whether a patient has a certain disease. Let M denote that a patient has a particular disease and M' that he does not. These two cases are mutually exclusive and exhaustive, so $P(M') = 1 - P(M)$. A medical test produces some data D

which might be a simple positive/negative test result, or it could be a more extensive set of blood values, heart rate measurements, etc. We would like to use these data to find out how probable it is that the patient has the disease. That is, we would like to find $P(M|D)$. As $P(M'|D) = 1 - P(M|D)$, we automatically find out the probabilities of both models. We are doing model comparison between two complementary models.

From the fundamental laws of probability (section 1.2.1)

$$
\begin{aligned}
P(D) &= P(D, M) + P(D, M') \\
&= P(D|M)P(M) + P(D|M')P(M')
\end{aligned}
\tag{3.1}
$$

where $P(D)$ is the probability (or probability density)[1] of having observed this particular piece of data at all (under either model). We are interested in $P(M|D)$. This is related to the above quantities via Bayes' theorem (section 1.2.4)

$$
P(M|D) = \frac{P(D|M)P(M)}{P(D)}.
\tag{3.2}
$$

Substituting equation 3.1 into this gives

$$
P(M|D) = \frac{P(D|M)P(M)}{P(D|M)P(M) + P(D|M')P(M')}.
\tag{3.3}
$$

Dividing by the numerator we can write this as

$$
P(M|D) = \frac{1}{1 + \frac{1}{R}}
\tag{3.4}
$$

where

$$
R = \frac{P(D|M)P(M)}{P(D|M')P(M')}
\tag{3.5}
$$

is the *posterior odds ratio* of the two models.[2] We see from equation 3.2 that this also equals the ratio of the posterior probabilities for the two models, $P(M|D)/P(M'|D)$ (because $P(D)$ cancels).

In order to determine the posterior probability that the model is true given the data, $P(M|D)$, we therefore need three quantities

- $P(D|M)$, the probability of measuring D when M is true;
- $P(D|M')$, the probability of measuring D when M is not true;
- $P(M)$, the probability that M is true, independent of the data – the *prior probability*.

This last quantity tells us how probable the model is – how likely the patient has the disease – prior to obtaining the data, i.e. independent of the test result. Typically we have some information about this, such as from medical histories of the disease. But if we really had no information to tell us that M is more or less likely than M', then we could set

---

[1] M is a discrete proposition, so $P(M)$ and $P(M|D)$ are actual probabilities, whereas D could be a real variable (scalar or vector), in which case $P(D)$ and $P(D|M)$ would be probability densities. This mixing is neither a conceptual nor a practical difficulty. The laws of probability hold and the dimensions of the equations agree. As the nature of the variable will determine whether we are talking about a probability or a probability density, for simplicity I will just refer to P as a probability.

[2] Be aware that the term "odds ratio" also has other definitions in statistics.

Result D	Model M true?	
	yes	**no**
positive	true positive $P(D\|M)$	false positive $P(D\|M')$
negative	false negative $P(D'\|M)$	true negative $P(D'\|M')$

Table 3.1 The meaning of terms used to describe test results. Columns must sum to 1. Rows generally do not.

$P(M) = P(M')$. These terms then cancel in the odds ratio, leaving what is known as the *Bayes factor*

$$BF = \frac{P(D|M)}{P(D|M')}. \qquad (3.6)$$

This is the ratio of the probabilities of the data under each model.

If we have many different models $\{M_i\}$ (perhaps these are different diseases) then equation 3.1 has to be replaced by

$$P(D) = \sum_i P(D|M_i)P(M_i) \qquad (3.7)$$

(this is equation 1.15) where the sum is over *all* models, i.e. $\sum_i P(M_i) = 1$. This requires that the set of models be mutually exclusive (no disease is a subclass of another) and exhaustive (one of the models has to be the absence of all the other diseases). It follows from equation 3.2 that the posterior probability of any one model, say M_1, is

$$P(M_1|D) = \frac{P(D|M_1)P(M_1)}{\sum_i P(D|M_i)P(M_i)}. \qquad (3.8)$$

If we do not have a complete set of models then we cannot compute these posterior probabilities. However, we can still compute the odds ratio or Bayes factor between any two models. In practice we would calculate these against a common baseline model (e.g. the absence of all the other diseases).

Let's now put this theory into practice with an example.

3.2.2 Does a positive test result mean you have a disease?

A certain test for breast cancer is 90% reliable, which means that if the person has breast cancer she will test positive with a probability of 0.9. The probability of testing positive in absence of the disease (i.e. by mistake) is 0.07. Among women aged 40 to 50 showing no symptoms, 8 in 1000 have breast cancer.[3] A 40-something woman tests positive. What is the probability that she has breast cancer? Make an intuitive guess before reading on.

---

[3] These numbers have been taken from Gigerenzer (2002).

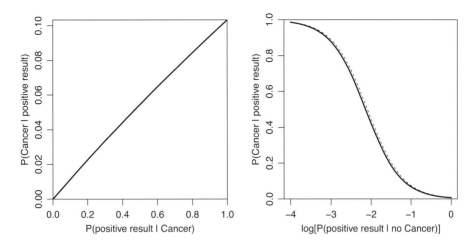

Fig. 3.1 The variation of the probability of having cancer given a positive test result $P(M|D)$ as a function of (left) the true positive rate $P(D|M)$ and (right) the false positive rate $P(D|M')$, in each case keeping the other quantities fixed. The dashed line in the right panel is for $P(D|M) = 1$. Note the log (base 10) scale on the horizontal axis in the right panel.

We first need to understand the information given in the question. Let D be the test result (data), which is either positive or negative. Let M be the hypothesis (model) that the woman has cancer. We are told that the true positive rate, $P(D|M)$, is 0.9. The figure of 0.07 is the false positive rate, $P(D|M')$. (Table 3.1 may be a useful reference for these terms.) The final statement is the *base rate* – the prior probability that M is true – which tells us that $P(M) = 0.008$.

We can now use equation 3.4 to compute $P(M|D)$, the posterior probability that the woman has cancer given that she has tested positive. The answer is low, just 0.09. Many people are surprised by this. Some people confuse the quantity we want with the reliability $P(D|M)$, and therefore expect an answer of 0.9. Others incorrectly interpret the false positive rate $P(D|M')$ as a statement about the disease, rather than a statement about the test. They erroneously equate it with $P(M'|D)$, and then conclude the probability of having the disease, which is $P(M|D) = 1 - P(M'|D)$, to be 0.93. Bayes theorem' shows that it is neither of these. The correct answer is much smaller in this case because it depends on the small base rate. Even though a positive test result is quite probable when the woman has the disease, it is very unlikely that she does have the disease. And although a positive test result in the absence of the disease is quite improbable, it is very likely that she doesn't have the disease in the first place. The first possibility – positive result, disease present – is quantified by $P(D, M) = P(D|M)P(M)$. The second possibility – positive result, disease absent – is quantified by $P(D, M') = P(D|M')P(M')$. It is the ratio of these which is decisive. This ratio is R (equation 3.4) and this determines $P(M|D)$ (equation 3.5).

We can understand this better by investigating how the result varies as a function of the assumed probabilities (you can use the R code at the end of this section to do this). The left panel of figure 3.1 shows how the probability of having cancer given a positive test result, $P(M|D)$, varies as a function of the true positive rate $P(D|M)$, keeping the false positive rate and base rate constant. It is virtually linear because the odds ratio R (equation 3.5) is much less than one (< 0.12 for all $P(D|M)$), so it follows from equation 3.4 that $P(M|D) \simeq R = \text{constant} \times P(D|M)$. If you redo the plot with a larger base rate you will see nonlinearity.

The right panel of figure 3.1 shows how $P(M|D)$ varies as a function of the false positive rate $P(D|M')$, keeping the true positive rate and base rate constant at their original values. It shows that the false positive rate has to be quite low before the test gives reliable results. For example, it must drop below 0.0072 ($\log P(D|M') = -2.14$) before the probability of having the disease given a positive test result rises above 0.5. To get a correct detection of the disease with a probability of at least 0.9 would require a false positive rate of less than 8×10^{-4}.

Improving the true positive rate will not help this test. Even when it is 1.0 the curve in the right panel of figure 3.1 hardly changes (the dashed line). The base rate in this example is $P(M) = 0.008$ or $\log P(M) = -2.10$. Inspection of the equations shows that only once the false positive rate drops significantly below the base rate does the test start to become useful. As you probably have no control over the base rate, getting a better test means lowering the false positive rate.[4]

Here is a useful rule of thumb: assuming a test has a high reliability (true positive probability), a positive result only means you are likely to have the disease if the false positive rate is lower than the base rate.

If you test positive for a rare disease, do the above analysis. Better still, do it to help you decide whether or not to take the test in the first place. Given that false positives can have detrimental consequences (such as worry and unnecessary or harmful treatment), it may be that for some people, some tests are not worth doing.

We shall look in more detail at the problem of model comparison in chapter 11, including for the case that the models have free parameters.

R file: `cancer_test.R`

```
##### Apply Bayes theorem to infer cancer probability given test result

# Return P(M|D) given P(D|M), P(D|M'), P(M)
post <- function(p_d_m, p_d_nm, p_m) {
  p_nm   <- 1-p_m
  oddsr <- (p_d_m * p_m) / (p_d_nm * p_nm)
  p_m_d <- 1/(1+1/oddsr)
  return(p_m_d)
}

# Vary reliability of test P(D|M)
```

---

[4] We are assuming here that we know the base rate exactly. This is often not the case, and an uncertainty in this will propagate into an uncertainty in our value of $P(M|D)$. In the current case, where $P(M|D) \propto P(D|M)$, this propagation is linear: a 10% error in the base rate gives approximately a 10% error in the posterior, for example.

```
p_d_m   <- seq(0.0, 1.0, 0.01) # prob. true positive
p_d_nm <- 0.07                  # prob. false positive
p_m     <- 8/1000               # prior probability of m
p_m_d   <- post(p_d_m, p_d_nm, p_m)
pdf("cancer_test_1.pdf", 4, 4)
par(mfrow=c(1,1), mgp=c(2.0,0.8,0), mar=c(3.5,3.5,1,1), oma=0.1*c(1,1,1,1))
plot(p_d_m, p_m_d, type="l", lwd=2, xaxs="i", yaxs="i",
     xlab="P(positive result | Cancer)", ylab="P(Cancer | positive result)")
dev.off()

# Vary false positive probability P(D|!M)
p_d_m   <- 0.9                   # prob. true positive
p_d_nm <- 10^seq(-4,0,0.02) # prob. false positive
p_m     <- 8/1000               # prior probability of m
p_m_d   <- post(p_d_m, p_d_nm, p_m)
pdf("cancer_test_2.pdf", 4, 4)
par(mfrow=c(1,1), mgp=c(2.0,0.8,0), mar=c(3.5,3.5,1,1), oma=0.1*c(1,1,1,1))
plot(log10(p_d_nm), p_m_d, type="l", lwd=2, ylim=c(0,1), yaxs="i",
     xlab="log[P(positive result | no Cancer)]",
     ylab="P(Cancer | positive result)")
lines(log10(p_d_nm), post(p_d_m=1.0, p_d_nm, p_m), lty=2)
dev.off()
```

3.2.3 Thinking in terms of frequencies may help

Probabilities are not frequencies. A frequency refers to how often something occurs when repeated in some way, whereas probability refers more broadly to knowledge and certainty (or lack thereof) gained from data. Nonetheless, it can often help our understanding to use frequencies to represent probabilities.

Let's consider the breast cancer example again, but with a 100% true positive rate, $P(D|M) = 1$. As before the false positive rate is 0.07 and the base rate is $8/1000$. Imagine a set of 10 000 women in the defined group who undergo the test. Of these, a fraction $8/1000$ have breast cancer, which is 80, and the remaining 9920 do not. Of the 80 with cancer, all will test positive. Of the 9920 without cancer, $0.07 \times 9920 = 694$ will test positive. In total $694 + 80$ have tested positive. Thus the probability of having the disease following a positive test result is

$$\frac{80}{80 + 694} \simeq 0.10. \tag{3.9}$$

This is of course the same calculation as before, but now carried out more intuitively.

3.3 Data modelling with parametric models

As discussed earlier in this chapter, the generative model is the theoretical entity that predicts observable data from the model parameters. The model in the previous section had no (explicit) parameters: the model was either true or false. Most models have one or more parameters, which I will denote by θ. A simple example of a generative model with parameters is the straight line

$$f(x; a, b) = a + bx. \tag{3.10}$$

The parameters of this are the intercept a and the slope b. The measurements y will differ from this due to noise, so we write

$$y = f(x; a, b) + \epsilon \tag{3.11}$$

where ϵ denotes a number drawn from a random distribution, the noise model (measurement model), which typically also has parameters.

Given a set of data – in the line example a set of measurements $D = \{y\}$ at specified values of $\{x\}$ – we want to infer the values of the parameters of the generative model (and perhaps of the measurement model too). In some cases one might be interested in finding just the single "best" set of parameters that predict the data. But as the data are noisy, there is no unique solution. So more generally we want to know what range of parameters is supported by the data (and its noise). That is, we would like to find the probability distribution over the parameters. The broader the distribution the less well constrained are the parameters and the larger our uncertainty. We can quantify this with the standard deviation, for example. There may also be multiple regions of high probability density, corresponding to multiple solutions. These could never be characterized by a single "best" solution.

The probability density function we are after is $P(\theta|D, M)$. This is the PDF over the parameters given both the data D and what we know already about the model M. An example of such prior knowledge might be that we know on physical grounds that the slope cannot be negative. The PDF $P(\theta|D, M)$ is called the parameter *posterior* PDF. We use Bayes' theorem to express this in terms of other quantities which, as we will see, are readily available. The difference with respect to the previous section is that we now apply Bayes' theorem to the model parameters, rather than to the model itself, to give

$$P(\theta|D, M) = \frac{P(D|\theta, M)P(\theta|M)}{P(D|M)}. \tag{3.12}$$

Once we have determined the three terms on the right side of the above equation then we have determined the posterior PDF over the parameters. Recall that all four terms in this equation are probability density functions in whatever quantity appears before the bar "|". (In the previous section we dealt with discrete variables – test positive or negative; model true or false – so all the quantities there were actual probabilities.) Let us now examine these four terms to understand what they mean and how we can determine them. I will then use a simple example to illustrate how we apply this in practice. More advanced examples follow in later chapters.

3.3.1 The likelihood

The first term in the numerator of equation 3.12, $P(D|\theta, M)$, is the *likelihood*. This is the key function in data modelling because it describes both the phenomenon and the measurements. It tells us the probability of getting the data we measured given some value of the parameters. Although this is the probability of D conditional on θ (and M), once the data have been measured they are fixed, so we are more interested in the likelihood's dependence on the parameters. Some authors therefore refer to the "likelihood of the parameters". Yet it remains a PDF in the data with units D^{-1}, so we should really refer to the "likelihood of the data" (as "likely" is just a synonym for "probable").

M specifies both the measurement model and the generative model. In terms of our straight line fitting example, the generative model is the equation for the line $f(x; a, b)$, and the measurement model tells us how the measurements of y at a given x differ from $f(x; a, b)$ on account of the noise. The measurement model describes the distribution of ϵ in equation 3.11. An example is a Gaussian with standard deviation σ. Here σ is the parameter of the measurement model. The likelihood for any one point y is then

$$P(y\,|\,\theta, M) \;=\; \frac{1}{\sigma\sqrt{2\pi}} \exp\left[-\frac{[y - f(x; a, b)]^2}{2\sigma^2}\right]. \tag{3.13}$$

This tells us that the measurement has a Gaussian distribution about the true value. Note that $\theta = (a, b, \sigma)$ is the union of parameters from both the generative model and measurement model.

3.3.2 The prior

The other term in the numerator of equation 3.12 is the *prior* $P(\theta\,|\,M)$, which is a PDF over the model parameters. It encapsulates the information we have, independent of the data, about the possible values of the model parameters. In general this covers the parameters of both the generative and measurement models. In the case of the line fitting example the prior may tell us the physically permitted ranges of the slope and intercept, or more generally, what their relative probabilities are. This will be informed by our background knowledge of the problem. It is called the prior because it is the information we have prior to obtaining the data.

Sometimes our prior information is vague and/or it may be difficult to express as a probability distribution. Different people may have different information or different opinions on what prior information is relevant. This is not a weakness of inference. It just reflects the reality that we do not only use our immediate measurements to reach scientific conclusions. I shall say more about priors in section 3.5 and in later chapters – in particular in section 5.3 – so I will defer further discussion until then.

3.3.3 The posterior

The posterior $P(\theta\,|\,D, M)$ is the PDF over the model parameters given the data and the background information on the model. Equation 3.12 tells us that the posterior is proportional to the product of the likelihood and the prior. The denominator in that equation, $P(D\,|\,M)$, is independent of θ so we can think of it as just a normalization constant for now. Thus once we have defined the prior and likelihood, and we've measured some data, we get the posterior probability density to within a normalization constant Z, i.e.[5]

$$P(\theta|D, M) = \frac{1}{Z} P(D|\theta, M)P(\theta|M). \qquad (3.14)$$

Conceptually, at least, inference really is that straightforward. Following my convention (section 1.2.6) I write the unnormalized posterior using an asterisk

$$P^*(\theta|D, M) = P(D|\theta, M)P(\theta|M). \qquad (3.15)$$

Both the posterior and the prior are PDFs over the model parameters. The difference is that whereas both are conditional on background information, the posterior is conditional on the data as well. This additional dependence is provided by the likelihood. Thus we can see Bayesian inference as a process of improving our knowledge of the parameters by using the data: we update the prior using the likelihood to obtain the posterior. In section 5.1 I will illustrate how this updating works and how it combines the information about the parameter contained in the data and in the prior. We will see another illustration of this in section 9.1.6.

This posterior PDF tells us everything we want to know about the relative probabilities of different parameters. Often we will want to summarize both the typical value of the parameters, for example using the maximum (mode) or mean, and their range, for example using the (co)variance or quantiles. We will discuss summaries of the posterior further in section 5.5.

Sometimes we will adopt a prior that is uniform over the possible range of parameters. In that case the posterior is directly proportional to the likelihood. But we must not forget that the posterior is a PDF over the parameters, not the data. So even if the likelihood is a common function of the data (like a Gaussian), it and the posterior will generally be a different function of the parameters. It follows that even though the likelihood is a normalized function *of the data*, the product of the likelihood with a uniform prior will not generally be a normalized function *of the parameters*.

Certain choices of the likelihood and prior can result in a convenient form for the posterior PDF. In particular, there are cases in which the posterior is in the same "family" of functions as the prior, in which case the prior and posterior are called *conjugate distributions*. For example, if the likelihood is Gaussian (in the data x) and the parameter of interest is its mean μ (the variance is fixed), then the likelihood is also a Gaussian in the mean (inspect equation 1.60). So if the prior on μ is also chosen to be a Gaussian then the posterior over μ is likewise Gaussian, because the product of two Gaussians (likelihood and prior) is another Gaussian. Conjugate priors are often used because they make the mathematics easier. We shall encounter various conjugate distributions in chapter 5.

I will have a lot more to say about the parameter posterior. Indeed, chapters 5 to 9 are concerned primarily with how, in practice, we estimate the posterior given data and priors.

---

[5] The normalization constant of the prior is also independent of θ, so we could absorb it into the proportionality too. The normalization constant of the likelihood, in contrast, is generally dependent on θ. We must therefore use a normalized likelihood when computing the posterior.

3.3.4 The evidence

The denominator in equation 3.12 is called the *evidence*. It gives the probability, assuming model M to be true, of observing the data D at all, for any values of θ. I referred to the evidence earlier as "just" a normalization constant for the posterior, because it is the integral of the numerator of equation 3.12 – the product of the likelihood with the prior – over all θ

$$P(D|M) = \int P(D|\theta, M)P(\theta|M)\, d\theta. \tag{3.16}$$

This can be thought of as the integral of the likelihood (a PDF in D) over the prior (a PDF in θ), so it is also called the *marginal likelihood*. We shall see in chapter 11 that the evidence plays a key role in model comparison.[6]

The evidence, as a normalization constant, is important if we want to compute certain quantities from the posterior. If we are only interested in the shape of the posterior or the relative probabilities of solutions, or certain statistics like the mode and full-width at half-maximum, then we don't need to normalize the posterior. But if we want to find the mean, standard deviation, quantiles, or anything else that requires us to integrate the posterior, then we need to normalize the posterior.[7]

Sometimes the posterior has a nice functional form, in which case its integral may be analytic or given by a standard integral. This is the case for all the PDFs introduced in section 1.4 and we will encounter more cases in chapter 6. But for many real-world problems we will have to resort to a numerical integration. If the posterior is univariate and non-zero over only a finite range (x_1, x_2), we can do a brute force numerical integration: we evaluate the function on a dense grid and compute the area as the sum of a series of narrow rectangles each of width δx and height x_i, i.e.

$$\int_{x_1}^{x_2} P(x)\, dx \simeq \sum_{i=1}^{N} P(x_i)\, \delta x \tag{3.17}$$

where $\delta x = (x_2 - x_1)/N$ and the $\{x_i\}$ are at the centers of the rectangles.[8] This is illustrated in figure 3.2 and is known as a Riemann sum.

Better approximations than this rectangle method are the trapezium rule and Simpson's

---

[6] We already used the evidence for model comparison in the cancer test example in section 3.2.2, where the two models were having cancer and not having cancer. The four probabilities listed in table 3.1 are mathematically equivalent to evidences, but as that problem didn't involve a model with parameters I did not call them evidences.

[7] To see this, let x be a real number constrained to lie between 0 and 1, and let $P(x) = 2x$ be a normalized PDF. This has a mean $\int_0^1 2x^2 dx = 2/3$. If we now multiply $P(x)$ by a constant ($\neq 1$) it is no longer normalized, and we would calculate a different (incorrect) mean. The mode, in contrast, is unchanged. My statement about the need to normalize the posterior assumes we are making computations using the functional form of the posterior PDF. If we instead have a set of samples drawn *from* the PDF, then because these are characteristic of the distribution, you can calculate the mean, standard deviation, quantiles, etc., from these samples without evaluating the normalization constant. We will see why this is when we discuss sampling from distributions in chapter 8.

[8] In order to plot a distribution we also just evaluate it on a dense grid and plot it at these as points. If we use some software to plot it as lines, then the software does some kind of interpolation between the point evaluations.

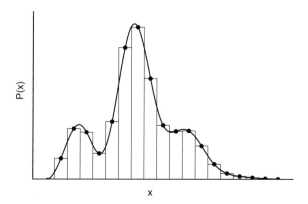

P(x)

x

Fig. 3.2   Illustration of the rectangle method for integrating a one-dimensional function.

rule. These use linear and quadratic interpolation (respectively) between neighbouring points – as opposed to constant interpolation in the case of the rectangle method – and so obtain higher accuracy for a given number of evaluations of $P(x)$ (although if $P(x)$ is fast to compute we can just use the rectangle method with narrower spacings). These approaches can be generalized to two dimensions using a double summation, but they become computationally intractable in higher dimensions. This is because the number of function evaluations required to achieve a given precision grows exponentially with the number of dimensions. This problem is known as the *curse of dimensionality* (an unusually evocative name for the field of statistics). For more complex functions in one or two dimensions, more sophisticated methods such as Gaussian quadrature may be required to do integration efficiently. Challenges are presented by higher dimensional functions, and by functions with probability concentrated in a very narrow region (as gridding techniques may miss the peak). For these we can use Monte Carlo methods, which we will turn to in chapter 8.

3.4 Making decisions

In section 3.2.2 we computed the posterior probability of having a disease given a positive medical test result. From the point of view of data analysis, this is the end of the problem. But if we need to decide whether or not to have treatment for the disease, we need to go further. To do this we need to know both the net benefit of the treatment for the case that we do have the disease, and the net cost of the treatment for the case that we do not have the disease. In an abstract sense the net benefit of a decision is called its *utility*, denoted U. The net benefit of having treatment T when we have the disease M we write as $U(T, M)$. (We assume for simplicity that the treatment is always successful.) The utility of this treatment when we do not have the disease is $U(T, M')$; this will be a negative number due to the negative side-effects of unnecessary treatment. If we test positive – measure data D – and have the treatment, then from the defintion of expectation (section 1.3) the expected utility of the treatment is

$$E[U(T)] = U(T, M)P(M \mid D) + U(T, M')P(M' \mid D). \qquad (3.18)$$

Typically we would elect to have the treatment only if its expected utility is positive. To reach this decision we have to decide what the two stated utilities are; these are likely to vary from person to person depending on how they value the negative side-effects of treatment. We could apply the same analysis to the utility of not having the treatment to find $E[U(T')]$ (and would presumably then decide for T or T' according to which gives the more positive utility).

This approach to decision making can also be applied when our model involves a continuous parameter θ, which the utility depends on, but where we don't know θ with certainty: our knowledge of it is described by the posterior PDF. To take an example, suppose we want to produce the strongest possible steel alloy by selecting one of a number of possible manufacturing processes d_k $(k = 1 \ldots K)$. Each of these processes results in an amount θ of a solute being added to the alloy. There is a degree of uncertainty in all of these processes: the amount of solute added is described by the PDF $P(\theta \mid d_k)$. If too much or too little solute is added then the alloy will be weak; the exact variation depends on the process. The strength of the resulting alloy is our utility, and as it depends on both d_k and θ we write it as $U(d_k, \theta)$. For a given process d_k the expected utility is

$$E[U(d_k)] = \int U(d_k, \theta)P(\theta \mid d_k) \, d\theta. \qquad (3.19)$$

If we compute this for each d_k we can then select the process that gives the highest expected utility.

The difference between inference and decision making is that in the latter we are not concerned with the parameter posterior per se, but rather with how this translates into utility for different possible courses of action.

3.5 Estimating one parameter from one data point

In chapters 5, 6, and 9 we will look at finding the posterior over one or more model parameters from a set of data using both exact analytic and approximate numerical methods. Here we look at a simpler problem – estimating a single parameter from a single measurement – and use it to gain some insight into inference.[9]

Distances to stars can be measured geometrically via their parallaxes. As the Earth orbits the Sun, the position of a star observed from the Earth will appear to move relative to more distant background stars (see figure 3.3). The nearer the target star, the larger this apparent movement. The size of the angular displacement is known as the *parallax*, ϖ. As all stars are very far away compared to the size of the Earth's orbit (150 million km),

---

[9] This section is based on my more detailed, yet still didactic, analysis of the problem in Bailer-Jones (2015). R code for producing the figures in this section is provided in the file `parallax.R`, available online.

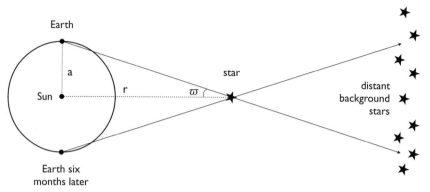

Fig. 3.3 The parallax ϖ of a star is the apparent angular displacement of that star (relative to distant background stars) due to the orbit of the Earth about the Sun. More precisely, the parallax is the angle subtended by the Earth's orbital radius a as seen from the star. As parallaxes are extremely small angles ($\varpi \ll 1$), $\varpi = a/r$ to a very good approximation. When ϖ is 1 arcsecond, r is defined as the *parsec*, which is about 3.1×10^{13} km. In this sketch the size of the Earth's orbit has been greatly exaggerated compared to the distance to the star, and the distance to the background stars in reality is orders of magnitude larger again.

their parallaxes are very small. The largest is of order one arcsecond, which is $1/3600$ of a degree. Such a parallax would be produced by a star that is about 3.1×10^{13} km away, an amount which is used as a distance unit in astronomy, the *parsec*. From simple geometry (and the fact that parallaxes are very small), if the distance to a star is r parsec, its parallax is $1/r$ arcseconds.[10] These units are implicit in everything that follows.

Given this definition, we might think that if we measure a parallax we can just invert it to determine the distance. But this is only the case if we have no measurement errors in ϖ. As we always have measurement errors, determining the distance given a parallax becomes an inference problem. Here we will investigate how to do this and see why inverting a noisy parallax is not the right thing to do.

For a star at true distance r, its true but unknown parallax is $1/r$. The measured parallax ϖ is a noisy measurement of $1/r$. Let us assume[11] that ϖ has a Gaussian distribution with unknown mean $1/r$ and known standard deviation σ_ϖ. That is, we assume ϖ has been drawn from the distribution

$$P(\varpi|r) = \frac{1}{\sigma_\varpi \sqrt{2\pi}} \exp\left[-\frac{1}{2\sigma_\varpi^2}\left(\varpi - \frac{1}{r}\right)^2\right] \quad \text{where} \quad \sigma_\varpi > 0, \quad (3.20)$$

which is Gaussian in ϖ, but of course not in r. This is the likelihood: the probability density

---

[10] Some readers will realize that because the distant background stars are not infinitely far away, measuring a parallax in this way would only give a relative parallax. More sophisticated procedures involving observing stars all over the sky can be used in order to obtain absolute parallaxes.

[11] This model is actually used in practice, with the standard deviation estimated from a noise model for the instrument.

of the data ϖ given the parameter r. Equation 3.20 has a finite probability for negative parallaxes, and this probability gets larger with increasing *fractional parallax uncertainty*, $f = \sigma_\varpi/\varpi$. Negative parallaxes arise because the angular measurements used to derive a parallax are noisy: it is possible that the measured displacement of the star is in the opposite direction of that expected from the movement of the observer along the baseline. A negative parallax does *not* correspond to a negative distance, because by definition $r \geq 0$. Negative parallaxes instead suggest that the star is probably quite distant, so the true parallax is small and noise made it negative. The measurement of a negative parallax therefore contains information.

3.5.1 Misleading intuition

Suppose we have a parallax measurement $\varpi \pm \sigma_\varpi$ and want to estimate the distance and its uncertainty. The intuitive approach is to report $1/\varpi$ as the distance estimate and to use a first-order Taylor expansion (section 2.8) to give σ_ϖ/ϖ^2 as the uncertainty. From the definition of the Gaussian likelihood (equation 3.20), the two-sigma intervals are

$$1/r = [\varpi - 2\sigma_\varpi, \varpi] \quad \text{and} \quad 1/r = [\varpi, \varpi + 2\sigma_\varpi], \tag{3.21}$$

each of which includes a fraction $0.954/2 = 0.477$ of the total probability of the distribution $P(\varpi|r)$. The transformation from $1/r$ to r is monotonic and so preserves the (integrated) probability. Thus the intervals

$$r = [1/(\varpi - 2\sigma_\varpi), 1/\varpi] \quad \text{and} \quad r = [1/\varpi, 1/(\varpi + 2\sigma_\varpi)] \tag{3.22}$$

each also contain a fraction 0.477 of the total probability over the distance. But whereas these intervals are equal-sized in $1/r$ (the Gaussian is symmetric), they are not equal-sized in r. The uncertainties do not transform symmetrically because of the nonlinear transformation from $1/r$ to r. For example, with $\varpi = 0.1$ and $\sigma_\varpi = 0.02$, the above intervals are $r = [16.67, 10]$ and $r = [10, 7.14]$ respectively. This is shown in figure 3.4. The black line in the left panel plots the likelihood as a function of $1/r$ for this case. As the likelihood is invariant when swapping ϖ and $1/r$, this plot is a Gaussian. The solid black line in the right panel shows the same likelihood, but now plotted against r. The asymmetry of the tranformation is clear. The first-order Taylor expansion suggested above corresponds to estimating the distribution in distance as a Gaussian with mean $1/\varpi$ and standard deviation σ_ϖ/ϖ^2. This is plotted as the black dashed line in the right panel. We see it is quite a poor approximation, even for this relatively small fractional uncertainty of $f = 1/5$.

What if the uncertainties are larger, say $f = 1/2$, so $\varpi = 0.1$ and $\sigma_\varpi = 0.05$? This likelihood is shown as the grey line in figure 3.4. The upper distance interval is now $r = [\infty, 1/\varpi]$. If f is even larger then this interval becomes undefined and we seem to "lose" some of the probability. (Remember, r cannot be negative.) As the Gaussian distribution has infinite support for all values of ϖ and $\sigma_\varpi > 0$, some finite amount of probability in the likelihood function will always correspond to an undefined distance, which is obvious for the grey line in the left panel of figure 3.4.

The problem here is that we are trying to make a probability statment about r using just

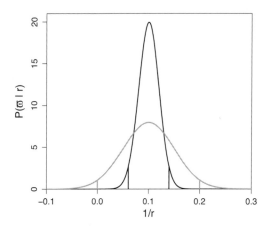
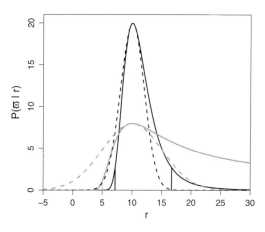

Fig. 3.4 The solid lines show the likelihood (equation 3.20) in the parallax problem for a measured parallax of $\varpi = 0.1$ as a function of $1/r$ in the left panel (a Gaussian) and as a function of r in the right panel. Note that the functions in the right panel are not PDFs over r, so they are not normalized. The black line is for $\sigma_\varpi = 0.02$ (so $f \equiv \sigma_\varpi/\varpi = 1/5$) and the grey line is for $\sigma_\varpi = 0.05$ ($f = 1/2$). The vertical lines denote the upper and lower 2σ limits around $1/\varpi$; the upper limit for the grey curve in the right panel is at $r = \infty$. The dashed lines in the right panel correspond to a Gaussian with mean $1/\varpi$ and standard deviation σ_ϖ/ϖ^2. Each of these Gaussians has been multiplied by the ratio of its standard deviation to that of the likelihood, $(\sigma_\varpi/\varpi^2)/\sigma_\varpi = 100$, in order to put them on the same vertical scale as the likelihood.

equation 3.20, yet this is not a probability distribution over r. The solution is to pose the problem correctly, as an inference problem.[12]

3.5.2 The inference problem

We tackle inference problems by writing down Bayes' theorem to give the posterior PDF over the model parameter (here r) given the data (here ϖ). This is

$$P(r\,|\,\varpi) \;=\; \frac{1}{Z}\,P^*(r\,|\,\varpi) \;=\; \frac{1}{Z}\,P(\varpi\,|\,r)\,P(r) \tag{3.23}$$

where $P^*(r\,|\,\varpi)$ is the unnormalized posterior and Z is the normalization constant. The likelihood is equation 3.20. What do we adopt as the prior? This should embody our knowledge of – or assumptions about – the distance, independent of the parallax we have measured. Some people object to priors on philosophical grounds (*How can science depend on assumptions?*), others on practical grounds (*How can I know the prior if I haven't yet mea-*

---

12   We cannot solve this problem through a simple change of variables (section 1.9.1), because that would give a transformation between the true distance and the noisy parallax, i.e. it would assume the true distance is the inverse of the noisy parallax, which it is not.

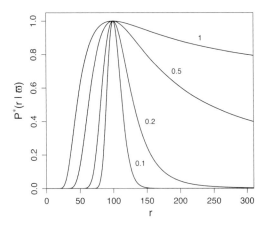

Fig. 3.5 The unnormalized posterior using the improper uniform prior (equation 3.25) for $\varpi = 1/100$ and four values of $f = 0.1, 0.2, 0.5, 1.0$. The unnormalized posteriors have been scaled to all have their mode at $P^*(r|\varpi) = 1$. Figure reproduced from Bailer-Jones (2015).

sured any distances?). The latter is a valid protest and will be discussed in later chapters. Yet without a prior we run into the problems we just saw in the previous section.

3.5.3 An improper uniform prior

A common approach is to adopt a uniform prior on the grounds that this does not prefer one value over another. This would not solve the problem, however, as we would end up with a posterior looking like the right panel of figure 3.4. The least we should do is accept physical reality and use

$$P^*(r) = \begin{cases} 1 & \text{if } r > 0 \\ 0 & \text{otherwise} \end{cases} \tag{3.24}$$

so as to introduce the uncontroversial assumption that distances must be positive. Because it extends to infinity, this prior cannot be normalized. Such priors are referred to as *improper*. In this case the (unnormalized) posterior $P^*(r|\varpi)$ is just the likelihood but now considered as a function of r rather than ϖ, and subject to the additional constraint $r \geq 0$, i.e.

$$P^*(r|\varpi) = \begin{cases} P(\varpi|r) & \text{if } r > 0 \\ 0 & \text{otherwise.} \end{cases} \tag{3.25}$$

Examples of this posterior are shown in figure 3.5 for $\varpi = 1/100$ and various values of f. Notice how the posterior gets broader for larger uncertainties in the data. It also becomes increasingly skew, as discussed in section 3.5.1. Conversely, the smaller the uncertainty, the more concentrated the posterior becomes around the value predicted purely by the data.

Inspection of equation 3.25/3.20 shows that

$$\lim_{r \to \infty} P^*(r|\varpi) = \text{constant}. \tag{3.26}$$

As the posterior does not converge to zero it has an infinite area and so cannot be normalized: it is improper. Consequently it has no mean, no standard deviation, no median, and no quantiles. The only plausible estimator of the distance is the mode of the posterior, which we see from figure 3.5 is well-defined for all values of f, and is equal to $1/\varpi$. Yet this estimator is invalid for non-positive parallaxes,[13] and it ignores the measurement uncertainty. Moreover, numerical experiments show that once f grows above about 0.2, this prior for distance estimation gives very large errors in terms of both bias and variance (terms which are defined in a modelling context in section 4.8). Improper posteriors are a bad thing.

3.5.4 A proper uniform prior

An obvious improvement on the above is to truncate the prior at some value. The prior, which can then be normalized, is

$$P(r) = \begin{cases} \dfrac{1}{r_{\text{lim}}} & \text{if } 0 < r \leq r_{\text{lim}} \\ 0 & \text{otherwise} \end{cases} \tag{3.27}$$

where r_{lim} is the largest distance we expect for any star in our survey. The unnormalized posterior is the same as in figure 3.5 but set to zero for $r > r_{\text{lim}}$, so

$$P^*(r|\varpi) = \begin{cases} P(\varpi|r) & \text{if } 0 < r \leq r_{\text{lim}} \\ 0 & \text{otherwise.} \end{cases} \tag{3.28}$$

This is shown for different values of f in the left panel of figure 3.6. The normalization can easily be done by integrating on a dense grid, as explained in section 3.3.4. The normalized posteriors are shown in the right panel of figure 3.6 (and make quite a different impression). The mean, standard deviation, and quantiles are all defined, and not just the mode. If we nonetheless use the mode as our distance estimator it is

$$r_{\text{mode}} = \begin{cases} \dfrac{1}{\varpi} & \text{if } 0 < \dfrac{1}{\varpi} \leq r_{\text{lim}} \\ r_{\text{lim}} & \text{if } \dfrac{1}{\varpi} > r_{\text{lim}} \\ r_{\text{lim}} & \text{if } \varpi \leq 0. \end{cases} \tag{3.29}$$

Another consequence of using a proper, normalized prior is that negative parallaxes now also produce plausible posterior PDFs (the dashed line in the plot), although this prior has the undesirable property of giving them all the same mode ($r = r_{\text{lim}}$). This is nonetheless

---

[13] The maximum is at infinity. You can get an idea of what the posterior looks like for negative parallaxes by plotting, e.g.

```
r <- seq(from=-10, to=10, by=0.01)
plot(r, exp(-0.5*(-1-1/r)^2), type="l")
```

although at negative values of r the posterior is set to zero by the prior.

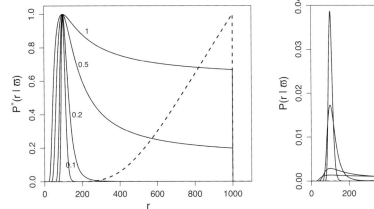 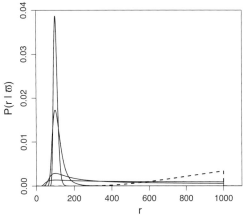

Fig. 3.6 Left: the unnormalized posterior using the truncated uniform prior with $r_{\mathrm{lim}} = 1000$ (equation 3.28). The four solid lines are for $\varpi = 1/100$ and $f = 0.1, 0.2, 0.5, 1.0$. These unnormalized posteriors have been scaled to all have their mode at $P^*(r\,|\,\varpi) = 1$. The dashed line is for a negative parallax, $\varpi = -1/100$, and $|f| = 0.25$. The right panel shows the same posteriors, but now normalized. Figure reproduced from Bailer-Jones (2015).

an improvement over inferring arbitrarily large distances for arbitrarily small, but noisy, parallaxes.

3.5.5 What is a good prior?

The uniform distance prior, even when truncated, may seem attractive because it does not prefer one distance over the other. It appears to be "uninformative". But there is nothing fundamental about a prior that is uniform in r. Why not uniform in, say, $\log r$? Is this less "fundamental"? Worse, a prior that is uniform in r corresponds to assuming that the volume density of stars in space varies as $P(V) \propto 1/r^2$, as seen from the observer.[14] Not only does this assign the observer a privileged position in the Galaxy (because the density falls off in all directions from us), but it assumes that the distribution of stars in the Galaxy is scale independent, which it demonstrably is not: the Galaxy has a characteristic length scale. This apparently harmless uniform prior is actually making strong assumptions that violate our astrophysical knowledge. There is in fact no such thing as an "uninformative" prior; just different priors that convey different information.

In the parallax example an improved prior might be

$$
P(r) = \begin{cases} \dfrac{3}{r_{\mathrm{lim}}^3}\, r^2 & \text{if } 0 < r \leq r_{\mathrm{lim}} \\ 0 & \text{otherwise} \end{cases} \tag{3.30}
$$

---

[14] $V \propto r^3$, so using $P(V)dV = P(r)dr$ to make a change of variables, we get $P(V) \propto P(r)/r^2$.

which corresponds to a uniform volume density of stars out to distance $r_{\lim}$. We could replace the hard cut-off with an exponential decrease in the volume density, $P(V) \propto \exp(-r/L)$, for which the prior is

$$
P(r) = \begin{cases} \dfrac{1}{2L^3} r^2 e^{-r/L} & \text{if } r > 0 \\ 0 & \text{otherwise} \end{cases} \tag{3.31}
$$

where $L > 0$ is a length scale which could be chosen based on current knowledge of the size of the Galaxy.[15]

Priors that are physically more acceptable necessarily make more physical assumptions. This may run into objections from those who claim that priors should play no role in inference; yet priors are all about introducing relevant contraints into the inference. By ignoring priors and just selecting the maximum of the likelihood (which here is $1/\varpi$), we can get nonsensical answers, and very poor estimates if the uncertainties are large. One may be tempted to circumvent this problem by discarding all data with large uncertainties, but that may result in discarding much of the hard-won data. Worse, it will bias scientific results on a population of objects, because by discarding poorer data we preferentially discard the more distant and fainter stars.

Opinions vary considerably on how strong priors should be and on what additional information we should use in our inference.[16] In the distance example one could make a case for using the best model we currently have of the distribution of stars in the Galaxy. But others will say that this influences the results too much, and we should do something more minimal. What one person calls a bias, another may call relevant information. Often priors cannot be avoided, even in principle (e.g. section 3.2.2), and inference is impossible without some assumptions. I generally advocate using a minimal prior consistent with the physical constraints, but even this will rarely result in a unique specification. A good approach is to try out different plausible minimal priors, and if these give very different results for the data available, this is just an indication that the data are not very informative. Your data are poor, so your results unavoidably depend more strongly on the prior. If you can't live with this uncertainty, you'll need to get better data.

We'll discuss further the issue of assigning priors, in particular in section 5.3.

3.6 An inference story

One of the key messages of this chapter is that inference needs to take into account all relevant information. This may be more than meets the eye, quite literally, as the following story demonstrates.

During World War Two, aircraft undertook bombing raids that were sometimes fatal for

[15] I will use this prior and this problem in section 12.7 to illustrate numerical optimization and bootstrap estimates.

[16] Readers with an astronomical background will recognise that we often have colour or even spectral information on stars. This can give fractional distance uncertainties as small as of order 10%, depending on the type of star. But the issues discussed still apply.

the aircrews. If the planes could be better armoured they would be more likely to survive. But armour is heavy, so it wasn't possible to put more armour everywhere. It was observed that while the returning planes often had damage on the wings and fuselage, they hardly ever had any damage on the cockpit or engines. It was concluded that the planes should receive more armour in these damaged areas in order to decrease aircraft losses. But assuming that planes could be hit anywhere with equal probability, why wasn't damage seen in certain areas of the returning planes? The answer is in the question: the observations are made of the *returning* planes. Presumably those planes that were hit on the cockpit or fuselage rarely returned. These are therefore the more vulnerable areas of the aircraft. Thus the conclusion based on all the data – i.e. also on what damage is *not* seen – is the opposite of the initial conclusion: more armour is required where damage is not seen.[17]

---

[17] This observation is part of a more detailed analysis undertaken by Abraham Wald (1943).

## 4

Linear models, least squares, and maximum likelihood

Here we shall learn the classical approach to fitting linear models via least squares and maximum likelihood, starting with the example of fitting a straight line in one dimension. We will see how this can easily be extended to higher dimensions and fitting nonlinear functions of the data. In the course of this we shall encounter the concepts of degrees of freedom, and the bias-variance decomposition. We will take the full Bayesian approach to this problem in chapter 9.

4.1 One-dimensional linear regression[1]

If we have two data points (x_1, y_1) and (x_2, y_2) then there is a unique straight line that fits these data. But what if we have three points or more, $\{x_i, y_i\}$?

There is no unique solution to this problem when stated in this way (unless the points are co-linear). We can imagine many different fits; but what is a good one?

Let the model for the straight line be $f(x; \theta)$, where θ denotes the parameters of the model, and $f_i = f(x_i; \theta)$ are the model predictions. A good fit in one sense will be a line for which the residuals of the fit, $\delta_i = y_i - f_i$, of the $i = 1 \ldots N$ points are small. We can combine these into a single metric by taking the sum of their squares (as squaring gets rid of the sign)

$$\text{SS}_{\text{res}} = \sum_i \delta_i^2 = \sum_i (y_i - f_i)^2. \tag{4.1}$$

This is the Euclidean distance between the data and the fit in an N-dimensional space, and is called the *sum of squared residuals* or the *residual sum of squares* (RSS). If we minimize this metric as a function of the parameters of the fit θ, then we can find the "best" solution. This is called *least squares minimization*. The idea of minimizing some error function (also called a cost, loss, or objective function) with respect to the model parameters is a key idea in fitting models. Mathematically we set

$$\frac{\partial \text{SS}_{\text{res}}}{\partial \theta_j} = 0 \tag{4.2}$$

for each of the model parameters θ_j, and then solve this set of equations for these parameters. We will have as many equations as parameters.

---

[1] "One-dimensional" here refers to the x being a one-dimensional quantity.

Suppose our model is $f = a + bx$, which is linear in the parameters (see section 4.2 for a definition). We want to find the best fitting coefficients, a and b.

$$SS_{res} = \sum_i (y_i - a - bx_i)^2. \tag{4.3}$$

Differentiating with respect to a and setting to zero gives

$$\sum_i -2(y_i - a - bx_i) = 0$$

$$\bar{y} - a - b\bar{x} = 0 \tag{4.4}$$

where $\bar{x} = \frac{1}{N}\sum_i x_i$ and $\bar{y} = \frac{1}{N}\sum_i y_i$. Differentiating with respect to b gives

$$\sum_i -2x_i(y_i - a - bx_i) = 0$$

$$\overline{xy} - a\bar{x} - b\overline{x^2} = 0. \tag{4.5}$$

Eliminating a from these two equations gives

$$b = \frac{\overline{xy} - \bar{x}\,\bar{y}}{\overline{x^2} - \bar{x}^2} \tag{4.6}$$

$$= \frac{\widehat{Cov}(x, y)}{\widehat{Var}(x)} \tag{4.7}$$

where $\widehat{Cov}(x, y)$ and $\widehat{Var}(x)$ are the sample covariance and sample variance respectively (see sections 1.3 and 1.6.1).[2] We also get

$$a = \bar{y} - b\bar{x}. \tag{4.8}$$

The line goes through the mean of the data $(\bar{x}, \bar{y})$.

In the more general case of a linear model with J parameters, we will have a system of J linear equations that need to be solved (see sections 4.5 and 4.6).

4.1.1 Uncertainty estimates on the model parameters

Noting that

$$\overline{xy} - \bar{x}\,\bar{y} = \frac{1}{N}\sum_{i=1}^N x_i y_i - \frac{\bar{x}}{N}\sum_{i=1}^N y_i$$

$$= \frac{1}{N}\sum_{i=1}^N (x_i - \bar{x})y_i \tag{4.9}$$

we can write the estimate for the gradient (equation 4.6) as a weighted sum of the data points $\{y_i\}$

$$b = \frac{1}{NV_x}\sum_{i=1}^N (x_i - \bar{x})y_i \tag{4.10}$$

---

[2] We can write this in terms of sample estimates because a factor of $(N - 1)/N$ appears in both the numerator and denominator, which then cancels.

where[3]

$$V_x = \overline{x^2} - \overline{x}^2 = \frac{1}{N} \sum_{i=1}^{N} (x_i - \overline{x})^2. \tag{4.11}$$

Using the result in section 2.8.5, the variance of a linear combination of independently measured variables $\{y_i\}$ is

$$\text{Var}(b) = \sum_i \left(\frac{\partial b}{\partial y_i} \right)^2 \sigma_i^2 \tag{4.12}$$

where σ_i is the standard deviation (uncertainty) in measurement y_i.[4] The x_i are assumed to be error free. From equation 4.10

$$\frac{\partial b}{\partial y_i} = \frac{x_i - \overline{x}}{N V_x} \tag{4.13}$$

so

$$\text{Var}(b) = \sum_{i=1}^{N} \left(\frac{x_i - \overline{x}}{N V_x} \right)^2 \sigma_i^2. \tag{4.14}$$

If all data points have the same standard deviation, $\sigma_i^2 = \sigma^2$, we can write this (using also equation 4.11) as

$$\begin{aligned} \text{Var}(b) &= \sigma^2 \sum_{i=1}^{N} \left(\frac{x_i - \overline{x}}{N V_x} \right)^2 \\ &= \sigma^2 \frac{N V_x}{(N V_x)^2} \\ &= \frac{\sigma^2}{N V_x}. \end{aligned} \tag{4.15}$$

You can likewise show that the variance in the intercept a is

$$\text{Var}(a) = \frac{\sigma^2 \overline{x^2}}{N V_x} = \text{Var}(b) \overline{x^2} \tag{4.16}$$

and the covariance between the gradient and intercept is

$$\text{Cov}(a, b) = -\frac{\sigma^2 \overline{x}}{N V_x} = -\text{Var}(b) \overline{x}. \tag{4.17}$$

Unless $\overline{x} = 0$, there is a non-zero covariance between the parameters. We interpret this as a parameter degeneracy: we can compensate for an increase in the gradient – to achieve almost as good a fit to the data – by decreasing the intercept. To remove this covariance we can translate our data to have $\overline{x} = 0$ prior to model fitting. The y-axis then goes through the mean of the data, so a small change in the gradient does not compensate for any change in the intercept.

---

[3] V_x is not quite the sample variance, because this would involve a factor of $N - 1$ rather than N.
[4] Equation 4.12 is exact because b is just a linear function of y_i, so the higher order terms in the Taylor expansion are zero.

The above analysis shows that we need to have uncertainty estimates in the y_i measurements to calculate the uncertainties in the parameters. What if we don't have any uncertainty estimates? In that case we must estimate them from the data. We could assume that the residuals $y_i - f_i$ have a common variance σ^2, which we estimate as $\sum_i (y_i - f_i)^2/(N-2)$. This estimate follows from the definition of variance in which the expected value of the residual is zero.[5] This is what the R function lm does (see section 4.1.5). Of course, if we *really* have zero uncertainty in the y_i measurements, then there can be no uncertainty in the parameters either, and the variance is zero (as follows from the above equations). If we have individual uncertainty estimates, then we might want to do a weighted fit (see section 4.1.4).

4.1.2 Uncertainties on predictions

Having found the function coefficients and their variances and covariance, we can predict the value of the function at any new point along with its uncertainty. If f_j is the prediction at x_j, then

$$f_j = a + bx_j \quad \text{and} \tag{4.18}$$
$$\mathrm{Var}(f_j) = \mathrm{Var}(a + bx_j). \tag{4.19}$$

It follows from the result in section 2.8.3 (equation 2.21), that

$$\mathrm{Var}(f_j) = \mathrm{Var}(a) + \mathrm{Var}(bx_j) + 2\,\mathrm{Cov}(a, bx_j). \tag{4.20}$$

Assuming that x_j has zero variance, then $\mathrm{Var}(bx_j) = x_j^2\,\mathrm{Var}(b)$. Using the results from the previous section, and writing $x_j = \overline{x} + \Delta x_j$, we see that

$$
\begin{aligned}
\mathrm{Var}(f_j) &= \mathrm{Var}(a) + x_j^2\,\mathrm{Var}(b) + 2x_j\,\mathrm{Cov}(a,b) \\
&= \mathrm{Var}(b)\,(\overline{x^2} + x_j^2 - 2x_j\,\overline{x}) \\
&= \frac{\sigma^2}{NV_x}\left(\overline{x^2} - \overline{x}^2 + (\Delta x_j)^2\right) \\
&= \frac{\sigma^2}{N}\left(1 + \frac{(\Delta x_j)^2}{V_x}\right).
\end{aligned} \tag{4.21}
$$

This is nonlinear in Δx_j. However, at points far from the original data $(\Delta x_j)^2/V_x \gg 1$, and so $\sqrt{\mathrm{Var}(f_j)} \sim |\Delta x_j|$, i.e. the standard deviation in the prediction grows linearly with the distance from the mean of the data.

4.1.3 Measures for goodness of fit

We have now fit the best straight line in a least squares sense, but is the fit any good? One measure of this is the quantity we minimized, the residual sum of squares of equation 4.1. Analogous to this we can define the *explained sum of squares* (also called the regression

---

[5] This is $N-2$ rather than N because we have estimated two model parameters from the data, and so have used up two degrees of freedom, a concept I will define later in section 4.3.

sum of squares), which measures how much the fitted values deviate from the mean of the data

$$SS_{ess} = \sum_i (f_i - \bar{y})^2. \tag{4.22}$$

The *total sum of squares* measures the total spread of the data about its mean,

$$SS_{tot} = \sum_i (y_i - \bar{y})^2. \tag{4.23}$$

It may come as no surprise that

$$SS_{tot} = SS_{res} + SS_{ess}. \tag{4.24}$$

The *residual standard error* reported by the `lm` function in R (we'll get to this in section 4.1.5) is

$$RSE = \sqrt{\frac{SS_{res}}{\nu}} \tag{4.25}$$

where ν is the degrees of freedom (section 4.3), which is $N - 2$ for this straight line fit.

Another useful metric is the *r squared* value – also known as the *coefficient of determination* – r^2. This is the fraction of the total sum of squares that is explained by the fit

$$r^2 = \frac{SS_{ess}}{SS_{tot}} = 1 - \frac{SS_{res}}{SS_{tot}}. \tag{4.26}$$

As the fit cannot explain more spread than is in the data, r^2 ranges from 0 to 1, with 0 being no fit and 1 a perfect fit (zero residuals). You can show that for a straight line fit this is equal to the square of the correlation coefficient.

Possibly more useful is the *adjusted r squared* value, defined as

$$r^2_{adj} = 1 - \frac{RSE^2}{\widehat{Var}(y)} = 1 - \frac{SS_{res}/\nu}{SS_{tot}/(N-1)}. \tag{4.27}$$

This describes the degree of variance described by the fit compared to the sample variance in the data $\widehat{Var}(y)$, which is the variance with no fit at all. Here ν is the number of degrees of freedom in SS_{res} (i.e. after computing the fit) and $N - 1$ is the number of degrees of freedom in SS_{tot} (i.e. after computing the average).

4.1.4 Linear model fitting with measurement errors

Sometimes we have data with individual error estimates σ_i on the measurements y_i. In that case we expect (and tolerate) deviations from the fit of this order for each point. The sensible thing to do is therefore to minimize

$$\chi^2 = \sum_i \frac{(y_i - f_i)^2}{\sigma_i^2} \tag{4.28}$$

instead of SS$_{\text{res}}$. (For now χ^2 is just a symbol. We'll learn about the χ^2 distribution in section 10.5.) Differentiating χ^2 with respect to b and setting to zero gives

$$\sum_i \frac{-2x_i(y_i - a - bx_i)}{\sigma_i^2} = 0. \tag{4.29}$$

Comparing this to equation 4.5, we see that the solution for the coefficients will have the same form as before, but with the averaged quantities replaced by their inverse variance weighted quantities. Working through the maths we get

$$b = \frac{(\sum \sigma_i^{-2})(\sum \sigma_i^{-2} x_i y_i) - (\sum \sigma_i^{-2} x_i)(\sum \sigma_i^{-2} y_i)}{(\sum \sigma_i^{-2})(\sum \sigma_i^{-2} x_i^2) - (\sum \sigma_i^{-2} x_i)^2} \tag{4.30}$$

and

$$a = \left(\frac{\sum \sigma_i^{-2} y_i}{\sum \sigma_i^{-2}} \right) - b \left(\frac{\sum \sigma_i^{-2} x_i}{\sum \sigma_i^{-2}} \right) \tag{4.31}$$

where all the sums are of course over i. Comparing these to equations 4.6 and 4.8 we see that we have generalized the mean quantities involved to be the inverse variance weighted means. This follows because of the linearity of the model: we had essentially the same system of linear equations. The variances and covariance of the parameters are similarly modified by the inverse variance. We will see a generalized matrix formulation of this in section 4.5.1.

4.1.5 Using R: the `lm` function

R provides a flexible interface for doing linear modelling with `lm`. It uses a formula notation, so to fit a straight line model to $\{x_i, y_i\}$ we write `lm(y ~ x)`. The following shows what it provides.

```
x <- c(10.9, 12.4, 13.5, 14.6, 14.8, 15.6, 16.2, 17.5, 18.3, 18.6)
y <- c(24.8, 30.0, 31.0, 29.3, 35.9, 36.9, 42.5, 37.9, 38.9, 40.5)
plot(x, y)
model1 <- lm(y ~ x)
summary(model1)

Call:
lm(formula = y ~ x)

Residuals:
    Min      1Q  Median      3Q     Max
-4.2029 -1.3692 -0.6237  1.2761  5.8294

Coefficients:
            Estimate Std. Error t value Pr(>|t|)
(Intercept)   4.5972     5.8853   0.781 0.457213
x             1.9798     0.3815   5.190 0.000833 ***
---
Signif. codes:  0 '***' 0.001 '**' 0.01 '*' 0.05 '.' 0.1 ' ' 1

Residual standard error: 2.889 on 8 degrees of freedom
Multiple R-squared:  0.771,Adjusted R-squared:  0.7424
```

```
F-statistic: 26.93 on 1 and 8 DF,   p-value: 0.0008328

attributes(model1)
  $names
   [1] "coefficients"   "residuals"        "effects"       "rank"
   [5] "fitted.values" "assign"            "qr"            "df.residual"
   [9] "xlevels"        "call"             "terms"         "model"

  $class
   [1] "lm"
```

Under the heading Residuals we see the quartiles of the residuals about the fit. Coefficients shows the estimated values of the parameters and their standard deviations, computed as described above. The function also performs a t test to determine the significance of these parameters given their standard errors (the columns labelled t value and Pr(>|t|)). I will explain these in section 10.4 after we have learned about hypothesis testing. For now it is sufficient to consider smaller probabilities as suggesting that the parameter is more likely to be required in the fit. After this come several statistics, some of which I defined earlier in this chapter. The various quantities calculated by lm can be listed using attributes and can be accessed using the $ symbol, e.g. with model1$coefficients. Regardless of any formal statistical tests, we should always inspect the quality of the fit by plotting it over the data, and by looking for any structure in the residuals. This is done by the following code, which produces figure 4.1.

R file: linear_regression.R

```
##### Demonstration of lm

x <- c(10.9, 12.4, 13.5, 14.6, 14.8, 15.6, 16.2, 17.5, 18.3, 18.6)
y <- c(24.8, 30.0, 31.0, 29.3, 35.9, 36.9, 42.5, 37.9, 38.9, 40.5)
model1 <- lm(y ~ x)
model1$fitted.values
predict(model1) # lm has a predict method
xv <- 10:20
#yv <- predict(model1, xv) # does not work: need to name the variable 'x'
yv <- predict(model1, data.frame(x=xv))
pdf("linear_regression.pdf", 8, 4)
par(mfrow=c(1,2), mgp=c(2.0,0.8,0), mar=c(3.5,3.5,1,1), oma=0.1*c(1,1,1,1))
plot(x, y)
lines(xv, yv)
# lm has an abline method so you can instead do abline(model1)
plot(x, model1$residuals, ylab="residuals", pch=16)
abline(h=0, col="grey60")
dev.off()
```

In the above example we have no uncertainties on the y values. In this case, R just uses the residual standard error (equation 4.25) as the estimate of σ in equation 4.15, as explained earlier. In fact, this is the standard way to use lm. But what if we *do* have a value for σ? It seems that lm does not accommodate this, although we can easily write an R function to do so. If we have individual uncertainties on each data point, i.e. the σ_i are all different (heteroscedastic), then once again we can just use the equations above (such as equation 4.14 for the variance in b). However, we probably then want to give more weight

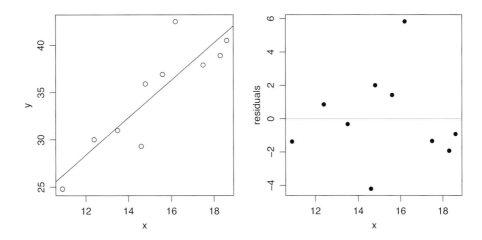

Using the R function `lm` to do linear regression. Left shows the data and fit. Right shows the residuals of the fit.

in the fit to points with smaller uncertainties, and thus do a weighted least squares fit. We can do this with `lm` by giving it a vector of weights via the parameter `weights`. A common thing to do with data from a Gaussian measurement model is *inverse variance weighting*, where each weight is σ_i^{-2}, as described in section 4.1.4.

4.2 Linear models

The model $f = a + bx$ is called linear because it is linear in the parameters a and b. It does not need to be linear in the data x. Thus a model $f = \sum_j \beta_j x^j$ (for any range of j) is a linear model. If $h_j(x)$ is some arbitrary function of the data with *fixed* parameters, such as $\sin(\omega_j x)$, then $f = \sum_j \beta_j h_j(x)$ is also a linear model, because it is linear in the unknown (or "free") parameters $\{\beta_j\}$. A model is linear if the unknown parameters only appear to the first power, i.e. there are no terms like β_j^2 or $\beta_j \beta_{j'}$. A linear model $f(\theta)$, where θ are the unknown parameters and a is a constant, has the properties

$$f(\theta_1 + \theta_2) = f(\theta_1) + f(\theta_2) \quad \text{and} \tag{4.32}$$
$$f(a\theta) = af(\theta). \tag{4.33}$$

Linear regression means doing regression with a linear model. We shall see in section 4.6 how to do linear regression with nonlinear functions of the data. Nonlinear regression, in contrast, means the model is nonlinear in the parameters. We shall encounter this in chapter 9.

4.3 Degrees of freedom

The *degrees of freedom* of an estimate is the number of independent measures that are used to make that estimate. Imagine we have N independent measurements. This represents N pieces of information, so N degrees of freedom. Suppose we are now told the mean of the numbers. Given this, the numbers are no longer independent: once we know the first $N - 1$ numbers, the final one is determined by the additional knowledge of the mean. By estimating one parameter from the data – here the mean – the remaining degrees of freedom goes down by one. If we now estimate the (sample) variance of the data, then the degrees of freedom that goes into this new calculation is $N - 1$. This is because the mean is also used in the calculation, so the variance is essentially estimated from just $N - 1$ independent numbers. The degrees of freedom in the estimates of the mean and variance (i.e. how many go into their calculation) are N and $N - 1$ respectively. Each subsequent independent estimate from the data reduces the remaining degrees of freedom by one. Hence, if we fit a linear model with p parameters to N data points, the number of degrees of freedom in the data after having made the fit (or just "in the fit") is $N - p$. A quadratic ($p = 3$) fit to seven data points, for example, would leave four degrees of freedom. We cannot fit a degree seven polynomial ($p = 8$) to seven data points because the degrees of freedom would be negative; fit parameters would remain undetermined. Generally speaking, a more robust fit is one which has more degrees of freedom, e.g. a quadratic curve can normally be better fit to 20 than to seven data points.

This simple relation between the degrees of freedom and the number of parameters only applies when the model is linear in its parameters. The idea that the degrees of freedom is reduced by one per parameter does not apply to nonlinear models. This is because nonlinear models with very few free parameters can be made to fit almost any data arbitrarily well, depending on the model. Consider the function $y = a \sin(\omega x + \phi)$. With suitable choices of the three parameters a, ω, and ϕ, we can fit *any* data set with any number of data points *perfectly*. The function may look ridiculously implausible, but χ^2 (section 10.5) will be zero, and the idea that the degrees of freedom is $N - p$ is meaningless.

4.4 Maximum likelihood and maximum posterior probability

In the previous chapter we learned that the Bayesian approach to finding the parameters of a parametric model is to write down the likelihood and prior, the product of which is proportional to the posterior PDF over the parameters. The result is an entire probability distribution over the parameters given the data. (We will see in chapter 9 how to do this for the line fitting problem.) From this we could then define an estimator, such as the mean, median, or mode of the distribution. In this chapter, in contrast, we have solved for a single set of parameters by minimizing the sum of squared residuals. Although finding the full posterior is normally preferable, there are situations were we are only interested in some "best" solution. We can define "best" as the maximum of the posterior and attempt to find this directly. If we want to use a uniform prior (and often we won't), then the maximum of the posterior coincides with the maximum of the likelihood. Finding this maximum with respect to the parameters is the procedure known as *maximum likelihood*.

We have a set of data $D = \{y_i\}$. The likelihood – probability density – of one data point for model M with parameters θ is $P(y_i | \theta, M)$. Assuming that the data have been measured independently, then from the rules of probability the likelihood of the whole data set is

$$P(\{y_i\} | \theta, M) = \prod_i P(y_i | \theta, M). \tag{4.34}$$

This is sometimes written as $L(\theta)$ (for brevity, or out of laziness), because we are interested in maximizing it with respect to the parameters for fixed data. But don't forget that it is still a PDF in the data! It is convenient to work with the log likelihood, as this converts the product into a sum

$$\ln L(\theta) = \ln \prod_i P(y_i | \theta, M) = \sum_i \ln P(y_i | \theta, M). \tag{4.35}$$

This is valid because the logarithm is a strictly monotonic function of its argument, so a maximum in $\ln L(\theta)$ is a maximum in $L(\theta)$. The maximum, which defines $\hat{\theta}$, is

$$\frac{d \ln L(\theta = \hat{\theta})}{d\theta} = 0. \tag{4.36}$$

If we're lucky this has a closed form and we can solve it analytically, as is the case with linear models. But often we have to solve numerically, something we will look at in section 12.6.

Having found the maximum likelihood solution, we still need to provide uncertainties; an estimate without an uncertainty is of limited value. One way to do this will be presented in section 7.1.

One of the positive features of maximum likelihood is that its solution is invariant under strictly monotonic transformations of the parameter, as we now see. Suppose the maximum of $L(\theta)$ is at $\hat{\theta}$. We transform the parameter via a strictly monotonic function $h(\theta)$. Such a transformation guarantees the function is invertible, i.e. $\theta(h)$ exists. The maximum likelihood solution is

$$0 = \frac{dL(\theta)}{d\theta} = \frac{L(\theta(h))}{dh} \frac{dh}{d\theta}. \tag{4.37}$$

As $dh/d\theta \neq 0$ (due to strict monotonicity), this equation is only satisfied when $dL(\theta(h))/dh = 0$, which gives a point estimate $\hat{h}$. But the above equation also tells us that this occurs at $\hat{\theta}$, so it follows that $\hat{h}(\theta) = h(\hat{\theta})$. Thus if we transform the parameter we can simply transform the maximum.

This invariance arises because the likelihood is not a *density* function in θ. Although we wrote the likelihood as $L(\theta(h))$ above, this is a shorthand for $P(D | \theta(h))$, a density function in D. Thus changing from θ to h does not change the density function. If we instead have some density function $P(\theta)$ in θ, then when we transform the parameter, the density function becomes $P(h) = P(\theta) d\theta/dh$. As $d\theta/dh$ is generally a function of θ, a maximum in $P(\theta)$ does not correspond to a maximum in $P(h)$. Thus, in general, the maximum of a density function is not invariant under a transformation of its parameter (or parameters). We already saw an example of this on page 34, when we transformed the coordinates of a bivariate density function from Cartesian to radial coordinates.

Note that it is only the maximum of the likelihood which is invariant under strictly monotonic transformations of θ. The shape of the likelihood can change, so other statistics like the mean or the variance (used to characterize uncertainties) are generally not invariant.

4.4.1 Maximum likelihood estimate of the weighted mean of Gaussian variables

We have measured a set of values $\{y_i\}$ with corresponding uncertainties $\{\sigma_i\}$. Assume these uncertainties are Gaussian. What is the maximum likelihood estimate of the mean of the data?

The question tells us each point has been drawn from a Gaussian distribution with $\mathcal{N}(\mu, \sigma_i)$, where μ is the mean we wish to estimate. Thus the likelihood L is

$$L(\mu) = \prod_i \frac{1}{\sigma_i \sqrt{2\pi}} \exp\left[-\frac{(y_i - \mu)^2}{2\sigma_i^2}\right] \tag{4.38}$$

$$\ln L(\mu) = -\sum_i \ln(\sigma_i \sqrt{2\pi}) - \sum_i \frac{(y_i - \mu)^2}{2\sigma_i^2}. \tag{4.39}$$

To get the maximum likelihood estimate we differentiate with respect to μ and set to zero

$$\sum_i \frac{y_i - \mu}{\sigma_i^2} = 0$$

$$\sum_i \frac{y_i}{\sigma_i^2} = \mu \sum_i \frac{1}{\sigma_i^2}$$

$$\mu = \frac{\sum_i y_i/\sigma_i^2}{\sum_i 1/\sigma_i^2}. \tag{4.40}$$

This justifies the use of an inverse variance weighted mean when we have errors on each data point (section 2.8.7): it is the maximum likelihood estimate. If the standard deviations are all the same, $\sigma_i = \sigma$, so we get $\mu = (1/N) \sum_i y_i$.

4.4.2 Least squares and χ^2 as a case of maximum likelihood

When doing linear regression (section 4.1) you may have been wondering why we do least *squares*. That is, why do we minimize $\sum_i (y_i - f_i)^2$? Why not minimize instead the sum of absolute deviations $\sum_i |y_i - f_i|$, for example? The answer is "maximum likelihood with Gaussian errors". The (log) likelihood (considered as a function of the model parameter μ) is given by equation 4.39, which we can write as

$$\ln L(\mu) = c - \frac{1}{2}\sum_i \frac{(y_i - \mu)^2}{\sigma_i^2} = c - \frac{1}{2}\chi^2 \tag{4.41}$$

for some constant c. Thus by minimizing the residual sum of squares (or χ^2 when we have errors on each point) we obtain the maximum likelihood solution. Gaussian error models are widespread, leading us to least squares.[6]

4.4.3 Minimizing L^1 and L^2 losses

The sum of squared residuals is also called the L^2 *loss function*, because it involves the residuals to the power of 2. (The L^2 *norm* or L^2 *distance* is the square root of this.) Suppose we have a normalized PDF $P(z)$. We would like to find the estimator $\hat{z}$ which minimizes the expected value of the squared residuals $E[(z - \hat{z})^2]$. From the definition of expectation (equation 1.23) this means that

$$0 = \frac{\partial}{\partial \hat{z}} \int (z - \hat{z})^2 \, P(z) \, dz$$

$$= -2 \int (z - \hat{z}) \, P(z) \, dz$$

$$\hat{z} = \int z \, P(z) \, dz = E[z] \tag{4.42}$$

as $\int P(z)dz = 1$. Thus provided $P(z)$ has finite zeroth, first, and second moments, the mean minimizes the L^2 loss function.

Which quantity minimizes the expected value of the absolute residuals $|z - \hat{z}|$ (the L^1 *loss function*)? We proceed in the same way. The function $|z - \hat{z}|$ looks like V, so its derivative with respect to $\hat{z}$ is -1 when $z - \hat{z} < 0$ and $+1$ when $z - \hat{z} > 0$. Hence we can write

$$0 = \frac{\partial}{\partial \hat{z}} \int |z - \hat{z}| \, P(z) \, dz$$

$$= -\int_{-\infty}^{\hat{z}} P(z) \, dz + \int_{\hat{z}}^{+\infty} P(z) \, dz. \tag{4.43}$$

The two integrals are equal when $\hat{z}$ is the median of $P(z)$ (by definition). Thus the median minimizes the L^1 loss function.

4.4.4 Maximum posterior

When we have a non-uniform prior $P(\theta)$, we cannot simply maximize the likelihood. We must instead maximize the posterior, the product of the likelihood with the prior. It's again easier to work with the logarithm. Bayes' theorem (equation 3.12; dependence on the model M is implicit) is

$$\ln P(\theta \,|\, D) = \ln L(\theta) + \ln P(\theta) - \ln Z(D) \tag{4.44}$$

---

[6] While it is sometimes fine to maximize the likelihood to find the best fitting parameters, we should not use this method to find the best model among a set of different models, e.g. to determine whether a polynomial or a sinusoidal model is the best curve. Why this is, and how we find the best model, will be explained in chapter 11.

where $Z(D)$ is the normalization constant (which is independent of θ, so will disappear when we differentiate). The maximum of this is often called the *maximum a posteriori* (MAP) estimate of the parameters. As the posterior is a density function in θ, the MAP is generally not invariant under a strictly monotonic transformation of θ (unlike the maximum likelihood, as we saw at the beginning of this section).

Of course, if the prior is uniform in θ, then the MAP will give the same solution as the maximum likelihood.[7] Yet once we introduce priors, the issue arises of how to represent the prior. If we say "uniform prior", do we mean uniform in θ or uniform in $\ln \theta$, for example? (Assigning priors is discussed in sections 3.5.5 and 5.3.) A prior uniform in θ is not uniform in $\ln \theta$. The shape of the prior is generally not invariant under transformations of the parameter ("reparametrizations"), and so the maximum of its product with the likelihood is also not invariant.

Likewise, other point estimates of the posterior as well as interval estimates – used to give uncertainties – are generally non-invariant under reparametrizations, so we usually will be concerned with the parameter representation. A general invariance is difficult to achieve (and may anyway not always be necessary). There are important types of invariance which can be attained, however, as we will see in section 5.3.

4.4.5 Maximum posterior estimate for Poisson data

We are given a set of N independent measurements $\{y_i\}$. We are told they have been drawn from a Poisson distribution with unknown parameter λ, but we do know a priori that $P(\lambda) \propto \exp(-\lambda/a)$ where a is known. What is the maximum posterior (MAP) estimate of λ?

It is instructive to first find the maximum likelihood solution. From the Poisson density function (equation 1.51) the likelihood is

$$L(\lambda) = \prod_{i=1}^{N} \frac{e^{-\lambda} \lambda^{y_i}}{y_i!} \tag{4.45}$$

the logarithm of which is

$$\ln L(\lambda) = \sum_{i=1}^{N} y_i \ln \lambda - \lambda - \ln(y_i!). \tag{4.46}$$

Differentiating with respect to λ and setting to zero we get the maximum likelihood solution,

$$\sum_{i=1}^{N} \left(\frac{y_i}{\lambda} - 1 \right) = 0$$

$$\lambda = \frac{1}{N} \sum_{i=1}^{N} y_i. \tag{4.47}$$

---

[7] For this reason maximum likelihood is often thought of as a special case of MAP. This isn't quite correct though, because the concept of priors does not exist in the maximum likelihood approach; so the issue of how to represent the parameter doesn't arise.

Let's now find the MAP solution. The normalized prior is

$$P(\lambda) = \frac{1}{a}e^{-\lambda/a} ,$$

(4.48)

from which we get

$$\frac{d \ln P(\lambda)}{d\lambda} = -\frac{1}{a}.$$

(4.49)

The log posterior (equation 4.44) is

$$\ln P(\lambda | \{y_i\}) = \ln L(\lambda) + \ln P(\lambda) + \text{constant}$$

(4.50)

where the constant is independent of λ. Differentiating with respect to λ and setting to zero gives

$$\sum_{i=1}^{N} \left(\frac{y_i}{\lambda} - 1 \right) - \frac{1}{a} = 0$$

$$\lambda = \frac{1}{N + 1/a} \sum_{i=1}^{N} y_i.$$

(4.51)

The effect of the prior is to reduce the estimate of λ relative to the maximum likelihood solution by an amount which depends on a. This is consistent with the prior, which favours smallers solutions. In the limit $a \to \infty$ the prior is flat (uniform improper prior), and the maximum posterior is just the maximum likelihood solution again.

4.5 Multi-dimensional linear regression

Let us now generalize linear regression to J dimensions. Borrowing language from machine learning, each "input" is a now a J-dimensional (column) vector $\mathbf{x}$, and the corresponding "output" is the scalar y. Given a set of $i = 1 \ldots N$ inputs and outputs, we want to fit a J-dimensional hyperplane[8] of y to $\mathbf{x}$. Our model for any one output y_i is

$$f_i = \beta_0 + \sum_{j=1}^{J} x_{i,j}\beta_j$$

$$= \mathbf{x}_i^{\mathsf{T}}\boldsymbol{\beta}$$

(4.52)

where in the second line I have absorbed the constant number 1 into the vector to form the $J + 1$ input vector $\mathbf{x}_i^{\mathsf{T}} = (1, x_{i,1}, \ldots, x_{i,j}, \ldots, x_{i,J})$. Here $\boldsymbol{\beta}$ is the $J + 1$ dimensional parameter vector in which β_0 is the first element. We can write the set of N inputs as an

---

[8] The prefix "hyper" is used to indicate the higher dimensional generalization of a geometric concept, such as a plane, sphere, or volume.

$N \times (J + 1)$ matrix X, in which each row is the vector of $J + 1$ "features" for input i, i.e.

$$X = \begin{pmatrix} \mathbf{x}_1^\mathsf{T} \\ \vdots \\ \mathbf{x}_N^\mathsf{T} \end{pmatrix}. \tag{4.53}$$

This is called the *design matrix*. The model can now be written

$$\mathbf{f} = X\beta \tag{4.54}$$

with the corresponding N outputs written as the vector $\mathbf{y}$.

Proceeding as with one-dimensional regression, we want to minimize

$$\begin{aligned} \mathrm{SS_{res}} &= \sum_{i=1}^{N}(y_i - f_i)^2 \\ &= \sum_i (y_i - \mathbf{x}_i^\mathsf{T}\beta)^2 \\ &= (\mathbf{y} - X\beta)^\mathsf{T}(\mathbf{y} - X\beta) \end{aligned} \tag{4.55}$$

with respect to β. This is quadratic in β so always has a turning point (and you can show that it is a minimum). Differentiating and setting to zero we get

$$\begin{aligned} 0 &= X^\mathsf{T}(\mathbf{y} - X\beta) \\ X^\mathsf{T}X\beta &= X^\mathsf{T}\mathbf{y} \\ \beta &= (X^\mathsf{T}X)^{-1}X^\mathsf{T}\mathbf{y} \end{aligned} \tag{4.56}$$

provided $X^\mathsf{T}X$ (the *information matrix*) is not singular. The model-predicted values of y are therefore

$$\begin{aligned} \mathbf{f} &= X\beta \\ &= X(X^\mathsf{T}X)^{-1}X^\mathsf{T}\mathbf{y} \end{aligned} \tag{4.57}$$

whereby the $N \times N$ matrix $X(X^\mathsf{T}X)^{-1}X^\mathsf{T}$ is sometimes called the *hat matrix*, because it put a hat on – makes an estimator out of – $\mathbf{y}$.

Note that for a Gaussian likelihood, the log likelihood is $-\frac{1}{2}\mathrm{SS_{res}}$, to within an additive constant. What is labelled above as the information matrix is a special case of a more general definition which will be encounter in section 5.3.2.

This method of getting a solution for the coefficients is known as *ordinary least squares*.

4.5.1 Generalized least squares

We can generalize line fitting further to the case of having different uncertainties on each of the N outputs $\{y_i\}$. We can even allow them to have correlations. Both of these are characterized by the $N \times N$ covariance matrix Σ, the elements of which are $c_{ij} = \mathrm{Cov}(y_i, y_j)$. The elements in the leading diagonal are the variances; the off-diagonal elements are the covariances (see section 1.6.2). Assuming a Gaussian error model, the likelihood of these data is an N-dimensional Gaussian

$$L(\{y_i\}) = \frac{1}{(2\pi)^{N/2}|\Sigma|^{1/2}} \exp\left(-\frac{1}{2}\chi^2\right) \quad \text{where}$$

$$\chi^2 = (\mathbf{y} - X\boldsymbol{\beta})^{\mathsf{T}}\Sigma^{-1}(\mathbf{y} - X\boldsymbol{\beta})$$

(4.58)

(cf. equation 4.28). By minimizing χ^2 with respect to $\boldsymbol{\beta}$ we maximize the likelihood (assuming Σ is fixed). This is a generalization of ordinary least squares. The residual sum of squares metric in equation 4.55 is a special case in which Σ is the identity matrix. Differentiating χ^2 with respect to $\boldsymbol{\beta}$ and setting to zero gives the solution

$$\boldsymbol{\beta} = (X^{\mathsf{T}}\Sigma^{-1}X)^{-1}X^{\mathsf{T}}\Sigma^{-1}\mathbf{y} .$$

(4.59)

4.6 One-dimensional regression with nonlinear functions

As discussed in section 4.2, the "linear" in "linear regression" refers to linear in the parameters, not in the data. Provided the model is linear in its parameters, then the metric we are minimizing, SS_{res}, is a quadratic function of the parameters (so its derivative is linear in the parameters), yielding a unique minimum we can find analytically. We can use linear regression to fit models which are nonlinear in the data, by doing multi-dimensional regression in which the other dimensions are nonlinear expression of the data. One way to do this is with the polynomial expansion up to the Jth order

$$f_J(x) = \sum_{j=0}^{J} x^j \beta_j .$$

(4.60)

Expressed in this way, we can use all the machinery from the previous section.

The R code below uses `lm` to investigate fitting a nonlinear relationship. `model1` is the linear expression $y = \beta_0 + \beta_1 x$, `model2` is $y = \beta_0 + \beta_1 x + \beta_2 x^2$, and `model2b` drops the linear term, $y = \beta_0 + \beta_2 x^2$. `model3` is $y = \beta_0 + \beta_1 x + \beta_2 x^2 + \beta_3 x^3$. The fits and residuals of the straight line and quadratic models are shown in figure 4.2. The code outputs are shown together with the code below. I suggest you work through it before reading on. The quadratic model gives a significantly better fit than the straight line one in terms of the residual standard error and the adjusted r-squared. The cubic model is only marginally better on this count. The t test (to be covered in chapter 10) suggests that neither the quadratic nor the cubic term is significant. This is not the case, however, because we see that the linear fit is much poorer than the quadratic fit. Both terms achieve low significances because this test only investigates what happens if we remove them one term at a time, in which case the other term fits much of the nonlinearity (this will be explained in section 10.4). The quadratic model without the linear term (`model2b`) fits the data much worse than the full quadratic model, as suggested by the residual standard error (it's even worse than the straight line model). If you plot the fitted curve, it even seems to bend the wrong way. The reason is that a model $y = \beta_0 + \beta_2 x^2$ always has its turning point at $x = 0$.

R file: `nonlinear_functions.R`

```
##### Demonstration of linear regression with nonlinear functions of the
##### data using lm()

x <- c(0.9, 2.1, 2.9, 3.8, 5.3, 6.0, 7.0, 8.4, 9.1, 9.8, 11.1, 11.9, 13.2,
       13.8, 14.8)
y <- c(19.5, 13.8, 16.6, 11.6, 8.3, 9.9, 7.6, 6.6, 6.2, 5.4, 5.9, 5.5, 5.6,
       4.4, 4.7)
par(mfrow=c(2,2), mgp=c(2.0,0.8,0), mar=c(3.5,3.5,1,1), oma=0.1*c(1,1,1,1))
plot(x,y) # looks nonlinear...


# Straight line model
model1 <- lm(y ~ x)
summary(model1) # r^2 is a bit low
abline(model1, col="blue", lw=2)
plot(x, model1$residuals, pch=20, col="blue", ylim=c(-5,5))
# we see structure in the residuals
abline(h=0)


# Quadratic model
model2 <- lm(y ~ x + I(x^2)) # note meaning of "+" and I()
summary(model2) # looks better than linear
xv  <- seq(from=0, to=30, by=0.1) # generate data for prediction,
                                  # as abline doesn't do curves
yv2 <- predict(model2, data.frame(x=xv))
plot(x,y)
lines(xv, yv2, col="red", lw=2)
plot(x, model2$residuals, pch=20, col="red", ylim=c(-5,5))
# we see less structure in residuals
abline(h=0)


# Cubic model
model3 <- lm(y ~ x + I(x^2) + I(x^3))
summary(model3)
# no evidence for cubic term; r^2 hardly drops compared to quadratic
yv3 <- predict(model3, data.frame(x=xv))
plot(x,y)
lines(xv, yv3, col="magenta", lw=2)
plot(x, model3$residuals, pch=20, col="magenta", ylim=c(-5,5))
# this looks no better than quadratic
abline(h=0)


# Quadratic model without the linear term
model2b <- lm(y ~ I(x^2)) # can we drop linear term in the quadratic model?
summary(model2b)
yv2b <- predict(model2b, data.frame(x=xv))
plot(x,y)
lines(xv, yv2b, col="brown", lw=2) # no we can't!
plot(x, model2b$residuals, pch=20, col="brown", ylim=c(-5,5))
abline(h=0)
```

```
# Plot linear and quadratic results together
pdf("nonlinear_functions.pdf", 8, 4)
par(mfrow=c(1,2), mgp=c(2.0,0.8,0), mar=c(3.5,3.5,1,1), oma=0.1*c(1,1,1,1))
plot(x,y)
abline(model1, lw=2)
lines(xv, yv2, lw=2, lty=2)
plot(x,   model1$residuals, pch=20, ylim=c(-4.5,4.5), ylab="residuals")
points(x, model2$residuals, pch=4)
abline(h=0, col="grey60")
dev.off()
```

Here is an edited version of the summaries produced by R for each of the four models.

```
lm(formula = y ~ x)
Residuals:
     Min      1Q  Median      3Q     Max
 -2.9938 -1.6293 -0.4736  1.3421  4.1089
Coefficients:
             Estimate Std. Error t value Pr(>|t|)
(Intercept)  16.2292     1.1652  13.929 3.43e-09 ***
x            -0.9312     0.1282  -7.264 6.33e-06 ***
Residual standard error: 2.136 on 13 degrees of freedom
Multiple R-squared:  0.8023,Adjusted R-squared:  0.7871
```

```
lm(formula = y ~ x + I(x^2))
Residuals:
      Min       1Q   Median       3Q      Max
 -2.00968 -0.65973  0.04368  0.73096  2.41122
Coefficients:
             Estimate Std. Error t value Pr(>|t|)
(Intercept) 20.68163    1.08735  19.020 2.51e-10 ***
x           -2.53278    0.31632  -8.007 3.73e-06 ***
I(x^2)       0.10133    0.01946   5.207 0.000219 ***
Residual standard error: 1.231 on 12 degrees of freedom
Multiple R-squared:  0.9394,Adjusted R-squared:  0.9292
```

```
lm(formula = y ~ x + I(x^2) + I(x^3))
Residuals:
      Min       1Q   Median       3Q      Max
 -2.02586 -0.50426 -0.05821  0.32114  2.68788
Coefficients:
              Estimate Std. Error t value Pr(>|t|)
(Intercept) 22.230245   1.602753  13.870 2.59e-08 ***
x           -3.566737   0.860392  -4.145  0.00163 **
I(x^2)       0.260299   0.124959   2.083  0.06137 .
I(x^3)      -0.006712   0.005215  -1.287  0.22451

Residual standard error: 1.199 on 11 degrees of freedom
Multiple R-squared:  0.9473,Adjusted R-squared:  0.9329
```

```
lm(formula = y ~ I(x^2))
Residuals:
     Min      1Q  Median      3Q     Max
 -3.2085 -2.6313 -0.5933  1.2637  6.6229
Coefficients:
             Estimate Std. Error t value Pr(>|t|)
(Intercept) 12.91773    1.19060   10.85 6.93e-08 ***
```

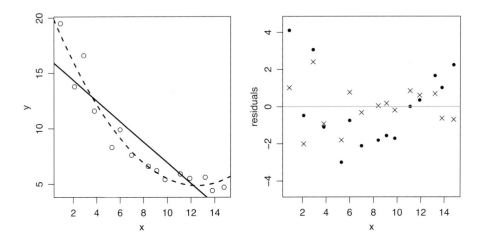

Fig. 4.2   Demonstration of linear regression with nonlinear functions of the data. Left shows the data (open circles) and fit for the straight line model (solid line) and quadratic model (dashed line). Right shows the residuals for the straight line model (filled circles) and quadratic model (crosses).

```
I(x^2)      -0.05017    0.01100   -4.56 0.000535 ***
Residual standard error: 2.979 on 13 degrees of freedom
Multiple R-squared:   0.6153, Adjusted R-squared:   0.5858
```

Using linear least squares on expansions of the data – so-called *basis functions* or *basis expansions* – is a powerful approach to defining and fitting nonlinear functions of data. We will look at these further in section 12.3.

4.7 Least squares with errors on both axes

So far we have considered the x variable (or $\mathbf{x}$ variables) as the "independent" variable and y as the "dependent" variable, because we have been fitting a model $f(x)$ or $f(\mathbf{x})$. This is because we think of x as being fixed, i.e. noise free. If y were instead noise free, then there is nothing to stop us from fitting a model $x = f^{-1}(y)$ instead. Least squares does not generally give the same solution – line in (x, y) space – because we are minimizing a different sum of squares.

But what do we do if we have errors in x as well as y? Errors in x must affect the solution. This is a so-called *errors in variables* problem. Let x' and y' represent the model-predicted (i.e. noise-free) values of x and y respectively. Our generative model is thus

$$y' = a + bx'. \tag{4.61}$$

Suppose we have the measurements $\{x_i\}$ and $\{y_i\}$ with corresponding uncertainties (Gaussian standard deviations) $\{\varsigma_i\}$ and $\{\sigma_i\}$ respectively. We now consider the residuals in both variables, $x - x'$ and $y - y'$. A plausible generalization of the χ^2 metric (equation 4.28) is the sum of squares of error-weighted residuals (Deming, 1943),

$$\text{SS}_{\text{res}} = \sum_i \left(\frac{y_i - y_i'}{\sigma_i} \right)^2 + \left(\frac{x_i - x_i'}{\varsigma_i} \right)^2, \tag{4.62}$$

as this now includes the residuals in x. If $\varsigma_i = \varsigma$ and $\sigma_i = \sigma$ for all i, then this can be written as

$$\text{SS}_{\text{res}} = \frac{1}{\sigma^2} \sum_i (y_i - a - b x_i')^2 + \lambda^2 (x_i - x_i')^2 \tag{4.63}$$

where $\lambda = \sigma / \varsigma$ is the ratio of the standard deviations. This has a (slightly complicated) analytic solution – for a, b, and the $\{x_i'\}$ – which depends not on the individual standard deviations, but only on λ. When $\lambda = 1$ the method is sometimes called *orthogonal least squares*, because it corresponds to measuring the residuals of the fit not parallel to the y-axis, but orthogonal to the line being fit.

This approach can be generalized for the case of non-zero covariance between the x and y variables. Let $\mathbf{z}_i = (x_i, y_i)^\mathsf{T}$ and $\mathbf{z}_i' = (x_i', y_i')^\mathsf{T}$ be the measured and true (unknown) positions of point i, respectively, and let their covariance be

$$\Sigma_i = \begin{bmatrix} \sigma_{x,i}^2 & \rho \sigma_{x,i} \sigma_{y,i} \\ \rho \sigma_{x,i} \sigma_{y,i} & \sigma_{y,i}^2 \end{bmatrix}. \tag{4.64}$$

Adopting a Gaussian measurement model, the likelihood for this single data point is a bivariate Gaussian (see equation 1.69)

$$L(a, b) = \frac{1}{2\pi |\Sigma_i|^{1/2}} \exp\left(-\frac{1}{2} (\mathbf{z}_i - \mathbf{z}_i')^\mathsf{T} \Sigma_i^{-1} (\mathbf{z}_i - \mathbf{z}_i') \right). \tag{4.65}$$

Solving this for the two model parameters is known as *total least squares*, and in general requires a numerical solution. We will see in section 9.4 how this can be generalized further for the case of arbitrary error models.

4.8 Bias-variance decomposition

I finish up this chapter with a useful way of looking at the sum of squared residuals.

Suppose we have a set of data $\{x\}$ (they could be scalar or vector) and corresponding values $\{y\}$, for which we assume a relationship

$$y = f(x) + \epsilon \tag{4.66}$$

exists, where $f(x)$ is an unknown deterministic function, and ϵ is a noise source with mean zero and variance σ^2. Fitting a model to these data means finding a good estimator for $f(x)$, which I label $\hat{f}(x)$. The best estimator in the least squares sense is the one which minimizes the sum of squared residuals $\sum_i (y_i - \hat{f}(x_i))^2$. I will now show that this metric can be written in terms of the bias and the variance of the estimator. First, because f is a deterministic function and the noise is additive, $E[f] = f$. Second, since $E[\epsilon] = 0$, it follows from the previous point and equation 4.66 that $E[y] = f$. Third, we can use the definition of variance (equation 1.30) to write

$$E[\hat{f}^2] = \text{Var}(\hat{f}) + E[\hat{f}]^2 \quad \text{and} \tag{4.67}$$
$$E[y^2] = \text{Var}(y) + E[y]^2 = \text{Var}(y) + f^2. \tag{4.68}$$

Using these we see that the expected squared residual is

$$\begin{aligned}
E[(y - \hat{f})^2] &= E[y^2] + E[\hat{f}^2] - 2E[y\hat{f}] \\
&= \text{Var}(y) + f^2 + \text{Var}(\hat{f}) + E[\hat{f}]^2 - 2fE[\hat{f}] \\
&= \text{Var}(y) + \text{Var}(\hat{f}) + (E[\hat{f}] - f)^2 \\
&= \text{Var}(y) + \text{Var}(\hat{f}) + E[\hat{f} - f]^2 \\
&= \sigma^2 + \text{Var}(\hat{f}) + \text{bias}^2. \tag{4.69}
\end{aligned}$$

The bias is defined as $E[\hat{f} - f]$, the expected difference between the estimator and the true function. Hence the expected squared residual is the sum of the estimator's variance and the square of the estimator's bias, plus an irreducible variance σ^2 due to the noise in the true relationship. A perfect model would have zero bias and zero variance. This is usually unachievable. Good models can be obtained either by reducing their bias, or their variance, or both. When we fit a model in practice, we usually have to trade off the bias against the variance: decreasing one often increases the other. We shall see this in chapter 12.

As we do not know the true model, $f(x)$, we cannot calculate this bias and this variance. The above is a way of thinking about the properties of the estimator (and perhaps evaluating them with simulations). But we can compute similar things from real data. Let $\delta_i = y_i - \hat{f}_i$ for $i = 1 \dots N$ be the residuals of a model fit. The root mean square (RMS) of these is

$$\text{RMS} = \sqrt{\frac{1}{N} \sum_{i=1}^{N} \delta_i^2}. \tag{4.70}$$

Contrast this with their standard deviation (SD) (neglecting the $N/(N-1)$ bias correction factor)

$$\text{SD} = \sqrt{\frac{1}{N} \sum_{i=1}^{N} (\delta_i - \bar{\delta})^2} \tag{4.71}$$

in which

$$\bar{\delta} = \frac{1}{N} \sum_{i=1}^{N} \delta_i \tag{4.72}$$

is a measure of the bias. Whereas RMS measures the total spread of the residuals including the bias, SD measures their spread about the bias. We see that

$$
\mathrm{SD}^2 \;=\; \frac{1}{N}\sum_{i=1}^{N}(\delta_i^2 + \bar{\delta}^2 - 2\bar{\delta}\delta_i)
$$

$$
\;=\; \mathrm{RMS}^2 + \bar{\delta}^2 - 2\bar{\delta}^2
$$

$$
\mathrm{RMS}^2 \;=\; \mathrm{SD}^2 + \bar{\delta}^2 \tag{4.73}
$$

which is analogous to equation 4.69, except that now the irreducible variance is part of the bias and standard deviation. It is possible to have estimators which have a small scatter in their residuals, i.e. a small SD, but still a large RMS due to a large bias.

## 5

Parameter estimation: single parameter

In chapter 3 I introduced the basic concepts of inference. In section 3.2 we used Bayes' theorem to interpret medical test results, and in section 3.5 we looked at the simple problem of estimating one parameter from one measurement. Here I take these ideas further to estimate the posterior probability density over a model parameter given a set of data, focusing on a single parameter problem and the use of so-called conjugate priors. We will see specifically how the prior and the likelihood combine to make the posterior, and how this depends on the amount of data available. I will also discuss assigning priors and summarizing distributions.

5.1 Bayesian analysis of coin tossing

We are given a coin and we toss it n times. It lands heads in r of them. Is the coin fair?

This question has no definitive answer; we can only answer it probabilistically. Let p be the unknown probability that the coin lands heads in any one toss. We can interpret the question to mean "what is the posterior PDF over p?", i.e. "what is $P(p|n, r, M)$?", where M describes our assumptions. Bayes' theorem (equation 3.12) tells us that the posterior is the product of the likelihood and prior, divided by a normalization constant.

While coin tossing is not a very scientific enterprise, this example stands in for any repeated process that has just two alternative outcomes, such as detecting or not detecting a type of source in a survey, or a disease in a test.

Let's first identify the likelihood, which is the probability of the data given the parameter. We set up a model M for the phenomenon and the measurement. This specifies that the coin lands heads in a single toss with probability p, that all tosses are independent, and that p is constant (does not change with the number of tosses already done). p is the only parameter of the model. The appropriate likelihood is the binomial distribution (section 1.4.1)

$$P(r|p, n, M) = \binom{n}{r} p^r (1-p)^{n-r} \tag{5.1}$$

where $r \leq n$. My notation here explicitly shows the conditioning on p, n, and M. Although n is part of the data we are given, the relevant likelihood describes the variation of r for given n, which is why n is on the right of the conditioning bar.[1] Thus in terms of Bayes'

---

[1] We could imagine a different problem in which r is given (fixed) and we are interested in the variation of n, e.g. how many times do we toss the coin until we achieve r heads. In that case we need the likelihood

theorem in equation 3.12, $\theta = p$, $D = r$, and n is an additional quantity which all terms are conditioned on. In this case Bayes' theorem is

$$P(p|r,n,M) = \frac{P(r|p,n,M)P(p|M)}{P(r|n,M)}.$$ (5.2)

I have removed the n in the conditioning of the prior $P(p|M)$ because the prior is independent of the number of coin tosses. In parameter estimation problems everything is conditioned on the model M, so to make the expressions less cumbersome authors sometimes remove this explicit conditioning. Sometimes n is removed too. It is important when manipulating probabilistic equations to know exactly what you are conditioning on. In this section I will retain all conditioning.

I will now adopt two different forms of the prior, derive and plot the posterior, and also work out the expectation value of p from this.

5.1.1 Uniform prior

Let us first adopt a uniform prior $\mathcal{U}(0,1)$ over p. In that case the posterior PDF is just proportional to the likelihood

$$P(p|r,n,M) = \frac{1}{Z} p^r (1-p)^{n-r}$$ (5.3)

where Z is a normalization constant which does not depend on p. I have absorbed the binomial coefficient into this normalization constant, since it depends only on n and r. Imagine we had $n = 20$, $r = 7$. We can plot the posterior PDF using the following code.

```
par(mfrow=c(1,1), mgp=c(2.0,0.8,0), mar=c(3.5,3.5,1,1), oma=0.1*c(1,1,1,1))
n <- 20
r <- 7
p <- seq(from=0,to=1,length.out=201)
plot(p, dbinom(x=r, size=n, prob=p), ylim=c(0, 0.2), xaxs="i", yaxs="i",
  xlab="p", ylab=expression(paste(P^symbol("*"), "(p | r,n,M)")), type="l",
  lwd=1.5)
```

The result is in figure 5.1. We see that the PDF peaks well away from a fair coin with $p = 1/2$, although a fair coin is not strongly disfavoured. The mode is at $r/n = 0.35$, which is probably most people's intuitive estimate for the value of p. Remember that this curve is *not* binomial in p (it is binomial in r). Furthermore, the posterior as plotted – as a function of p – is *unnormalized*; the integral over p is not unity. As normalization just rescales the whole curve by a scalar multiple, it does not change the peak (the mode) or relative probabilities of solutions. But normalization is necessary if we want to calculate expectation values, e.g. the mean or variance.[2] If we denote the unnormalized posterior PDF as $P^*(p|r,n,M)$ – which in this case is $p^r(1-p)^{n-r}$ – then the expectation value

---

$P(n|p,r,M)$, which is not the binomial distribution, but rather the negative binomial distribution (see equation 11.50).

[2] Normalization for this purpose will no longer be necessary if we were to *draw samples from* the distribution, as we shall see in chapter 8.

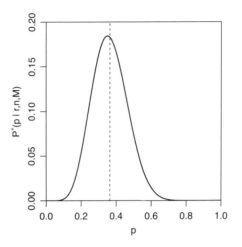

Fig. 5.1 The unnormalized posterior PDF for the probability p of a coin giving heads on a single toss, when $r = 7$ of $n = 20$ observed tosses are heads. A uniform prior over p is used. The mode is at $r/n = 0.35$. The dashed line is the mean, which is at 0.364.

of p is

$$E[p] = \int_0^1 p\,P(p|r,n,M)\,dp = \frac{1}{Z}\int_0^1 p\,P^*(p|r,n,M)\,dp \tag{5.4}$$

where $Z = \int_0^1 P^*(p|r,n,M)dp$. We can approximate these integrals with a sum. Here I use the rectangle rule (section 3.3.4), so for example

$$Z = \int_0^1 P^*(p|r,n,M)\,dp \simeq \sum_i P^*(p_i|r,n,M)\,\delta p \tag{5.5}$$

for some fixed (small) step size δp. Thus the normalized PDF is

$$P(p|r,n,M) \simeq \frac{1}{\sum_i P^*(p_i|r,n,M)\,\delta p}\,P^*(p|r,n,M). \tag{5.6}$$

When using such sums the step size appears in both the numerator and denominator of equation 5.4, so it cancels to leave

$$E[p] \simeq \frac{1}{\sum_i P^*(p_i|r,n,M)}\sum_i p_i\,P^*(p_i|r,n,M). \tag{5.7}$$

Although not used in the code below, we could estimate the variance (equation 1.30) in a similar way as

$$\mathrm{Var}(p) \simeq \frac{1}{\sum_i P^*(p_i|r,n,M)}\sum_i (p_i - E[p])^2\,P^*(p_i|r,n,M). \tag{5.8}$$

The following R code implements the mean as just described and overplots it in figure 5.1 with the dashed line.

```
pdense <- dbinom(x=r, size=n, prob=p) # unnormalized in p
p.mean <- sum(p*pdense)/sum(pdense)
abline(v=p.mean, lty=2)
```

It is instructive to repeat this example for $n = 20$ for a range of values of r. This is done by the following R code, which produces figure 5.2. The more heads we toss, the more our inference of p shifts towards larger values. But since we only have 20 tosses, a degree of uncertainty remains, as reflected by the finite width of the posterior PDF.

R file: coin1.R

```
##### Compute the posterior PDF for coin problem with a uniform prior for a
##### range of r

n <- 20
Nsamp <- 200 # no. of points to sample at
pdf("coin1.pdf", 9, 7)
par(mfrow=c(3,4), mgp=c(2,0.8,0), mar=c(3.5,3.5,1.5,1), oma=0.5*c(1,1,1,1))
deltap <- 1/Nsamp # width of rectangles used for numerical integration
p <- seq(from=1/(2*Nsamp), by=1/Nsamp, length.out=Nsamp) # rectangle centres
for(r in seq(from=0, to=20, by=2)) {
  pdense <- dbinom(x=r, size=n, prob=p)
  pdense <- pdense/(deltap*sum(pdense)) # normalize posterior
  plot(p, pdense, type="l", lwd=1.5, xlim=c(0,1), ylim=c(0,1.1*max(pdense)),
       xaxs="i", yaxs="i", xlab="p", ylab="P(p | r,n,M)")
  title(main=paste("r =",r), line=0.3, cex.main=1.2)
  p.mean <- deltap*sum(p*pdense)
  abline(v=p.mean, lty=2)
}
dev.off()
```

In the above code I divide the range 0–1 into Nsamp equal-sized intervals and compute the (at first unnormalized) posterior density pdense at the centre of each of these. Thus the first point in this grid is at 1/(2*Nsamp), the next is offset by $\delta p = 1/$Nsamp, etc. This grid is constructed by the following statement.

```
p <- seq(from=1/(2*Nsamp), by=1/Nsamp, length.out=Nsamp)
```

I then normalize the posterior – make the area under the curve equal to one – by dividing pdense by deltap*sum(pdense) (equation 5.6). It is then a probability *density* function, so sum(pdense) is not equal to one. As $\sum_i P(p_i|r, n, M)\delta p = 1$, it follows that sum(pdense) is just equal to 1/deltap, which is Nsamp.

5.1.2 Beta prior

In practice you are unlikely to have a personal prior for p which is uniform. Given a random coin, I suspect you are far more likely to believe that the coin is fair, or close to fair.[3] An appropriate density function for parameters constrained to lie in the range 0–1 is the beta

---

[3] If you were attending a magicians' convention you might well adopt a broader prior. Background information is important when analysing data. We do not interpret data independent of context.

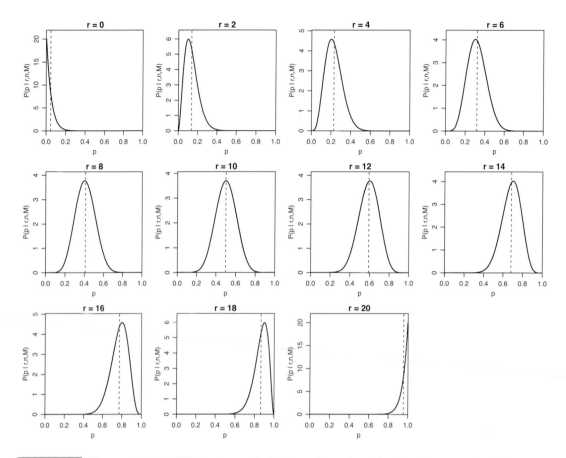

Fig. 5.2 The posterior PDF for the probability p of a coin giving heads on a single toss, when r of $n = 20$ observed tosses are heads. A uniform prior over p is used. The dashed line is the mean. Note the different vertical scales on the individual panels (the posterior is normalized).

distribution (section 1.4.3). As a reminder, this is described by two parameters, α and β. Its PDF is

$$P(p) = \frac{1}{B(\alpha, \beta)} p^{\alpha-1}(1-p)^{\beta-1} \quad \text{where} \quad \alpha > 0,\ \beta > 0,\ 0 \leq p \leq 1. \tag{5.9}$$

If $\alpha = \beta$ the function is symmetric, the mean and mode are 0.5, and the larger the value of α (when $\alpha \geq 1$) the narrower the distribution. An example is shown in figure 5.3. Multiplying this prior by the likelihood, and absorbing terms independent of p into the normalization constant Z, gives the posterior

$$P(p|r, n, M) = \frac{1}{Z} p^r(1-p)^{n-r} p^{\alpha-1}(1-p)^{\beta-1}$$

$$= \frac{1}{Z} p^{r+\alpha-1}(1-p)^{n-r+\beta-1}. \tag{5.10}$$

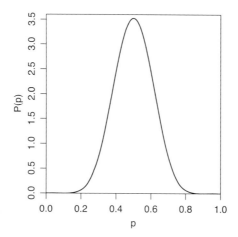

Fig. 5.3   The beta distribution prior with $\alpha = \beta = 10$.

The posterior has the same form as the prior (for this likelihood); it is also a beta distribution.[4] When this happens, we say that the prior and posterior are *conjugate distributions*. The prior is the *conjugate prior* for this likelihood function. This is not only convenient, but also instructive, because it helps us to appreciate how the data update the prior to produce the posterior. If we start off with a prior with parameters α_p and β_p, and then measure r heads from n tosses, the posterior is a beta distribution with parameters

$$\alpha = \alpha_p + r$$
$$\beta = \beta_p + n - r. \tag{5.11}$$

As larger values of these parameters correspond to narrower distributions,[5] we see that the data improve our knowledge of p by narrowing the posterior. The mean and mode of the posterior (from equation 1.4.3) are

$$\text{mean} = \frac{\alpha_p + r}{\alpha_p + \beta_p + n} \quad \text{and} \tag{5.12}$$

$$\text{mode} = \frac{\alpha_p + r - 1}{\alpha_p + \beta_p + n - 2} \tag{5.13}$$

respectively. We could use either of these as a single "best" estimate of p.

Equation 5.9 tells us that the uniform prior is a beta distribution with $\alpha = \beta = 1$. Thus with a uniform prior, the posterior (equation 5.3) is also a beta distribution, the mean and

---

[4] It follows from the definition of the beta distribution that the normalization constant in equation 5.10 is $Z = B(r + \alpha, n - r + \beta)$.

[5] If we set $\alpha = \beta$ into the expression for the variance of the beta distribution (equation 1.59) then we see that for large α, $\text{Var}(p) \sim 1/\alpha$.

mode of which are

$$\text{mean} = \frac{1+r}{2+n} \tag{5.14}$$

$$\text{mode} = \frac{r}{n}. \tag{5.15}$$

Given r heads from n tosses, what would *your* estimate of p be? The intuitive answer, r/n, is equal to the mode of the posterior when adopting a uniform prior. (This is also the maximum likelihood solution, which you can verify by differentiation; see section 4.4.) But if you tossed the coin three times and you got zero heads, would you really estimate p to be zero? Aren't you still more likely to think p is nearer to $1/2$ than to zero, given that the vast majority of coins you encounter are very close to fair?[6] Instead of the mode you could use the mean of the posterior as your estimator, which for zero heads from three tosses gives an estimate for p of 0.2 (and the standard deviation is 0.16, indicating there is considerable uncertainty). The nice thing about the mean in this case is that it gives sensible answers in the limit of little data. Indeed, even if $r = n = 0$ (no data, so the posterior is just the prior) the mean is $1/2$, whereas the mode is undefined (because the posterior is uniform). The standard deviation (of the uniform distribution) is $1/\sqrt{12} = 0.29$.

Equation 5.14 is sometimes called *Laplace's rule of succession*. The larger the number of coin tosses, the more peaked the posterior becomes, and the mean and mode both converge to r/n.

Let us now adopt a beta prior with $\alpha = \beta = 10$. This is plotted in figure 5.3. I think this is still a very conservative (broad) prior for a randomly encountered coin, but it is useful for illustrating the influence of the prior.[7] The following code performs the same experiment as before of varying r for $n = 20$, and produces figure 5.4.

R file: coin2.R

```
##### Compute posterior PDF for coin problem with a beta prior for a
##### range of r

n <- 20
alpha.prior <- 10
beta.prior  <- 10
Nsamp <- 200 # no. of points to sample at
pdf("coin2.pdf", 9, 7)
par(mfrow=c(3,4), mgp=c(2,0.8,0), mar=c(3.5,3.5,1.5,1), oma=0.5*c(1,1,1,1))
deltap <- 1/Nsamp # width of rectangles used for numerical integration
p <- seq(from=1/(2*Nsamp), by=1/Nsamp, length.out=Nsamp) # rectangle centres
for(r in seq(from=0, to=20, by=2)) {
  pdense <- dbeta(x=p, shape1=alpha.prior+r, shape2=beta.prior+n-r)
  plot(p, pdense, type="l", lwd=1.5, xlim=c(0,1), ylim=c(0, 6.5),
```

---

[6] If you still think inference should not depend on prior information, do this experiment with someone with a real coin, and bet real money on the outcome of the next toss of the coin. I'm sure that after three tails in three tosses you won't put a large amount of money on the next toss being tails.

[7] If $\alpha > 1$ and $\beta > 1$ the prior is zero at $p = 0$ and $p = 1$, meaning that the posterior will always be zero at these values no matter how much (finite) data we collect. It is generally a bad idea to assign a prior density of zero to a value of a parameter which is not actually impossible, even if it is highly implausible. On the other hand, with enough data, the posterior can be non-zero arbitrarily close to $p = 0$ and $p = 1$, so this is of little practical consequence (in this case).

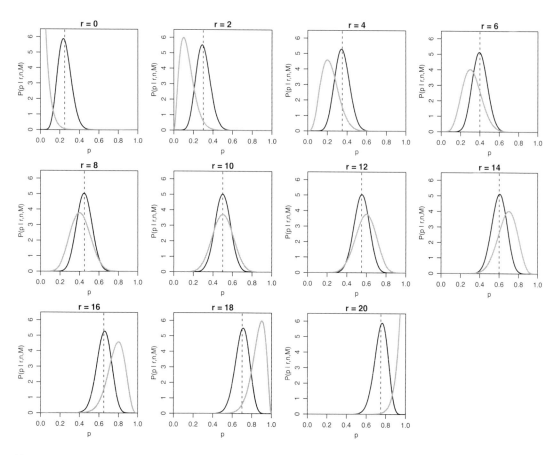

Fig. 5.4   As figure 5.2, but now using a beta prior on p with $\alpha = \beta = 10$. The black line is the posterior. For comparison, the grey line shows the posterior when using the uniform prior. Note that this extends beyond the top of the plots in the top left and bottom right panels.

```
        xaxs="i", yaxs="i", xlab="p", ylab="P(p | r,n,M)")
title(main=paste("r =",r), line=0.3, cex.main=1.2)
p.mean <- deltap*sum(p*pdense)
abline(v=p.mean, lty=2)
# overplot posterior obtained from a uniform prior
pdense.uniform <- dbinom(x=r, size=n, prob=p)
lines(p, pdense.uniform/(deltap*sum(pdense.uniform)), lwd=2,
      col="grey60")
# Can verify that pdense can also be found by direct calculation
#pdense2 <- dbinom(x=r, size=n, prob=p) *
#          dbeta(x=p, shape1=alpha.prior, shape2=beta.prior)
#pdense2 <- pdense2/(deltap*sum(pdense2)) # normalize posterior
#lines(p, pdense2, col="red", lty=2)
}
dev.off()
```

In the above code we no longer need to normalize pdense, because it is a beta distribution in p calculated by the R function dbeta, which is a normalized density function. The direct calculation of the posterior in the code (commented out) produces pdense2 which is not normalized, so we must normalize this manually.

It is instructive to plot the likelihood, prior, and posterior together, which is done by the following code. The result is shown in figure 5.5.

R file: coin3.R

```
##### Plot prior, likelihood, and posterior PDF for coin problem with a
##### beta prior for a range of r with n fixed

n <- 20
alpha.prior <- 10
beta.prior  <- 10
Nsamp <- 200 # no. of points to sample at
pdf("coin3.pdf", 9, 7)
par(mfrow=c(3,4), mgp=c(2,0.8,0), mar=c(3.5,3.5,1.5,1), oma=0.5*c(1,1,1,1))
deltap <- 1/Nsamp # width of rectangles used for numerical integration
p <- seq(from=1/(2*Nsamp), by=1/Nsamp, length.out=Nsamp) # rectangle centres
prior <- dbeta(x=p, shape1=alpha.prior, shape2=beta.prior)
for(r in seq(from=0, to=20, by=2)) {
  like  <- dbinom(x=r, size=n, prob=p)
  like  <- like/(deltap*sum(like)) # for plotting convenience only
  post  <- dbeta(x=p, shape1=alpha.prior+r, shape2=beta.prior+n-r)
  plot(p, prior, type="l", lwd=1.5, lty=2, xlim=c(0,1), ylim=c(0, 6.5),
       xaxs="i", yaxs="i", xlab="p", ylab="density")
  lines(p, like, lwd=1.5, lty=3)
  lines(p, post, lwd=1.5)
  title(main=paste("r =",r), line=0.3, cex.main=1.2)
}
dev.off()
```

We see from this plot how the (fixed) prior combines with the likelihood to form the posterior: the posterior (solid line) is just the product of the likelihood (dotted line) and the prior (dashed line), which is then renormalized. While the prior and posterior are automatically normalized by the R function dbeta, I explicitly normalize the likelihood before plotting. This doesn't really make sense, because it is not a PDF over p, but rather a PDF over r given p. I normalize it just to ease visual comparison (so the area under all curves is unity).

You might think that when the prior density is small, the posterior density must also be small. But the absolute value of the product of prior and likelihood is not important, because this product is renormalized to make the posterior. This is crucial. Even if the likelihood is small everywhere over the range of the prior, the posterior must still integrate to one.

As we get more and more (useful) data, the prior stays the same, but the likelihood becomes more peaked, so the posterior will be influenced more by the likelihood than by the prior. I demonstrate this using the same prior as above, a beta distribution with $\alpha = \beta = 10$. Let's start with $(r, n) = (2, 3)$, i.e. two heads and one tail. I will increase the amount of data in steps of factors of two, but keeping the proportion of heads to tails the

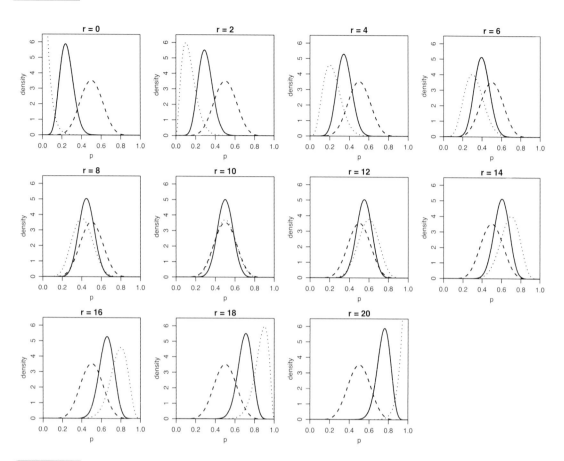

Fig. 5.5 As figure 5.4, but now showing in addition to the posterior (solid line), the likelihood (dotted line), and the prior (dashed line). The likelihood has been normalized to have unit area under the curve.

same, so $(r, n) = (2, 3), (4, 6), (8, 12), \ldots$. That is, I increase the signal-to-noise ratio in the data. The following code implements this; the results are shown in figure 5.6.

R file: coin4.R

```
##### Plot prior, likelihood, and posterior PDF for coin problem with a
##### beta prior for a range of r with ratio r/n fixed

alpha.prior <- 10
beta.prior  <- 10
Nsamp <- 200 # no. of points to sample at
pdf("coin4.pdf", 7, 7)
par(mfrow=c(3,3), mgp=c(2,0.8,0), mar=c(3.5,3.5,1.5,1), oma=0.5*c(1,1,1,1))
deltap <- 1/Nsamp # width of rectangles used for numerical integration
p <- seq(from=1/(2*Nsamp), by=1/Nsamp, length.out=Nsamp) # rectangle centres
prior <- dbeta(x=p, shape1=alpha.prior, shape2=beta.prior)
for(r in 2^(1:9)) {
```

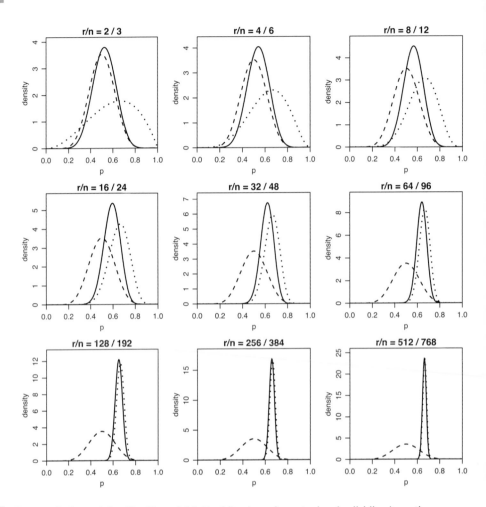

Fig. 5.6 The evolution of the likelihood (dotted line) and posterior (solid line) as the amount of data is increased, with r/n kept constant at $2/3$. The beta prior (dashed line) is kept fixed (with $\alpha = \beta = 10$). Note the different vertical scales on the individual panels.

```
n    <- (3/2)*r
like <- dbinom(x=r, size=n, prob=p)
like <- like/(deltap*sum(like)) # for plotting convenience only
post <- dbeta(x=p, shape1=alpha.prior+r, shape2=beta.prior+n-r)
plot(p, prior, type="l", lwd=1.5, lty=2, xlim=c(0,1),
     ylim=c(0,1.1*max(prior,post,like)), xaxs="i", yaxs="i",
     xlab="p", ylab="density")
lines(p, like, lwd=2, lty=3)
lines(p, post, lwd=1.5)
title(main=paste("r/n =",r,"/",n), line=0.3, cex.main=1.2)
}
dev.off()
```

The prior (dashed line) is the same in all panels (just the vertical scale changes). Initially we have few data, so the likelihood (dotted line) is broad, and the posterior (solid line) is dominated by the prior. As the amount of data increases, the posterior is increasingly dominated by the likelihood: the posterior shifts from the prior towards the likelihood. By the time we have a lot of data, the prior is basically irrelevant. This is all a consequence of the posterior being the product of the prior and likelihood (then renormalized).

From equation 5.11 we see that the effect of the prior is equivalent to adding α_{prior} successes to the actual observed number of successes, and adding β_{prior} failures to the actual observed number of failures.

In none of this have we considered the order of heads and tails. If we wanted to drop the assumption of the independence of the coin tosses, then we would need to take into account this ordering. This corresponds to a more complex (but legitimate) model for the coin, in which the binomial distribution is no longer the appropriate likelihood function.

5.2 Likelihoods can be arbitrarily small and their absolute values are irrelevant

We have seen in the previous section how the posterior is the product of the likelihood and prior, which is then renormalized. This renormalization is critical because the posterior must integrate to unity. Even if the prior and likelihood have little overlap, the posterior can still be strongly concentrated and have a large peak. We see this when comparing the panels labelled $r = 0$ and $r = 10$ in figure 5.5: the degree of overlap is very different, yet the posteriors have similar heights.

This highlights the fact that the absolute values of the likelihood are not relevant. What counts is their relative values. If we toss a coin twice, there is a probability of $1/4$ of getting a particular sequence of heads and tails. If we toss it n times, the probability is 2^{-n}. The larger the data set, the lower the likelihood of any particular outcome. Even if we have the true model with the true parameters, the probability (likelihood) of any *particular* data set can be very small, and for all but the smallest data sets, the likelihood is invariably very small for *any* data set. The same is true for continuous data when the likelihood is a probability density rather than an actual probability.

We can demonstrate this using the coin example. If we toss a fair coin n times, then the expected number of heads (the mean of the distribution) is $E[r] = n/2$. The probability that we observe r heads – the likelihood – is given by the binomial distribution. The left panel of figure 5.7 shows the variation of the logarithm of this likelihood as a function of $r/E[r]$ for four different fixed values of $E[r]$. The curve labelled $E[r] = 10$ means we toss the coin 20 times – for which we expect to get 10 heads – and shows the probability of getting 7, 8, ..., 13 heads. (The distribution is discrete, so this "curve" can only be plotted as a set of points.) I plot against $r/E[r]$ rather than r so that the curves are all centred on 1.0. Looking at the different curves we see how quickly the likelihood drops for larger sample sizes. For example, the probability of getting seven heads in 20 tosses (top curve

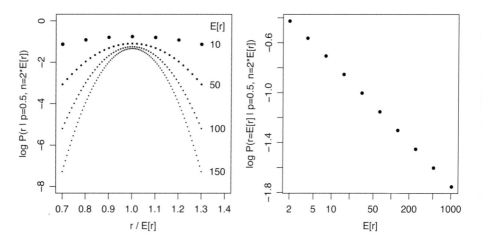

Fig. 5.7 The likelihood is smaller for larger data sets. This is shown here using the binomial likelihood with $p = 0.5$. Left: the variation of the log (base 10) likelihood as a function of $r/E[r]$, where $E[r]$ is the expected number of successes, which is half the sample size, $E[r] = n/2$. This is shown for four sample sizes, each plotted with a different point size and labelled with the value of $E[r]$. Right: the log (base 10) likelihood of the expected number of successes as a function of that number. This is the sequence of values at $r/E[r] = 1$ in the left panel, but now extended to larger sample sizes (the horizontal axis in the right panel is logarithmic).

for $r/E[r] = 7/10$ is 0.07 (log probability is -1.2), whereas the probability of getting 70 heads in 200 tosses (third curve down) is 6×10^{-6} (log probability is -5.2).

Yet the likelihood does not drop just for "unlikely" outcomes of the data. The right panel of figure 5.7 shows how the likelihood of the expected number of tosses varies with that expected number. We see that even at the expected value of the data (which is also the single most likely outcome: half the tosses are heads), the likelihood is small, and gets smaller the larger the data set.

The take-home message is that absolute values of the likelihood on their own tell us nothing about how good the model is.

This fact is central to the problem of frequentist hypothesis testing using p values, which we will cover in chapter 10. The appropriate p value in our example is the probability that the number of heads obtained in n tosses would be equal to or less than some particular threshold value, r. This is given by the cumulative probability of the binomial likelihood up to r

$$\text{p value} = \sum_{r'=0}^{r} \binom{n}{r'} p^{r'} (1-p)^{n-r'}. \tag{5.16}$$

In the frequentist paradigm, a hypothesis – in this case a fair coin, $p = 1/2$ – is rejected when the p value is small, because this indicates that the probability is small of getting as few heads as we did (r) or even fewer (i.e. even more "extreme" data). The problem with

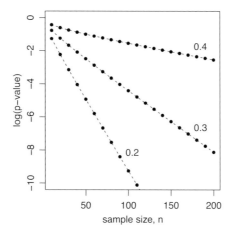

Fig. 5.8 In the coin tossing example the p value is the probability of getting r or fewer heads in n tosses, which is given by the cumulative binomial distribution (equation 5.16). The logarithm (base 10) of the p value is shown here as a function of n (in steps of 10) as the black points, for three different values of r/n: 0.4, 0.3, 0.2, for a fair coin (parameter $p = 1/2$; this "p" has nothing to do with the "p" in p value). These values and the step size were chosen to ensure that r is always an integer (the binomial is a discrete distribution). The dashed lines just connect the points for common r/n.

this idea, however, is that the larger the data set, the smaller the p value. We can see this by plotting the p value as a function of n for a fixed fraction of heads, r/n. That is, we ask ourselves what is the probability of getting, say, 30% or fewer heads in n tosses. This is shown in figure 5.8 for three values of r/n with $p = 1/2$, i.e. the true hypothesis. For example, if $n = 50$, then the probability of getting 15 or fewer heads (30%, the line marked 0.3) is 3.3×10^{-3} (log probability is -2.5), but for $n = 100$, the probability of getting 30 or fewer heads (also 30%) is 3.9×10^{-5} (log probability is -4.4). This is what we expect. But the usual approach in the frequentist paradigm is to reject a hypothesis once the p value drops below some fixed threshold, say 10^{-3}, without explicitly taking into account the size of the data set. Yet we see from figure 5.8 that whether or not the true hypothesis is rejected depends on the size of the data set. The larger the data set, the more likely we are to incorrectly reject the true hypothesis. This tells us that the p value is insufficient for deciding whether or not a hypothesis can be rejected. We must also take into account the sample size. Yet in real problems involving more complicated likelihood functions, it can be difficult to decide what threshold to use.

This problem with p values arises because it is using the absolute values of the likelihood, or sums (or integrals) thereof. We will see in chapter 11 how we can avoid this problem by instead considering the *ratios* of probabilities for different models.

The following two R scripts can be used to reproduce the experiments and plots in this section.

R file: `likelihoods_small.R`

```
##### Illustration with the binomial distribution that likelihoods are small

pdf("likelihoods_small.pdf", 8, 4)
par(mfrow=c(1,2), mgp=c(2.2,0.8,0), mar=c(3.5,3.5,0.5,1),oma=0.5*c(1,1,1,0))
expVal <- c(10, 50, 100, 150)
plot(0:1, 0:1, type="n", xlim=c(0.7, 1.4), ylim=c(-8, 0), xlab="r / E[r]",
     ylab=expression(paste(log," P(r | p=0.5, n=2*E[r])")))
for(i in 1:length(expVal)) {
  rVec <- seq(from=0.7*expVal[i], to=1.3*expVal[i], by=1)
  loglike <- (1/log(10))*dbinom(rVec, size=2*expVal[i], prob=0.5, log=TRUE)
  points(rVec/expVal[i], loglike, pch=20, cex=1/i)
}
text(1.31, -0.3, "E[r]", pos=4)
text(1.31, c(-1.18, -3.13, -5.25, -7.32), expVal, pos=4)
expVal   <- 2^(1:10)
loglike <- (1/log(10))*dbinom(expVal, size=2*expVal, prob=0.5, log=TRUE)
plot(expVal, loglike, log="x", pch=20, xlab="E[r]",
     ylab=expression(paste(log," P(r=E[r] | p=0.5, n=2*E[r])")))
dev.off()
```

R file: `pvalues_binomial.R`

```
##### Illustration with the binomial distribution that p-values depend on
##### sample size

pdf("pvalues_binomial.pdf", 4, 4)
par(mfrow=c(1,1), mgp=c(2.0,0.8,0), mar=c(3.5,3.5,1,1), oma=0.1*c(1,1,1,1))
sampSize <- seq(from=10, to=200, by=10)
fracLim  <- c(0.4, 0.3, 0.2)
plot(range(sampSize), c(-10,0), type="n", xlab="sample size, n",
     ylab="log(p-value)")
for(f in fracLim) {
  pVal <- pbinom(q=f*sampSize, size=sampSize, prob=0.5)
  points(sampSize, log10(pVal), pch=20)
  lines(sampSize,  log10(pVal), lty=2)
}
text(c(170, 170, 90), c(-1.8, -6.7, -8.2), fracLim, pos=4)
dev.off()
```

5.3 Assigning priors

Probabilistic inference provides answers to well-posed problems. But it does not define our models, or our priors, or tell us which data to collect or how to collect it. These are all important questions which affect the inference and thus the conclusions. Here we look at how to assign priors.

We saw in section 3.5 with the parallax problem that a prior can be essential to making any sensible inference; there was no general answer without one. Both this example and the coin example in section 5.1 show how the posterior PDF over a parameter depends on both the prior and the likelihood. As the data become more informative (e.g. the fractional parallax error is smaller or we make more coin tosses), the posterior becomes increasingly dominated by the likelihood, and the prior becomes less and less relevant. This is because as the data get better, the likelihood – when viewed as a function of the parameters – becomes narrower. So no matter what the shape of the prior, it will be approximately constant across this narrow stretch of likelihood (unless the prior has finite support and partially excludes the likelihood). In such cases we need not be too concerned with the exact specification of the prior. When the data are poor, on the other hand, the prior plays a more dominant role. As indeed it should: it is what we know about the parameter in the limit of no data. So when the data are not highly informative we do need to be concerned about the prior. How, then, do we assign a prior?

The prior should incorporate any relevant information we have about the problem. One invariably has some kind of prior information on the model parameters, just from the very fact of having chosen the model. Sometimes one knows the parameter cannot occupy a certain range by definition, e.g. the distance must be positive, or the probability of a coin landing heads lies between 0 and 1. These may seem trivial, but we already saw in section 3.5 how the positivity of distance enabled us to sensibly interpret arbitrarily small and negative parallaxes. Limits of measurement or observability may also provide bounds.

We often have information beyond such hard constraints. For the coin example in section 5.1, a flat prior on p for a coin chosen at random is very conservative. We saw this from the inferred posterior: with 20 tosses and seven heads, adopting a uniform prior led to a very broad posterior (figure 5.1). I suspect most people, given these data, would implicitly conclude on a much narrower posterior than this. This is because we actually have a strong prior that the coin is very close to fair. No coin is perfectly symmetric, but deviations from this are small. You would need a lot of data to convince you otherwise. The prior quantifies this.

We implicitly use priors all the time in every day life. If a brown animal dashes past you in the forest one day, but you didn't get a good look at it, are you more likely to think it was a deer or a dinosaur? Yet while few people seriously doubt the existence or importance of priors, it can be difficult to specify them exactly in practice. For the coin example we would presumably adopt a prior symmetric about $p = 0.5$, but how narrow should it be? There is no rule here, because priors depend on what you know, believe, and understand about the problem. Turning this notion into a probability distribution is hard and rarely unique.

Often we adopt standard distributions for priors, like the beta distribution, and we'll see more examples of using standard distributions below and in the next chapter. If the parameter has finite support – or if we want our prior to enforce this – then we could even define a discrete prior ourselves using a histogram.

There are some principles which can help us to adopt appropriate priors. One is the *principle of insufficient reason*, also called the *principle of indifference*. This essentially says that if we have a set of mutually exclusive outcomes, and we don't expect any one of them to be more likely to occur than the other, then we should assign them equal probabilities.

This is almost a tautology, but it can nonetheless help (it leads us to derive the binomial distribution, for example – see section 1.4.1).

However, if we are unable to decide on the set of outcomes, we will run into problems. Consider the proposition – which I'll call R – "Norah is wearing a red dress today". If we have no idea whether the dress is red or not, then we might conclude that $P(R) = 1/2$, and $P(R') = 1/2$, where R' is the complementary proposition "Norah is not wearing a red dress today". But if we then consider an additional proposition W – "Norah is wearing a white dress today" – do we now change our probabilities to $P(R) = 1/3$, $P(W) = 1/3$, and $P(\text{neither } R \text{ nor } W) = 1/3$? Our probabilities appear to change based on how we choose to enumerate the possibilities. How many colours are there? Are they equally probable a priori? Is she even wearing a dress?

In general we will want to adopt a prior which is as conservative (uninformative) as possible, consistent with the information we have. Below we will look at some simple but useful cases of this. There are principled approaches for determining priors given certain information, perhaps the best known being the principle of *maximum entropy*. The idea is to find the least informative (most entropic) distribution given certain information.[8]

Another approach is *empirical Bayes*, in which we estimate the priors from some general properties of the data (or perhaps a subset thereof). Note that we should not iterate the Bayesian procedure, using the posterior as the prior for a reanalysis of the same data, and then repeating this N more times in the hope of removing the effect of the original prior. This will result in a posterior which is proportional to $P(\theta)P(D|\theta)^{N+1}$. This is indeed dominated by the data and will make the prior irrelevant for large N, but only because we have have reused the same data $N + 1$ times by erroneously multiplying non-independent (in fact identical) likelihoods.

In contrast, we *can* take the posterior from one analysis to be the prior of the next analysis if these two analyses involve independent data. The final posterior will be identical to having combined the two data sets together with the original prior. This follows because the likelihood factorizes. Let D_1 and D_2 be the two independent data sets. Then

$$P(\theta|D_1, D_2) \propto P(D_1, D_2|\theta)P(\theta)$$
$$\propto P(D_2|\theta)P(D_1|\theta)P(\theta)$$
$$\propto \underbrace{P(D_2|\theta)}_{\text{likelihood for } D_2} \times \underbrace{P(\theta|D_1)}_{\text{posterior from } D_1} . \tag{5.17}$$

This highlights again that the prior and posterior are both PDFs on the parameters, but based on different information.

Another way of assigning priors is *hierarchical Bayes*. This involves selecting a form for the prior parameter distribution, but then putting a prior distribution over its parameters. We then use this to marginalize over the parameter prior PDF. That is, we introduce a higher level prior with fixed parameters to average over the lower level prior. This topic is covered by several text books, such as Kruschke (2015) and McElreath (2016).

A pragmatic approach to dealing with an inability to specify a unique prior is to test

[8] If only the mean and the variance of a probability distribution are known, then the principle of maximum entropy shows that the Gaussian is the least informative distribution.

how sensitive our results are to the choice of prior. We may vary the bounds selected, or the shape of the function. Given two or more priors, all of which seem equally reasonable, we can derive the posteriors and compare them. If they give results which we consider to be indistinguishable from the point of view of their interpretation, then we can consider the choice of prior as unimportant. If not, then this is because the data are insufficiently conclusive (by definition of what we have just considered to be reasonable priors). We should also not forget that the posterior PDF, just like the prior, is a probability *distribution*: it does not give certainty. In science, as in life, there is none.

5.3.1 Location and scale parameters

Suppose we have a model that specifies the location (in an abstract sense) x_0 of some quantity, and we have no prior knowledge of its value other than that it lies between some limits. We adopt a prior $P(x_0)$. Our result (posterior PDF) should be independent of the origin of the coordinate system we adopt for this location. This requires that the prior be invariant with respect to a linear translation of x_0. We can write this translation as $x_1 = x_0 + c$ for some constant c. This change of variables requires

$$P(x_o)\, dx_o \;=\; P(x_1)\, dx_1 \;=\; P(x_0 + c)\, dx_0. \tag{5.18}$$

This can only be satisfied if $P(x_0)$ is constant over the limits, and zero outside. This is the uniform prior, which is therefore the invariant distribution with respect to additive changes in a quantity.

Suppose now we wish to infer the size or scale of something. Complete ignorance here means that we know nothing about this scale, other than that it must be positive. Our prior PDF should therefore be invariant with respect to being stretched. Let this scale be a length, w_0. If we specify a prior over w_0 in units of metres, the prior should be the same if we then decide to express it in centimetres instead (as the units should not matter if we are ignorant of the scale). This transformation is $w_1 = aw_0$ for a positive constant a. This demands

$$P(w_0)\, dw_0 \;=\; P(w_1)\, dw_1 \;=\; P(aw_0)\, a\, dw_0. \tag{5.19}$$

This can only be satisfied if $P(w_0) \propto 1/w_0$, as then $P(aw_0) \propto 1/(aw_0)$. Thus a prior proportional to the inverse of a parameter is invariant with respect to rescalings of that parameter. Through a change of variables we see that this is equivalent to $P(\log w_0) = \text{constant}$ (see the example in section 1.9.1). This tells us that equal multiples of w_0 have the same probability (e.g. the probability between 1 and 10 is the same as the probability between 10 and 100), which is just what ignorance of scale means. This type of prior is often called a *Jeffreys prior* (but see the next section).

Location and scale parameters are common. If we want to infer the mean of a Gaussian, the above principle tells us to use a uniform distribution for the prior. When inferring its variance, we should use a Jeffreys prior. Note that

$$P(\log \sigma^2) = \text{constant} \quad \Rightarrow \quad P(\log \sigma) = \text{constant} \tag{5.20}$$

so it does not matter whether we talk about a Jeffreys prior in the variance or in the standard deviation.

We may have one of these priors on a transformed version of the parameter. For example, if we are trying to infer a length x from an angular measurement ϕ, and we assign the angle a uniform prior, then a transformation of variables shows us that the prior over $x \propto \tan \phi$ has a Cauchy distribution (section 1.4.7).

One disadvantage of these particular uniform priors is that they are improper (cannot be integrated). We must be careful when we use such priors: as the posterior is the product of the prior and likelihood, it is possible (depending on the data) that the posterior will also be improper, in which case it is meaningless. We saw with the parallax problem in section 3.5 that improper priors can lead to wild inferences if the data are poor. But often the likelihood – when seen as a function of the parameters – will drop rapidly enough beyond a finite-sized region so that the resulting posterior is proper. The likelihood may even be quite peaked, in which case the exact form of the prior will have little influence on the resulting posterior. However, improper priors are unusable when we do model comparison with the evidence, because then we must integrate the unnormalized posterior over the parameter (this topic will be covered in chapter 11). A pragmatic solution is to apply upper and lower limits to make it proper, and then to test the sensitivity of our results to these limits. Doing this for the Jeffreys prior between limits θ_{min} and θ_{max} and normalizing gives

$$P(\theta) = \frac{1}{\theta \ln(\theta_{max}/\theta_{min})}. \tag{5.21}$$

Setting priors for location and scale parameters are examples of setting priors that are invariant with respect to irrelevant reparametrizations of the problem. We will not always want such invariances, however. A more fundamental consideration is the nature of the problem. In the astronomical distance inference problem in section 3.5, for example, we know that the Galaxy has a characteristic scale length, so our prior should respect this. In general, priors should be independent of irrelevant information. A nice illustration of this is the *Bertrand paradox*, in which different – but apparently equally valid – assumptions about what constitutes "random" leads to different answers to a probabilistic problem. Jaynes (1973) describes this problem and shows that a proper consideration of the invariances in the problem leads to a unique solution.

5.3.2 Jeffreys prior

What I referred to in the previous section as the Jeffreys prior can be seen as a case of a more general approach to setting priors introduced by Harold Jeffreys (Jeffreys 1961; see also Robert *et al.* 2009).[9] His goal was to produce posteriors that are invariant under reparametrizations. As the likelihood is motivated by the problem context – the generative model and the measurement model – we can imagine that achieving this invariance will put conditions on the prior that depend (only) on the likelihood. Let $\boldsymbol{\theta}$ be a J-dimensional vector of parameters and $P(x|\boldsymbol{\theta})$ be the likelihood. The Jeffreys prior is defined to be

---

[9] In fact, Harold Jeffreys introduced various rules for specifying priors. The general rule, described in this section, does not always give the location and scale priors of the previous section, and there were situations where Jeffreys did not recommend the general rule. For a discussion see Kass & Wasserman (1996).

proportional to the square root of the determinant of the *Fisher information matrix* $\mathcal{I}(\boldsymbol{\theta})$,

$$P(\boldsymbol{\theta}) \propto \sqrt{|\mathcal{I}(\boldsymbol{\theta})|}. \tag{5.22}$$

The Fisher information matrix is the $J \times J$ matrix with elements $\mathcal{I}_{i,j}$ equal to the expectation of the second derivatives of the log likelihood[10]

$$\mathcal{I}_{i,j} = -E\left[\frac{\partial^2 \ln P(x\,|\,\boldsymbol{\theta})}{\partial\theta_i\partial\theta_j}\right] \tag{5.24}$$

where the expectation is taken with respect to the data x, i.e.

$$E\left[\frac{\partial^2 \ln P(x\,|\,\boldsymbol{\theta})}{\partial\theta_i\partial\theta_j}\right] = \int \frac{\partial^2 \ln P(x\,|\,\boldsymbol{\theta})}{\partial\theta_i\partial\theta_j}P(x\,|\,\boldsymbol{\theta})\,dx. \tag{5.25}$$

The second derivative is a measure of the curvature of a function. Thus if the log likelihood has a sharply peaked maximum (as a function of θ), the second derivative at this point will be large and negative, so the information will be large and positive.[11]

The origin of the Jeffreys prior for a one parameter problem can be understood as follows. If we make a smooth, monotonic change of variables $\theta = \theta(\psi)$, then the information for ψ is

$$\mathcal{I}(\psi) = -E\left[\frac{\partial^2 \ln P(x\,|\,\psi)}{\partial\psi^2}\right]$$

$$= -E\left[\frac{\partial^2 \ln P(x\,|\,\theta)}{\partial\theta^2}\right]\left|\frac{d\theta}{d\psi}\right|^2$$

$$= \mathcal{I}(\theta)\left|\frac{d\theta}{d\psi}\right|^2 \tag{5.26}$$

where $\theta(\psi)$ is independent of x so could be taken out of the expectation. This equation has the same form as the expression for the prior under the change of variables (section 1.9.1)

$$P(\psi) = P(\theta)\left|\frac{d\theta}{d\psi}\right|. \tag{5.27}$$

Thus if we set $P(\theta) \propto \sqrt{|\mathcal{I}(\theta)|}$ in this equation, we see that the transformation of this prior is still proportional to the square root of the information (because the Jacobian cancels). As the information depends only on the likelihood, this shows that the information contained in the prior is invariant under the transformation.

---

[10] By carrying out the differentiation and taking expectations you can show that this may also be written as

$$\mathcal{I}_{i,j} = E\left[\frac{\partial \ln P(x\,|\,\boldsymbol{\theta})}{\partial\theta_i}\frac{\partial \ln P(x\,|\,\boldsymbol{\theta})}{\partial\theta_j}\right]. \tag{5.23}$$

[11] If you return to section 4.5 you will recall that the log likelihood for multi-dimensional linear regression was $-\frac{1}{2}\mathrm{SS}_{\text{res}}$ to within an additive constant. The negative second derivative of this with respect to the parameters $\boldsymbol{\beta}$ is X^TX, the (Fisher) information matrix.

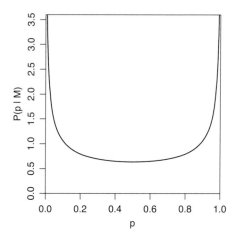

Fig. 5.9 The beta distribution prior with $\alpha = \beta = 1/2$ (also called the arcsine distribution).

Example: Jeffreys prior for a binomial likelihood

As an example we compute the Jeffreys prior for the binomial likelihood, $P(r|p, n)$ in equation 1.43, which has a single parameter p. Taking the logarithm of this likelihood and differentiating twice with respect to p gives

$$\frac{\partial^2 \ln P(r|p, n)}{\partial p^2} = -\frac{r}{p^2} - \frac{(n - r)}{(1 - p)^2}. \tag{5.28}$$

Using the definition of expectation for discrete distributions (equation 1.22), the information is

$$\mathcal{I}(p) = -\sum_{r=0}^{n} \frac{\partial^2 \ln P(r|p, n)}{\partial p^2} P(r|p, n). \tag{5.29}$$

As the Fisher information is linear in r, and $E[r] = \sum_r rP(r|p, n) = np$, this sum is easy to evaluate

$$\mathcal{I}(p) = \frac{np}{p^2} + \frac{(n - np)}{(1 - p)^2}$$

$$= \frac{n}{p(1 - p)}$$

$$P(p) \propto p^{-1/2}(1 - p)^{-1/2} \tag{5.30}$$

where in the last line I have used equation 5.22 and have absorbed n into the proportionality constant. This prior is a beta distribution with parameters $(\alpha, \beta) = (1/2, 1/2)$ (this is also called an arcsine distribution) and is shown in figure 5.9. It is symmetric about $p = 1/2$, of course, and goes to infinity at $p = 0$ and $p = 1$. Note that in this case the Jeffreys prior is also a conjugate prior.

Taking the same approach we can show that the Jeffreys prior for the standard deviation

σ of a Gaussian distribution is $1/\sigma$ (when the mean is known).[12] This is the same result which we derived in equation 5.19 to achieve the invariance of a parameter with respect to multiplicative scalings. This makes sense, as the standard deviation of a Gaussian acts like a scale parameter. You can likewise show that the Jeffreys prior for the Poisson rate parameter (λ in equation 1.51) is $1/\sqrt{\lambda}$.

5.3.3 Are priors subjective?

Some people are uncomfortable with using probabilistic inference (Bayesian statistics) because it seems to depend on the existence of subjective priors, and not just on the data. But if we think a bit more carefully, we will realise that the data are subjective in some sense too. What data did we choose to collect? What did we discard? How did we process them? We also make a decision when we choose the likelihood function, a decision that is based on how we understand the data to have been generated and what we assume their noise properties to be. This choice is independent of the measured values of the data, and different people could plausibly make different choices. Is the choice of the likelihood function any more "objective" than the choice of the prior? The data analysis and interpretation depend on more than just the data themselves. They also depend on what we know and what we assume, as examples in this and previous chapters have shown.

Probabilities reflect our own state of knowledge. Suppose I write "X" on one card, "Y" on another, and put each into a different sealed envelope. I then shuffle them and give one envelope to person A and the other to person B. Before opening anything, both A and B think there is a probability of 0.5 that A has card X. If B now opens her card and sees she has Y (but says nothing), then she knows (probability 1) that A has card X. But A's state of knowledge has not changed. Different people can correctly assign different probabilities to the same proposition because they have different knowledge. This may seem like an irrelevant example, but we have precisely this situation in science: the data we obtain give us incomplete knowledge of the true situation.

So in the sense that priors do not come from the data, they are indeed subjective. Yet so is the process of analysing and interpreting data, because people with different knowledge about a problem are justified to analyse and interpret the data differently. If I have information which tells me that some data can be rejected as outliers, then surely I should use that information. But if you do not have this information, then you should not start rejecting data in an ad hoc manner.

The important point about assigning priors is that two people with the same knowledge should assign the same prior. Given the same data and choice of the likelihood, they will then agree on the posterior.

Priors are unavoidable when doing inference. Many statistical approaches that pretend

---

[12] When the standard deviation is known, this general approach gives a uniform prior for the mean, which is the same result derived in equation 5.18. When both mean and standard deviation are unknown, however, the general approach gives the prior $P(\mu, \sigma) \propto \sigma^{-2}$. Jeffreys (1961) argues that this is inappropriate, and that we should instead adopt independent priors even when both parameters are unknown. This is what we will do in section 6.2.3.

not to employ them actually do so implicitly. The probabilistic approach allows one to express priors openly, to quantify them, and to investigate the sensitivity of results to them.

So the answer to the question posed by this section title is "yes", to which we should add "just like the rest of data analysis".

5.4 Some other conjugate priors

We saw in section 5.1.2 that the conjugate prior for the binomial likelihood is a beta distribution. Many likelihood distributions have conjugate priors. The key to finding the conjugate prior is to identify a distribution which, when multiplied by the likelihood, will have the same form of dependence on the model parameters. If we know what the conjugate prior is, then finding out the parameters of the posterior just involves some algebra. We will look at a couple more examples here.

Poisson likelihood

Comparing the Poisson distribution, equation 1.51, with the gamma distribution, equation 1.61, we see that both have the same dependence on λ if we set $x = \lambda$ in the latter. If we measure N data points $\{r_i\}$ independently from a process described by a Poisson distribution with rate parameter λ, then the likelihood is the product of N Poisson distributions. Dropping terms independent of λ we can write this as

$$P(\{r_i\}|\lambda) \propto \lambda^{N\bar{r}} e^{-N\lambda} \tag{5.31}$$

where $\bar{r}$ is the mean of the data. Likewise the gamma distribution prior can be written as

$$P(\lambda|k,\theta) \propto \lambda^{k-1} e^{-\lambda/\theta}. \tag{5.32}$$

Taking their product to get the (unnormalized) posterior, we can easily see that the result is a gamma distribution with parameters k' and θ' in which

$$k' - 1 = k - 1 + N\bar{r} \quad \text{so}$$
$$k' = k + N\bar{r} \quad \text{and} \tag{5.33}$$
$$\frac{1}{\theta'} = \frac{1}{\theta} + N \quad \text{so}$$
$$\theta' = \frac{\theta}{1 + N\theta}. \tag{5.34}$$

In the limit of a large amount of data, the posterior has parameters $k' \simeq N\bar{r}$ and $\theta' \simeq 1/N$, which are independent of the prior, as we would expect.

Gaussian likelihood

Consider now a Gaussian likelihood in the variable x, $\mathcal{N}(\mu, \sigma)$, for which σ is known and we want to infer μ. If we adopt a Gaussian prior $\mathcal{N}(m, s)$ on μ, then because the Gaussian is invariant under an exchange of its variable with its mean, the posterior is proportional to the product of two Gaussians in μ, i.e.

$$P(\mu|x,\sigma,s,m) \propto \frac{1}{\sigma s} \exp\left[-\frac{(x-\mu)^2}{2\sigma^2}\right] \exp\left[-\frac{(m-\mu)^2}{2s^2}\right]. \qquad (5.35)$$

I have written the posterior conditioned on the known parameters of the prior and likelihood to emphasise that μ is the only unknown quantity. You can show in a few lines of algebra that this product is in fact another Gaussian in μ, with

$$\text{mean} = \left(\frac{x}{\sigma^2} + \frac{m}{s^2}\right)\left(\frac{1}{\sigma^2} + \frac{1}{s^2}\right)^{-1} \qquad (5.36)$$

$$\text{standard deviation} = \left(\frac{1}{\sigma^2} + \frac{1}{s^2}\right)^{-1/2}. \qquad (5.37)$$

This looks plausible: the mean is the inverse variance weighted average of the terms in the prior and likelihood (which are treated equally). If we had a set of N independent measurements $\{x\}$ drawn from this likelihood, then the expressions for the mean and standard deviation of the posterior are similar except with x replaced by the mean of the data, $\bar{x}$, and σ replaced by $\sigma/\sqrt{N}$ (see section 6.2.1). In the limit $s \to \infty$, the Gaussian prior becomes an (improper) uniform prior, and the posterior is equal to the likelihood, but now seen as a function of μ rather than x.

If, for the Gaussian likelihood, we instead know μ and want to infer σ^2, then the conjugate prior on σ^2 (not on σ) is the inverse gamma distribution. We shall come back to this distribution in section 6.2.3.

Conjugate priors are convenient, but they are only useful if our prior knowledge really can be represented by the corresponding distribution.

5.5 Summarizing distributions

The posterior PDF over a parameter is *the* answer to an inference problem. One should never forget that we always have a distribution. Nonetheless, it is convenient – and often necessary – to summarize it with a few numbers. For some problems the posterior will be a standard distribution, such as a Poisson or Gaussian, which is defined entirely by its parameters (the variance for the Poisson, the mean and variance for the Gaussian). This will occur when we use a conjugate prior for the likelihood, as we just saw. Such cases are rare in practice, unfortunately, so we must think carefully about appropriate distribution summaries.

We already looked at some summary metrics in chapters 1 and 2. The mean and variance are common choices. However, these both require that the posterior is normalizable, which may not be the case if we used an improper prior (for example). We have also seen examples where the mean is an inefficient estimator – for the uniform distribution (see section 2.5) – and an inconsistent estimator – for the Cauchy distribution (see section 2.3.2). The standard deviation also does not make sense if the distribution does not have infinite

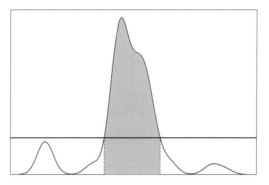

Fig. 5.10 The highest density interval (HDI) for a specified probability p is the span of the variable that encloses the region of highest probability density, the integral of which is p. It can be found by lowering a horizontal line (or plane in higher dimensions) which is parallel to the parameter axis until the grey region shown contains probability p.

support, such as when it is constrained to be positive, as was the case when estimating distances from parallaxes (see section 3.5): it is meaningless to summarize a quantity as 2 ± 5 if we know it cannot be negative.[13]

Considering first one-dimensional distributions, a more robust choice for the location of a distribution may be the median or the mode. The mode of a theoretical distribution can be found by differentiation, but in many cases this will not have an analytic solution, in which case we will have to use a numerical method (see section 12.6). The distribution may have more than one substantial mode, in which case we will want to report all of them (quoting a mean or median in this case would make little sense). If we have a sample of discrete data described as real variables, then each value occurs exactly once (finite numerical precision aside). To calculate the mode from samples we first need to make a density estimate, a procedure I will explain in section 7.2.

A robust choice for the scale (width) of a distribution may be the full-width at half-maximum (FWHM), although this alone does not tell us how much probability is enclosed within this width. Note that the mode and FWHM may be defined even if the probability distribution is improper. When the distribution is skew, or a quantity has a natural bound (e.g. must be positive), then we might be better off reporting quantiles of the distribution (see section 1.5). If we want to summarize such a distribution with two numbers, then the 5% and 95% quantiles (for example) may be more meaningful than the location and scale parameters. The difference between these define a 90% confidence interval. If this distribution is a posterior PDF, this tells us there is a 90% probability that the parameter lies within this interval.[14]

A $p \times 100\%$ confidence interval is not uniquely defined: there are an infinite number of

---

[13] Even 2 ± 1 makes no sense if – as is usually the case – we interpret these numbers as the mean and standard deviation of a Gaussian, because negative deviations beyond 2σ would be impossible.

[14] I am referring here to Bayesian confidence intervals, which are sometimes called credible intervals to distinguish them from frequentist confidence intervals (which we will encounter in section 10.2.3).

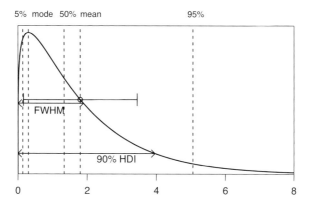

5% mode 50% mean   95%

FWHM

90% HDI

0   2   4   6   8

Fig. 5.11 Various metrics for summarizing a distribution. The vertical dashed lines are labelled by the metric names, the numbers in percentages being the respective quantiles. The 5% and 95% quantiles together form the 90% equal-tailed confidence interval. The error bar shows the range of the mean ± one standard deviation. The upper double-headed arrow shows the full-width at half-maximum (FWHM), the lower one the 90% highest density interval (HDI).

choices of bounds that enclose a fraction p of the probability ($0 \le p \le 1$). The one just mentioned is the equal-tailed interval. While intuitive and easy to compute, it may not be suitable for a skew distribution; it could even exclude the mean or mode, for example. Another common choice is the *highest density interval* (HDI), which – as the name suggests – encloses the regions of highest probability density, including the mode. If the distribution is unimodal, then it can be found by lowering a horizontal line parallel to the parameter axis until the region it defines by interception with the distribution includes the fraction p of the probability (see figure 5.10). If the distribution is multimodal then we can do the same thing, but we must consider what to do if we include a second (local) maximum. For a unimodal distribution, the HDI is also the shortest confidence interval corresponding to the specified value of p. This follows because it includes the regions of highest density. The HDI and several other summaries of a distribution are illustrated in figure 5.11.

Summaries of multivariate distributions necessarily involve more numbers. The mean and mode are now vectors and the variance is replaced by the covariance matrix. Quantiles are no longer uniquely defined, and it is not trivial to identify the equivalent of equal-tailed intervals. The multivariate generalization of the HDI is the highest density region. It is still uniquely defined for a unimodal distribution, and it is common practice to use it to define volumes that include a fraction p of the probability. For a two-dimensional distribution this corresponds to finding contours that include a fraction p of the probability. Such regions could be complicated and must normally be found numerically in practice.

Which summaries of a distribution are most appropriate in any situation depends on the shape of the distribution and on our objectives. Visual inspection is always recommended, but this will be impractical if we have a lot of data sets, and will provide limited insight if the distribution is more than two-dimensional. Whenever you read summaries of a posterior, always ask yourself what the full posterior PDF might look like, and thus what the limitations of the summaries might be.

How easy it is to calculate the above summaries depends on how we calculate the posterior. If we have samples *drawn from* the posterior – as opposed to densities *calculated at* particular values of it – then, as we will see in section 8.5.3, it is straightforward to calculate the mean, variance and higher moments, as well as quantiles for one-dimensional PDFs.

## 6

Parameter estimation: multiple parameters

In the previous chapter we looked at single parameter inference problems, as well as problems with conjugate priors that had analytic solutions. Now we move on to problems with multiple parameters, and will learn how to compute the joint, conditional, and marginal distributions. We will see how some well-known results concerning Gaussian distributions arise from the inference process.

6.1 Conditional and marginal distributions

When we have more than one parameter, the full posterior PDF becomes a multivariate distribution. Yet often we still want to find a one-dimensional distribution over one of its parameters. As introduced at the end of section 1.6.2, two such distributions are of particular interest: the conditional distribution and the marginal distribution.

Suppose a model has two parameters (a, b). If we fix one of the parameters, say $b = b_0$, then we can infer the one-dimensional posterior PDF for a, $P(a\,|\,b = b_0, D)$. From Bayes' theorem this is

$$P(a\,|\,b = b_0, D) \; = \; \frac{1}{Z_a} P(D\,|\,a, b = b_0) P(a\,|\,b = b_0) \qquad (6.1)$$

where Z_a is a normalization constant. This is a *conditional posterior distribution* because it is conditioned on a fixed value of b. It corresponds to taking a slice through the full distribution at $b = b_0$. If the priors on a and b are independent then $P(a\,|\,b) = P(a)$.

If, on the other hand, b is not fixed and so must also be determined from the data, then we infer the two-dimensional distribution $P(a, b\,|\,D)$ from the data and the priors on a and b using Bayes' theorem

$$P(a, b\,|\,D) \; = \; \frac{1}{Z_{ab}} P(D\,|\,a, b) P(a, b) \qquad (6.2)$$

where Z_{ab} is a normalization constant. If we then want to get the posterior of just a, we *marginalize* (integrate) over b to give

$$P(a\,|\,D) \; = \; \int P(a, b\,|\,D)\, db \qquad (6.3)$$

$$= \; \frac{1}{Z_{ab}} \int P(D\,|\,a, b) P(a, b)\, db. \qquad (6.4)$$

If the priors are independent then $P(a, b) = P(a)P(b)$. Marginalizing is like projecting the distribution along an axis.

Note the fundamental difference between $P(a|b, D)$ and $P(a|D)$. The former includes more given information, namely the value of b. Assuming that b determines the data to some degree (i.e. it is not an irrelevant parameter), then $P(a|b, D)$ must be narrower than $P(a|D)$, because by fixing b the data are used entirely to constrain a.

Marginalization is a powerful feature of probability analysis because it allows us to include parameters which are an essential part of the model, but which we may not actually be interested in. We marginalize over them to get the posterior PDF for the parameters of interest.

It is worth noting in this context the distinction between accuracy and precision (which I defined in section 2.7) in the context of posterior distributions. The width of the posterior PDF over a parameter is a measure of the *precision* of the estimate of that parameter. A narrower distribution means a higher precision. *Accuracy*, in contrast, is a measure of how close a point estimate of the PDF (e.g. the mode) is to the true value. We can have accurate but imprecise results: the point estimate of our PDF is close to the truth, but the PDF itself is very broad. And we can have precise but inaccurate results: the PDF is very narrow, but centred on a value that lies far from the truth. As we don't normally know the truth (that's why we're doing the inference), the posterior only tells us about the precision.

6.2 Inferring the parameters of a Gaussian

We turn now to a two-parameter problem, namely the inference of the mean and standard deviation of a Gaussian distribution. We will find the conditional, joint, and marginal posteriors for these two parameters.

We have a set of N data points $D = \{x_i\}$ drawn independently from a Gaussian with mean μ and standard deviation σ. The likelihood for these data is therefore

$$
\begin{aligned}
P(D|\mu, \sigma) &= \prod_{i=1}^{N} \frac{1}{(2\pi)^{1/2}\sigma} \exp\left[-\frac{(x_i - \mu)^2}{2\sigma^2}\right] \\
&= \frac{1}{(2\pi)^{N/2}\sigma^N} \exp\left[-\frac{1}{2\sigma^2} \sum_{i=1}^{N}(x_i - \mu)^2\right].
\end{aligned} \tag{6.5}
$$

Given only D, we would like to infer one or both of the parameters of the Gaussian. The physical situation might be that we have made N measurements of some quantity, whereby μ is its unknown value and σ is the (possibly also unknown) measurement error. The maths will be easier if we write the summation in the likelihood as

$$\sum_{i=1}^{N} (x_i - \mu)^2 = N(\overline{x} - \mu)^2 + NV_x \quad \text{where} \tag{6.6}$$

$$\overline{x} = \frac{1}{N} \sum_{i=1}^{N} x_i \quad \text{and}$$

$$V_x = \frac{1}{N} \sum_{i=1}^{N} (x_i - \overline{x})^2$$

which can be shown in a few lines of algebra. V_x is approximately the variance in the data (see equation 1.33). We will now look at three different cases according to what is known.

6.2.1 Standard deviation known

Suppose we know σ and want to get the posterior over μ. This is the conditional posterior, equation 6.1. As μ is a location parameter, I adopt an improper uniform prior (see section 5.3.1). The posterior is therefore just proportional to the above likelihood, so

$$P(\mu \,|\, D, \sigma) \propto \exp \left[-\frac{1}{2\sigma^2} \left(N(\overline{x} - \mu)^2 + NV_x \right) \right]$$

$$\propto \exp \left[-\frac{(\overline{x} - \mu)^2}{2\sigma^2/N} \right] \tag{6.7}$$

where I have absorbed terms that are independent of μ into the proportionality. This posterior is a univariate Gaussian with mean $\overline{x}$ and standard deviation $\sigma/\sqrt{N}$. This is the well-known result that the "best" estimate of μ is the mean of the data, and its uncertainty is $\sigma/\sqrt{N}$. (This is not equal to the standard error in the mean – equation 2.13 – because that arises when we estimate σ from the data, whereas here σ is known.) Compare and contrast this with what the central limit theorem (section 2.3) says. It gets the same result, but for large N and for data drawn independently from any distribution (with finite mean and variance). Here we instead have data drawn from a Gaussian distribution, but the resulting posterior is Gaussian for any N.

6.2.2 Mean known

Suppose we now know μ and want to get the posterior over σ (admittedly not so common in practice). We saw in section 5.3.1 that a suitable prior for a scale parameter is the Jeffreys prior, so I adopt $P(\sigma) \propto \sigma^{-1}$. Multiplying this prior by the likelihood we get the univariate posterior

$$P(\sigma \,|\, D, \mu) \propto \frac{1}{\sigma^{N+1}} \exp \left[-\frac{1}{2\sigma^2} \left(N(\overline{x} - \mu)^2 + NV_x \right) \right]. \tag{6.8}$$

I defer discussion of this distribution because we will encounter a rather similar one in the general case.

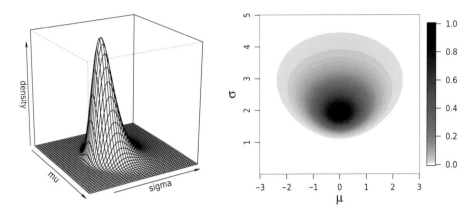

The two-dimensional posterior $P(\mu, \sigma \,|\, D)$ in equation 6.9, which is for the parameters of a Gaussian using a uniform prior on μ and a Jeffreys prior on σ. This distribution is show in the left panel as a three-dimensional perspective mesh. The right panel indicates the value of the posterior density using a grey scale which has been been scaled so that its maximum is one. The distribution is shown in both cases for $\overline{x} = 0$, $V_x = 2^2$, and $N = 10$.

6.2.3 Neither standard deviation nor mean known

We now turn to the general case of neither μ nor σ known. We adopt the same priors as before (they are independent of each other), so $P(\mu, \sigma) \propto \sigma^{-1}$. The bivariate posterior is therefore

$$P(\mu, \sigma \,|\, D) \propto \frac{1}{\sigma^{N+1}} \exp\left[-\frac{1}{2\sigma^2} \left(N(\overline{x} - \mu)^2 + NV_x \right) \right] \qquad (6.9)$$

to within a constant that does not depend on the parameters. This is a non-trivial function of both μ and σ, and is plotted in figure 6.1 for the data $\overline{x} = 0$, $V_x = 2^2$, $N = 10$. Note that the individual measurements do not appear in the posterior: these three statistics are sufficient. The code to make the plots is at the end of this section. You can use it to investigate how the shape of the posterior changes when changing the data, in particular V_x and N. Increasing N while keeping V_x constant, for example, results in a more compact PDF, i.e. we achieve a more precise determination of μ and σ, as we would expect.

If we now fix μ we get the conditional posterior over σ, equation 6.8. That this has exactly the same form as equation 6.9 follows from $P(\mu, \sigma \,|\, D) = P(\sigma \,|\, D, \mu)P(\mu)$ with $P(\mu)$ as a delta function. Note that I have not normalized either posterior. The normalized posteriors will have different units: $P(\sigma \,|\, D, \mu)$ has units σ^{-1}, whereas $P(\mu, \sigma \,|\, D)$ has units $(\mu\sigma)^{-1}$.

If we instead fix σ then we get the conditional posterior of μ, which is equation 6.7. We can visualize the conditional distribution as that obtained when taking a slice through the two-dimensional posterior. Equation 6.7 tells us that any horizontal slice through the right panel of figure 6.1 is a Gaussian.

Marginal posterior for μ

When both μ and σ are unknown, the posterior for one of the parameters is found by marginalizing over the other (equation 6.4). Let us first find $P(\mu|D)$ by marginalizing over σ. This can be thought of as projecting the two-dimensional posterior onto the μ-axis. Using the change of variables $\sigma = 1/z$ $(d\sigma = -dz/z^2)$ and the standard integral

$$\int_0^\infty z^n \exp(-\beta z^m)\, dz = \frac{\Gamma(\gamma)}{m}\beta^{-\gamma} \quad \text{where} \quad \gamma = \frac{n+1}{m} \quad \text{and} \quad (m, n, \beta) > 0$$
(6.10)

(Γ is the gamma function defined by equation 1.62) we get

$$P(\mu|D) = \int_0^\infty P(\mu, \sigma|D)\, d\sigma$$
$$\propto \int_0^\infty z^{N-1} \exp\left[-z^2\left(\frac{N(\bar{x}-\mu)^2 + NV_x}{2}\right)\right] dz$$
$$\propto \left[1 + \frac{(\bar{x}-\mu)^2}{V_x}\right]^{-N/2}$$
(6.11)

where I have absorbed factors independent of μ into the proportionality constant in the second line, and have absorbed an additional factor of $(NV_x/2)^{-N/2}$ in going to the third line. This density function is called a *Student's t distribution* (or just "t distribution"). The standard, normalized form for this distribution is

$$P(t) = \frac{\Gamma\left(\frac{\nu+1}{2}\right)}{\sqrt{\nu\pi}\,\Gamma\left(\frac{\nu}{2}\right)}\left(1 + \frac{t^2}{\nu}\right)^{-\frac{\nu+1}{2}} \quad \nu > 0$$
(6.12)

where ν, the *degrees of freedom*, is its sole parameter. Examples of the distribution are shown in figure 6.2. It is symmetric about $t = 0$ and becomes increasingly like a standardized Gaussian for larger values of ν. Comparing equation 6.12 with equation 6.11, we see that our marginal posterior is a t distribution with $\nu = N - 1$ degrees of freedom and

$$t = \frac{(\bar{x}-\mu)}{\sqrt{V_x/(N-1)}}.$$
(6.13)

Here t is the difference between the sample mean and the true mean, scaled by a quantity which, from reference to equation 2.13, is the standard error in the mean $\hat{\sigma}/\sqrt{N}$, where $\hat{\sigma} = \sqrt{V_x N/(N-1)}$ is an estimate of the standard deviation of the data. The best estimate of μ is thus $\bar{x}$ (which is the mean, mode, and median of the posterior), and its uncertainty is the standard error in the mean. We stated this classic statistical result in section 2.4, but now we have arrived at it via an inference procedure for data drawn from a Gaussian likelihood when we adopt Jeffreys priors on the mean and standard deviation. Previously we did not state what the distribution over the mean was; now we have shown it to be a t distribution.

As is apparent from figure 6.2, the t distribution has heavier tails than the standardized Gaussian. This reflects the fact that when σ is unknown, the data are not as informative about the mean as they are when σ is known (in which case the posterior is a Gaussian,

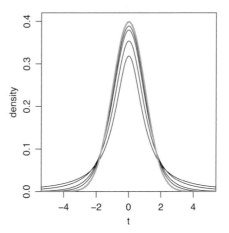

Fig. 6.2 The Student's t distribution for 1, 2, 5, and 10 degrees of freedom (dof). The larger the dof, the higher the central peak. The thick grey line shows a standardized Gaussian for comparison, to which the t distribution converges for infinite degrees of freedom.

as we saw in section 6.2.1). The more data we have, the better the determination of the parameters, the closer the t distribution gets to a Gaussian, and $\hat{\sigma}$ asymptotes to $\sqrt{V_x}$.

If we adopted a uniform prior on σ, $P(\sigma) \propto \sigma^0$, rather than a Jeffreys prior ($\propto \sigma^{-1}$), we see from inspection of the above equations that the marginal posterior $P(\mu\,|\,D)$ would again be a t distribution, but with $N - 2$ degrees of freedom. For large N the distribution hardly depends on the value of N, so the difference between these priors then becomes irrelevant. But given that σ cannot be negative, it would be illogical to put a uniform prior on it.

We will encounter the t distribution again in section 10.2.2.

Marginal posterior for σ and σ^2

The marginal posterior for σ is found by integrating equation 6.9 over μ

$$P(\sigma\,|\,D) = \int_{-\infty}^{\infty} P(\mu, \sigma\,|\,D)\,d\mu$$

$$\propto \frac{1}{\sigma^{N+1}} \exp\left[-\frac{NV_x}{2\sigma^2}\right] \int_{-\infty}^{\infty} \exp\left[-\frac{N(\bar{x} - \mu)^2}{2\sigma^2}\right] d\mu$$

$$\propto \frac{1}{\sigma^N} \exp\left[-\frac{NV_x}{2\sigma^2}\right] \tag{6.14}$$

as the Gaussian integral is just $\sqrt{2\pi\sigma^2/N}$. Plots of this for $V_x = 2^2$ and various values of N are shown in figure 6.3. Note that equation 6.14 is similar to the conditional posterior for σ in equation 6.8 if we set $\bar{x} = \mu$. They then only differ by a factor of σ, with the marginal distribution being slightly broader on account of the mean not being known.

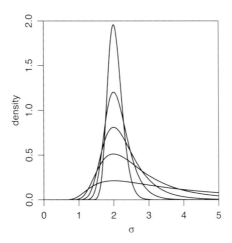

Fig. 6.3 The posterior $P(\sigma\,|\,D)$ (equation 6.14) for $V_x = 2^2$ and $N = 2, 5, 10, 20, 50$, whereby the higher peaks correspond to the larger N.

We can write equation 6.14 in terms of the variance σ^2 using a transformation of variables (section 1.9.1) $P(\sigma\,|\,D) = 2\sigma P(\sigma^2\,|\,D)$. This gives

$$P(\sigma^2\,|\,D) \propto \frac{1}{(\sigma^2)^{(N+1)/2}} \exp\left[-\frac{NV_x}{2\sigma^2}\right] \tag{6.15}$$

which is an *inverse gamma distribution*. Examples of this are shown in figure 6.4. For a variable z it is usually written in terms of the two parameters, the shape α and scale β,

$$P(z) = \frac{\beta^\alpha}{\Gamma(\alpha)} z^{-(\alpha+1)} e^{-\beta/z} \quad \text{where} \quad \alpha > 0, \ \beta > 0, \ z > 0. \tag{6.16}$$

In our application $z = \sigma^2$, $\alpha = (N-1)/2$, and $\beta = NV_x/2$. To satisfy the condition on α the distribution is only defined for $N \geq 2$. And to satisfy the condition for β there need to be at least two *different* data points, to ensure $V_x > 0$. This makes sense. From a logical point of view we don't expect to be able to make an inference on σ with just one data point, because we used improper priors. From a mathematical point of view the distribution is not defined for $N = 1$, because it cannot be normalized. As we used improper priors there is no guarantee that we would get a proper (normalizable) posterior, and here for $N = 1$ we do not. As soon as we have at least two (finite and different) data points, the likelihood ensures we get convergence of the posterior distribution. For $N = 0$ we have no data, so the posterior is identical to the prior anyway.

I wrote in section 5.4 that the inverse gamma distribution was the conjugate prior for the variance of a Gaussian likelihood of *known* mean μ. This is now evident from the form of the dependence on $z = \sigma^2$ of equation 6.16, which is

$$P(\sigma^2) \propto (\sigma^2)^{-(\alpha+1)} e^{-\beta/\sigma^2}. \tag{6.17}$$

Adopting this as our prior and multiplying it by the likelihood in equation 6.5, which we

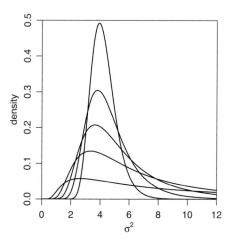

Fig. 6.4 The inverse gamma distribution of equation 6.16 with $V_x = 2^2$ and $N = 2, 5, 10, 20, 50$, whereby the higher peaks correspond to the larger N. The mode is at $V_x(N/N-1)$, so as $N \to \infty$ the mode tends towards V_x.

can write as

$$P(D|\mu, \sigma^2) \propto (\sigma^2)^{-N/2} \exp\left[-\frac{1}{2\sigma^2} \sum_{i=1}^{N} (x_i - \mu)^2\right], \tag{6.18}$$

we see from inspection that the posterior has the same functional form as the prior, with shape parameter $\alpha + N/2$ and scale parameter $\beta + (1/2) \sum_i (x_i - \mu)^2$.

The following code produces the density plot in the right panel of figure 6.1. I compute the (unnormalized) two-dimensional posterior on a dense grid, then use the function image.plot in the fields package to plot this as a greyscale image along with a bar to indicate the scale. The palette is made with the RColorBrewer package, which of course produces colour scales too (replace Greys with YlOrRd to produce a red/yellow scale, for example). To produce the mesh plot in the left panel of figure 6.1 use

```
persp(x=mu, y=sigma, z=postDen, phi=20, theta=60, d=5,
   xlab=expression(mu), ylab=expression(sigma), zlab="density")
```

but use only 50 elements in mu and sigma (i.e. set length.out=50 in their constructions) so that the shading works out nicely.

R file: 2D_gaussian_posterior.R

```
##### Plot 2D posterior over Gaussian mu and sigma
##### for uniform prior on mu and Jeffreys prior on sigma

library(fields) # for image.plot
library(RColorBrewer) # for colorRampPalette
mypalette <- colorRampPalette(brewer.pal(9, "Greys"), space="rgb",
                              interpolate="linear", bias=2.5)
mycols <- mypalette(64)
```

```
# Define function to return the unnormalized posterior
post <- function(mu, sigma, xbar, Vx, N) {
   (1/sigma^(N+1))*exp( (-N/(2*sigma^2)) * ((xbar-mu)^2 + Vx) )
}

# Define data and calculate posterior density on a dense grid
xbar   <- 0
Vx     <- 2^2
N      <- 10
mu     <- seq(from=-3,   to=3, length.out=1e3)
sigma <- seq(from=0.01, to=5, length.out=1e3)
postDen <- matrix(data=NA, nrow=length(mu), ncol=length(sigma))
for(i in 1:length(mu)) {
   for(j in 1:length(sigma)) {
      postDen[i,j] <- post(mu=mu[i], sigma=sigma[j], xbar=xbar, Vx=Vx, N=N)
   }
}
postDen <- postDen/max(postDen) # scale so maximum is one

pdf("2D_gaussian_posterior.pdf", 5, 4)
par(mfrow=c(1,1), mar=c(3.5,3.5,0.5,1), oma=c(0.1,0.1,0.5,0.1),
    mgp=c(2.2,0.8,0), cex=1.0)
image.plot(z=postDen, x=mu, y=sigma, nlevel=1024, xlab=expression(mu),
           ylab=expression(sigma), col=mycols, cex.lab=1.5)
dev.off()
```

6.3 A two-parameter problem: estimating amplitude and background

We now turn to a parameter estimation problem for which the posterior does not have a convenient functional form.[1]

A spectrograph is a device for spreading out light along a detector as a function of wavelength. We use this to measure the intensity – the number of photons d – as a function of the wavelength x. Suppose we know that this spectrum comprises a single emission line on top of a constant background. We would like to determine the amplitude a of the emission line in the presence of both measurement noise and an unknown background level b. Our model for the line is that it has a Gaussian shape centred at $x = x_0$ with standard deviation w. Thus the expected signal (number of photons) at any position x is

$$s = t \left[a \exp \left(-\frac{(x - x_0)^2}{2w^2} \right) + b \right] \tag{6.19}$$

where t is proportional to the exposure time, and so is proportional to the expected number of photons collected. If we expose for longer we expect to get more signal from both source and background, but without changing the shape of either. This model is shown as the grey

---

[1] This example is based on one in Sivia & Skilling (2006).

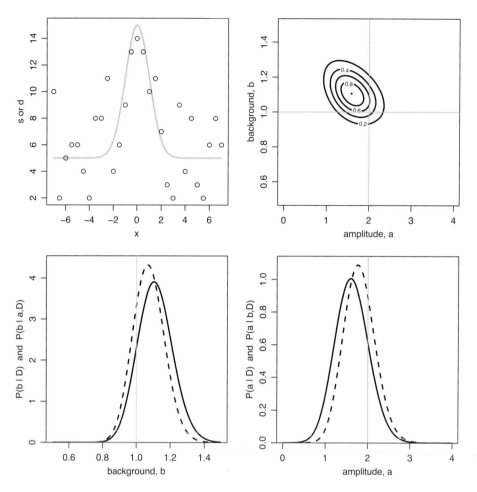

Estimating the amplitude and background of a signal. Top left: true model (grey curve) and observed data (open circles). Top right: two-dimensional unnormalized posterior, scaled to have a maximum of unity. The contours are equal spacing in probability density, labelled by the density relative to the maximum. Bottom two panels: the marginalized posteriors (solid lines), and conditional posteriors (dashed lines) using the true values of the parameters. The vertical grey lines indicate the true parameters.

line in the top-left panel of figure 6.5 for $x_0 = 0, w = 1, t = 5, a = 2, b = 1$. These are the true values of the model parameters which I will use to simulate the data in this example.

We measure this signal at a number of different positions $\{x_i\}$ on the detector. If the expected number of photons at any of these positions is s (which is not generally an integer), then the number of photons actually observed, d, follows a Poisson distribution with mean

s, i.e.

$$P(d|s) = \frac{s^d e^{-s}}{d!}. \tag{6.20}$$

One possible realization of the measurements, and the one which I will use for the subsequent inference, is shown as the open circles in the top-left panel of figure 6.5. There are 29 data points measured uniformly from $x = -7w$ to $x = +7w$ in steps of $0.5w$. Denoting the set of measurements as $D = \{d_i(x_i)\}$, the likelihood of the data is

$$P(D|x_0, w, t, a, b) = \prod_i \frac{s_i^{d_i} e^{-s_i}}{d_i!} \tag{6.21}$$

where $s_i = s(x_i)$ from equation 6.19 and so introduces the dependence on the model parameters.

The model has five parameters, but let us assume that x_0, w, and t are known. We therefore want to infer $P(a, b|D, M)$ from the data, where the model M expresses that we know the shape of the line and the values of the fixed parameters (and that the noise model is Poisson). Let us adopt the minimalistic prior that a and b cannot be negative. I set no upper limit, so this is an improper prior. Provided the posterior drops to very low densities at large a and b, we can later truncate it at some larger values of a and b in order to normalize it. This is equivalent to the prior dropping to zero at these values, although by construction this has no relevant impact on the posterior.[2] The prior $P(a, b)$ is constant when both a and b are positive, and zero otherwise. The posterior is therefore

$$P(a, b|D, M) = \begin{cases} \dfrac{1}{Z} \displaystyle\prod_i \frac{s_i^{d_i} e^{-s_i}}{d_i!} & \text{if } a \geq 0 \text{ and } b \geq 0 \\ 0 & \text{otherwise} \end{cases} \tag{6.22}$$

and the log posterior is

$$\ln P(a, b|D, M) = \begin{cases} \displaystyle\sum_i d_i \ln(s_i) - s_i + \text{constant} & \text{if } a \geq 0 \text{ and } b \geq 0 \\ -\infty & \text{otherwise} \end{cases} \tag{6.23}$$

where the constant absorbs terms which do not depend on a or b (as they do not effect the shape of the function). Given the $\{x_i\}$ we calculate the expected counts $\{s_i\}$ from the generative model, equation 6.19, which gives the dependence on the parameters. Equation 6.23 has a nonlinear dependence on the parameters, so solving for the maximum, full-width at half-maximum, etc., of the posterior is not analytically simple. But we can plot the posterior just by calculating it on a grid of values of $\{a_j, b_k\}$. This is done by the R code in the following section. I use a regular grid of size $K \times K$ with $K = 100$ over a pre-defined range with spacing δa in a and δb in b. I found the best range by trial and error. The contour function in R is then used to find and plot smooth contours of constant

---

[2] We must be careful to ensure that the probability mass in the posterior beyond the truncated range really is negligible. This can be hard to know in more complex or higher dimensional problems. Of course if the likelihood – seen as a function of the parameters – drops to zero within a finite region, or if it asymptotes fast enough to have a finite integral, then this truncation is not necessary.

probability density. The result is shown in the top-right panel of figure 6.5. We see a clear anticorrelation between a and b. This means we can get almost equally good models of the data if we simultaneously increase b and decrease a slightly (or vice versa). This makes sense, because if we attribute more of the measured signal to the background, we must attribute less to the spectral line.

The bottom two panels of figure 6.5 show the marginal posterior PDFs of the two parameters. These are formally found by integration (equation 6.3). But as I have a grid, I approximate them by simply summing over the other parameter. For a this is

$$P(a_j \mid D) \simeq \delta b \sum_{k=1}^{K} P(a_j, b_k \mid D), \tag{6.24}$$

and similarly for b. The posterior is normalized in the program via the rectangle rule using this grid, in the same way as done in section 5.1.1.

We can use these grid evaluations also to estimate the mean μ_a and variance σ_a^2 of a (and likewise for b), as well as the covariance $\mathrm{Cov}(a, b)$ and correlation coefficient ρ (see section 1.6.1). We must ensure that the grid evaluations extend well into the tails of the posterior to ensure that we capture essentially all of the probability. It follows from the definitions of these quantities that

$$\mu_a = \int a P(a \mid D) da \simeq \delta a \sum_{j=1}^{K} a_j P(a_j \mid D) \tag{6.25}$$

$$\sigma_a^2 = \int (a - \mu_a)^2 P(a \mid D) da \simeq \delta a \sum_{j=1}^{K} (a_j - \mu_a)^2 P(a_j \mid D) \tag{6.26}$$

$$\mathrm{Cov}(a, b) = \iint (a - \mu_a)(b - \mu_b) P(a, b \mid D) \, da \, db$$

$$\simeq \delta a \delta b \sum_{j=1}^{K} \sum_{k=1}^{K} (a_j - \mu_a)(b_k - \mu_b) P(a_j, b_k \mid D) \tag{6.27}$$

$$\rho = \frac{\mathrm{Cov}(a, b)}{\sigma_a \sigma_b}. \tag{6.28}$$

Strictly speaking I should have a factor of $K/(K-1)$ in equations 6.26 and 6.27 because the means μ_a and μ_b have been estimated from the data. But with $K = 100$ this introduces a negligible error on top of the approximation from gridding. For the data shown in the top-left panel of figure 6.5 the resulting estimates of the parameters (mean $\pm$ standard deviation) are $a = 1.63 \pm 0.40$ and $b = 1.11 \pm 0.10$, and the correlation coefficient is $\rho = -0.40$. Note that we cannot use the R functions mean, sd, cov, and cor to calculate the above, because these functions assume the samples they operate on have equal probability density. That is, they assume the samples have been drawn from the posterior PDF. But that is not the case here: we have evaluated the PDF at pre-defined points. We will learn in chapter 8 how we can sample from arbitrary probability density functions.

If we knew the value of a or b, then we could compute the conditional posteriors (equation 6.1). These are plotted as dashed lines in the lower two panels of figure 6.5, using the true values of the parameters in each case. Note how the conditional posteriors are slightly narrower and higher than the marginal posteriors. This just reflects the fact that when we fix some parameters, more of the data can be used to determine the other parameters. This requires that the data depend on the fixed parameters, of course. If the fixed parameters were irrelevant, then fixing them would not improve our inference about the other parameters.

6.3.1 R code for fitting the amplitude and background

The code below performs all the computations described above and produces the plots in figure 6.5. The code is documented and should be reasonably self-explanatory. Equation 6.19 is evaluated by the function `signal`, and `logupost` gives the (natural) logarithm of the unnormalized posterior (equation 6.23). By default, functions in R return whatever is evaluated in their last line, so you don't have to use the `return` command for this. The posterior is sampled on a uniform grid of size Nsamp $\times$ Nsamp. Each dimension is sampled in the same way as in `coin1.R` in section 5.1.1. Note that some data points can have a very low likelihood and therefore a very low posterior density. As explained below (section 6.3.3) these – in particular `exp(z)` – can become numerically identical to zero. If, as you experiment with changing values (see the suggestions after the code), you produce a posterior which is very narrow, and you use a sampling which is too coarse, then you may find that all grid evaluations give zero posterior density. This would cause a divide by zero when you try to normalize the posteriors. It would be straightforward to catch these errors, but this is not implemented in this code.

R file: `signal_background_estimation.R`

```
##### Infer posterior PDF over amplitude and background parameters

# Define function to return true signal at position x (generative model)
signal <- function(x, a, b, x0, w, t) {
  t*(a*exp(-(x-x0)^2/(2*w^2)) + b)
}

# Define function to return (natural) log posterior over (a,b).
# Prior on a and b: P(a,b) = const if a>0 and b>0, = 0 otherwise.
# Likelihood for one point is Poisson with mean d(x), so total
# likelihood is their product. Unnormalized posterior is product of these.
# d and x are equal length vectors (or scalars). The rest are scalars.
logupost <- function(d, x, a, b, x0, w, t) {
  if(a<0 || b <0) {return(-Inf)} # the effect of the prior
  sum(dpois(d, lambda=signal(x, a, b, x0, w, t), log=TRUE))
}

# Set model parameters (true and fixed)
x0    <- 0 # centre of peak
w     <- 1 # sd of peak
atrue <- 2 # amplitude
btrue <- 1 # background
t     <- 5 # scale factor (exposure time -> sets SNR)

# Simulate some data (by drawing from the likelihood)
```

```r
set.seed(205)
xdat   <- seq(from=-7*w, to=7*w, by=0.5*w)
strue <- signal(xdat, atrue, btrue, x0, w, t)
ddat   <- rpois(length(strue), strue)

# Define sampling grid to compute posterior (will be normalized
# over this range too). uniGrid spans the range 0-1 with Nsamp
# points. This is then scaled to cover the ranges alim and blim.
alim   <- c(0.0, 4.0)
blim   <- c(0.5, 1.5)
Nsamp <- 1e2
uniGrid <- seq(from=1/(2*Nsamp), to=1-1/(2*Nsamp), by=1/Nsamp)
delta_a <- diff(alim)/Nsamp
delta_b <- diff(blim)/Nsamp
a <- alim[1] + diff(alim)*uniGrid
b <- blim[1] + diff(blim)*uniGrid

# Compute log unnormalized posterior, z = ln P^*(a,b|D), on a regular grid
z <- matrix(data=NA, nrow=length(a), ncol=length(b))
for(j in 1:length(a)) {
  for(k in 1:length(b)) {
    z[j,k] <- logupost(ddat, xdat, a[j], b[k], x0, w, t)
  }
}
z <- z - max(z) # set maximum to zero

# Compute normalized marginalized posteriors, P(a|D) and P(b|D)
# by summing over other parameter. Normalize by gridding.
p_a_D <- apply(exp(z), 1, sum)
p_a_D <- p_a_D/(delta_a*sum(p_a_D))
p_b_D <- apply(exp(z), 2, sum)
p_b_D <- p_b_D/(delta_b*sum(p_b_D))

# Compute mean, standard deviation, covariance, correlation, of a and b
mean_a <- delta_a * sum(a * p_a_D)
mean_b <- delta_b * sum(b * p_b_D)
sd_a   <- sqrt( delta_a * sum((a-mean_a)^2 * p_a_D) )
sd_b   <- sqrt( delta_b * sum((b-mean_b)^2 * p_b_D) )
# To calculate the covariance I need to normalize P(a,b|D) = exp(z).
# I do it here by brute force with two loops (there are better ways in R).
# The normalization constant is Z = delta_a*delta_b*sum(exp(z)).
# This is independent of (a,b) so can be calculated outside of the loops.
# The factor delta_a*delta_b will just cancel in the expression for
# cov_ab, so I omit it entirely.
cov_ab <- 0
for(j in 1:length(a)) {
  for(k in 1:length(b)) {
    cov_ab <- cov_ab + (a[j]-mean_a)*(b[k]-mean_b)*exp(z[j,k])
  }
}
cov_ab <- cov_ab / sum(exp(z))
rho_ab <- cov_ab / (sd_a * sd_b)
cat("  a = ", mean_a, "+/-", sd_a, "\n")
cat("  b = ", mean_b, "+/-", sd_b, "\n")
cat("rho = ", rho_ab, "\n")
```

```
# Compute normalized conditional posteriors, P(a|b,D) and P(b|a,D)
# using true values of conditioned parameters. Vectorize(func, par)
# makes a vectorized function out of func in the parameter par.
p_a_bD <- exp(Vectorize(logupost, "a")(ddat, xdat, a, btrue, x0, w, t))
p_a_bD <- p_a_bD/(delta_a*sum(p_a_bD))
p_b_aD <- exp(Vectorize(logupost, "b")(ddat, xdat, atrue, b, x0, w, t))
p_b_aD <- p_b_aD/(delta_b*sum(p_b_aD))

# Make plots

pdf("signal_background_estimation.pdf", 7, 7)
# Plot true model and data
par(mfrow=c(2,2), mgp=c(2,0.8,0), mar=c(3.5,3.5,1,1), oma=0.1*c(1,1,1,1))
xplot <- seq(from=min(xdat), to=max(xdat), by=0.05*w)
splot <- signal(xplot, atrue, btrue, x0, w, t)
plot(xplot, splot, ylim=range(c(splot, ddat)), xlab="x", ylab="s or d",
     type="l", col="grey", lwd=2)
points(xdat, ddat)
# Plot unnormalized 2D posterior as contours.
# Note that they are labelled by posterior density relative to peak,
# NOT by how much probabilty they enclose.
contour(a, b, exp(z), nlevels=5, labcex=0.5, lwd=2, xlab="amplitude, a",
        ylab="background, b")
abline(v=2,h=1,col="grey")
# Plot the 1D marginalized posteriors
plot(b, p_b_D, xlab="background, b", yaxs="i",
     ylim=1.05*c(0,max(p_b_D, p_b_aD)), ylab="P(b | D)   and   P(b | a,D)",
     type="l", lwd=2)
lines(b, p_b_aD, lwd=2, lty=2)
abline(v=btrue, col="grey")
plot(a, p_a_D, xlab="amplitude, a", yaxs="i",
     ylim=1.05*c(0,max(p_a_D, p_a_bD)), ylab="P(a | D)   and   P(a | b,D)",
     type="l", lwd=2)
lines(a, p_a_bD, lwd=2, lty=2)
abline(v=atrue, col="grey")

dev.off()
```

6.3.2 Suggested experiments

It is instructive to experiment with changing both the simulated data and the values of the parameters in the above example. Here are some suggestions.

(1) Change the exposure time used to generate the data. Recall from the telescope example on page 19 that increasing the number of photons gathered will increase the signal-to-noise ratio. Try values of t of 0.5, 1, 2, 10, 50, 100. You will need to adjust the range over which the posterior is calculated (alim, blim) in order to sample it adequately (i.e. to cover its full range and to sample it at high enough resolution). If you set the limits to less than 0 you will see that the posterior is correctly truncated by the prior. A larger t corresponds to more photons. You should find that this gives a more accurate and precise determination of the parameters (and a smaller t a lower accuracy and precision).

(2) Vary the sampling resolution of x used to generate the data, but keeping the sampling range the same (so the number of data points will change). The samples are defined as follows.

```
xdat <- seq(from=-7*w, to=7*w, by=0.5*w)
```

In the above we have a resolution of $0.5w$, so try changing this to $0.1, 0.25, 1, 2, 3$ times w, for example. A better sampling of the spectral line will result in a more accurate and precise determination of its amplitude.

(3) Vary the sampling range (keep it centred on x_0), but keep the sampling resolution the same (so the number of data points will change). I suggest values of $\pm 3w$, $\pm 10w$, $\pm 20w$, $\pm 50w$. If a larger fraction of the grid evaluations are dominated by the background, is the accuracy and precision with which we determine the amplitude reduced? What about the background? Is the covariance affected?

(4) Vary the sampling range (keep it centred on x_0) between $0.5w$ and $2w$, but now with a sampling resolution of $0.1w$. In these cases we are barely sampling across the whole line. What happens to the posterior PDFs? If we set the range to $\pm 0.5w$ but have high resolution, say $0.01w$, do we improve things?

(5) Change the ratio a/b used to simulate the data (keeping both positive in accordance with the prior). The smaller this is, the less prominent the line. Is the amplitude then less accurately determined?

6.3.3 A note on computation with finite precision

When you experiment with the above example you will begin to realise (if you haven't already) that solving continuous mathematical problems on discrete computers involves dangers. Functions cannot be sampled with infinite resolution, nor can numbers be represented with infinite precision. In particular, numbers cannot be arbitrarily small on computers. This becomes a problem when sampling likelihoods or posteriors. As the likelihood for N independent data points is the product of N numbers that are often less than one, the likelihood can become arbitrarily small for larger data sets. This is the case even at the true parameter values (see section 5.2). While the absolute (small) value of the likelihood does not matter in principle, it does matter in practice, because with finite precision the likelihood (and therefore posterior) may be truncated numerically to zero for all sampled parameters. This will lead to a zero normalization constant, resulting in divide by zero and possibly other errors. You can run into this problem quicker than you realise, so you should be careful to trap such errors in code which manipulates likelihoods.

One way to address this is to work with log likelihoods, because this provides us with a much larger dynamical range: the number 10^{-1000} may be truncated by the computer representation to be numerically identical to zero, but the number -1000 will not be. In some situations we must calculate actual likelihoods, however. This is the case when we calculate the marginal likelihood (equation 3.16) which we will need to do in chapter 11. We can mitigate numerical problems to some extent with the following method.

As we will see in section 8.3, the marginal likelihood may be approximated by the average of the likelihoods computed at a set of parameter values drawn from the prior. Let $\{\log L_l\}$ be this set of N log likelihoods. Suppose that some or all of these values are so negative (because the likelihoods are so small) that $10^{\log L_l}$ is truncated by the computer representation to be identical to zero. The estimated marginal likelihood

$$E = \frac{1}{N} \sum_{l=1}^{N} 10^{\log L_l} \tag{6.29}$$

would then be incorrect (underestimated), and in the worst case zero. To overcome this we add a constant h to every $\log L_l$ term. Provided the range of the set $\{L_l\}$ is less than the computer's dynamic range,[3] we can choose h to ensure that $10^{(h + \log L_l)}$ is not truncated for any l. When we work with these we will compute the marginalized likelihood

$$E' = \frac{1}{N} \sum_{l=1}^{N} 10^{(h + \log L_l)}$$
$$= 10^h E$$
$$\log E' = h + \log E \tag{6.30}$$

from which we can calculate the quantity we actually want, E. A good approach is to set the largest values of the exponent to be zero, as we may not care about just the smallest of the set of likelihoods being numerically indistinguishable from zero. In that case an appropriate choice for h is $-\max(\log L_l)$.

---

[3] This is about 630 orders of magnitude on my machine when using double precision.

## 7

Approximating distributions

Posterior distributions can be quite complicated. In this chapter I will first look at how we can approximate a distribution using a Taylor expansion. I will then introduce the method of kernel density estimation, which is used to estimate a distribution given samples drawn from it. A well-known example of this is histograms. This technique will be indispensable when we come to using Monte Carlo sampling methods in the next two chapters.

7.1 The quadratic approximation

7.1.1 One dimension

In the real world there are relatively few problems for which either the full posterior or a marginalization thereof is a standard distribution. Posterior PDFs are sometimes quite complicated and difficult to summarize, as we shall see in chapter 9. Nonetheless, if the PDF is dominated by a single mode, and that mode is quite "peaky", then it is often sufficient to approximate the posterior around that peak. The more informative the data, the greater the amount of probability concentrated around the peak, and the better such an approximation will be. Here we look at a convenient approximation for such cases.

To do this, I use the logarithm of the posterior PDF

$$\Phi = \ln P(\theta|D) \tag{7.1}$$

where, for now, θ is a scalar. As the logarithm is a strictly monotonic function of its argument, turning points in the PDF correspond to turning points in its logarithm. Whereas, for informative data, the probability density will span many order of magnitudes, its logarithm will vary more slowly with θ. This suggest that a reasonable approximation for Φ can be achieved by expanding it around its mode $\hat{\theta}$ with a second-order Taylor expansion:

$$\Phi \simeq \Phi(\hat{\theta}) + (\theta - \hat{\theta})\frac{d\Phi}{d\theta}\bigg|_{\hat{\theta}} + \frac{1}{2}(\theta - \hat{\theta})^2\frac{d^2\Phi}{d\theta^2}\bigg|_{\hat{\theta}}$$

$$\text{where} \quad \frac{d\Phi}{d\theta}\bigg|_{\hat{\theta}} = 0 \quad \text{so}$$

$$\Phi - \Phi(\hat{\theta}) \simeq \frac{1}{2}(\theta - \hat{\theta})^2\frac{d^2\Phi}{d\theta^2}\bigg|_{\hat{\theta}}. \tag{7.2}$$

Taking the exponential of this and using equation 7.1 we get

$$P(\theta \mid D) \simeq A \exp\left(\frac{1}{2}(\theta - \hat{\theta})^2 \frac{d^2\Phi}{d\theta^2}\bigg|_{\hat{\theta}}\right) \tag{7.3}$$

which is a Gaussian with mean $\hat{\theta}$ and variance

$$\sigma^2 = \left(-\frac{d^2\Phi}{d\theta^2}\bigg|_{\hat{\theta}}\right)^{-1} \tag{7.4}$$

as the term $A = P(\hat{\theta} \mid D)$ is independent of θ. The second derivative is negative at a mode (maximum), ensuring that the variance is positive.

The quadratic approximation approximates a general PDF as a Gaussian, which is convenient because this has nice properties. For example, if we have a posterior which we cannot (or do not want to) normalize, then we cannot calculate its standard deviation to use as an uncertainty measure. But with the quadratic approximation we can calculate this relatively easily. Whether this is a good approximation naturally depends on how accurate the second-order Taylor expansion is in the case at hand.

Example: exponential distribution

Suppose we have a likelihood function

$$P(x \mid \theta) = \frac{1}{\theta} \exp(-x/\theta) \tag{7.5}$$

for $x \geq 0$ and $\theta \geq 0$, and a prior $P(\theta)$ which is uniform for $\theta \geq 0$ and zero otherwise. We measure N data points $\{x_i\}$. What is the quadratic approximation of the posterior PDF for θ?

Assuming the data are measured independently, the likelihood of the data is the product of the individual likelihoods. The posterior is proportional to the product of this with the prior, so the posterior is

$$P(\theta \mid \{x_i\}) \propto \frac{1}{\theta^N} \exp(-N\bar{x}/\theta) \tag{7.6}$$

where $\bar{x}$ is the mean of the data and the (missing) normalization constant is independent of the model parameter θ. Taking the natural logarithm and differentiating twice with respect to θ we get

$$\Phi = -N \ln\theta - \frac{N\bar{x}}{\theta} + \text{constant} \tag{7.7}$$

$$\frac{d\Phi}{d\theta} = -\frac{N}{\theta} + \frac{N\bar{x}}{\theta^2} \tag{7.8}$$

$$\frac{d^2\Phi}{d\theta^2} = \frac{N}{\theta^2} - \frac{2N\bar{x}}{\theta^3}. \tag{7.9}$$

From this we can see that the maximum (zero first derivative) is $\hat{\theta} = \bar{x}$, and the second derivative at this value is $-N/\bar{x}^2$. There is only one maximum. Hence we can approximate this posterior around its maximum as a Gaussian with mean $\bar{x}$ and variance $\bar{x}^2/N$.

7.1.2 Two dimensions

The quadratic approximation, like the Taylor expansion, generalizes to higher dimensions. Suppose we have the two-dimensional posterior $P(a, b|D)$. The second-order Taylor expansion at the maximum $(\hat{a}, \hat{b})$ of the log posterior (so the first derivatives are zero) is

$$\Phi \simeq \Phi(\hat{a}, \hat{b}) + \frac{1}{2}\left[(a - \hat{a})^2\frac{\partial^2\Phi}{\partial a^2} + (b - \hat{b})^2\frac{\partial^2\Phi}{\partial b^2} + 2(a - \hat{a})(b - \hat{b})\frac{\partial^2\Phi}{\partial a\partial b}\right] \quad (7.10)$$

where all the derivatives are evaluated at the maximum. The quantity in the square brackets can be written as a matrix multiplication

$$R = (a - \hat{a} \ \ b - \hat{b})\begin{pmatrix} v_{aa} & v_{ab} \\ v_{ab} & v_{bb} \end{pmatrix}\begin{pmatrix} a - \hat{a} \\ b - \hat{b} \end{pmatrix} \quad (7.11)$$

where

$$v_{aa} = \frac{\partial^2\Phi}{\partial a^2}, \quad v_{bb} = \frac{\partial^2\Phi}{\partial b^2}, \quad v_{ab} = \frac{\partial^2\Phi}{\partial a\partial b}. \quad (7.12)$$

Using the same logic as with one parameter, the approximation to the posterior around its maximum can be written as

$$P(a, b|D) \simeq A\exp\left(\frac{1}{2}R\right) \quad (7.13)$$

which, given the definition of R, is the equation for a two-dimensional Gaussian in (a, b) with mean $(\hat{a}, \hat{b})$ and covariance matrix

$$\mathrm{Cov}(a, b) \equiv \begin{pmatrix} \sigma_a^2 & \sigma_{ab}^2 \\ \sigma_{ab}^2 & \sigma_b^2 \end{pmatrix} = -\begin{pmatrix} v_{aa} & v_{ab} \\ v_{ab} & v_{bb} \end{pmatrix}^{-1}$$

$$= \frac{1}{v_{aa}v_{bb} - v_{ab}^2}\begin{pmatrix} -v_{bb} & v_{ab} \\ v_{ab} & -v_{aa} \end{pmatrix}. \quad (7.14)$$

From this we can read off the values of the variances σ_a^2 and σ_b^2 of the two parameters, and their covariance σ_{ab}^2.

7.1.3 Higher dimensions

The quadratic approximation can be extended to higher dimensions. Consider the J-dimensional parameter vector $\boldsymbol{\theta}$ with maximum $\hat{\boldsymbol{\theta}}$, i.e. $\nabla\Phi(\boldsymbol{\theta} = \hat{\boldsymbol{\theta}}) = 0$. Then

$$\Phi(\boldsymbol{\theta}) \simeq \Phi(\hat{\boldsymbol{\theta}}) + \frac{1}{2}(\boldsymbol{\theta} - \hat{\boldsymbol{\theta}})^\mathsf{T}[\nabla\nabla\Phi(\hat{\boldsymbol{\theta}})](\boldsymbol{\theta} - \hat{\boldsymbol{\theta}}). \quad (7.15)$$

The posterior is a multivariate Gaussian with covariance matrix equal to the negative of the inverse of the matrix of second derivatives of the logarithm of the posterior. This is quite a mouthful and is more easily expressed as

$$\mathrm{Cov}(\boldsymbol{\theta})_{\hat{\boldsymbol{\theta}}} = -[\nabla\nabla\Phi(\hat{\boldsymbol{\theta}})]^{-1}. \quad (7.16)$$

The J-dimensional square matrix $\nabla\nabla\Phi(\hat{\boldsymbol{\theta}})$ is called the *Hessian matrix*, the elements of which are $\partial^2\Phi/\partial\theta_i\partial\theta_j$ evaluated at $\boldsymbol{\theta} = \hat{\boldsymbol{\theta}}$. This is a symmetric matrix (as the covariance is symmetric).[1] We see that the width of the distribution – the uncertainty in the parameters – is determined by the second derivatives of the logarithm of the posterior, which are a measure of its curvature.

7.2 Density estimation

Up until this point we have been able to plot arbitrary PDFs simply by evaluating them on a pre-defined grid. But as we shall see in the next chapter this it not always practicable. So in preparation for the following two chapters consider the following situation: given a sample of N data points $\{x_i\}$ drawn from an unknown PDF, how can we find (or rather estimate) that PDF?

One approach is to adopt a parametrized model for the PDF, for example a Gaussian or Poisson distribution. We saw in section 4.4 how we could then maximize the product of probability densities with respect to the model parameters in order to find the best fitting parameters.

Unfortunately, in many situations we will not be able to find a parametrized model that approximates the data sufficiently well. We then need to take a non-parametric approach. Suppose the data are univariate. If they are also discrete, and we have many samples at each value, then we could just stack up the data into a histogram. If the data are real, each value is unique: our data set is essentially a string of delta functions. We can still construct a histogram by averaging over narrow bins of the data, although there are issues with this, as we shall see below. A histogram is in fact a simple case of a more general method called *density estimation*, which we shall investigate below.

Sometimes we do have an equation for the density distribution. In section 6.3 this distribution was a bivariate posterior, and we plotted it and calculated summary statistics simply by evaluating it on a grid. But if the distribution has more dimensions or a complex variation, then evaluation on a grid will be computationally expensive and may not capture all the variations. In that case it is preferable to draw samples from the distribution and then use density estimation to approximate the distribution. We will see how to do this sampling in chapter 8 using Monte Carlo methods.

I use several R packages below. `density` and `persp` are all in the base R packages. `image.plot` is in the package `fields`. Venables & Ripley (2002) introduce the package `MASS`. This contains the R methods `truehist` and `kde2d` as well as the data set `geyser`, all of which I will use below. My example R scripts are also based on scripts in Venables & Ripley.

7.2.1 Histograms

A histogram estimates the density by aggregating data points into bins of pre-specified width and location. These bins are adjacent, equal-width (usually), uniform functions over

---

[1] This is a symmetric matrix only if the order of differentiation does not matter, for which the functions must be continuous functions. But they must be continuous for the Taylor expansion to be valid.

the variable x. Examples are shown in figure 7.1. Let the bin width be λ. The number of points that fall into each bin is summed up and this is taken as the frequency (counts) of data over that bin; call this c_i for bin i. We can normalize the histogram by dividing the counts in each bin by the area under the histogram (which is $\lambda \sum c_i$) so that the histogram now integrates to one. Each bin then reports the density – the fraction of objects per unit x, which is $f_i = c_i/(\lambda \sum c_i)$ – under the assumption that the density is constant over the bin. Both representations are of course discontinuous at the bin boundaries. We can appreciate the difference between counts and density by considering a uniform distribution. If we half the bin width, the counts c_i will half, but the densities f_i will stay constant. The density will still integrate to one, because although there are now twice as many bins, they are half as wide.

Histograms are quick and easy, and useful when we have a lot of data such that we have well-populated bins. However, they can be sensitive to the placement and width of the bins, as we see in the following example.

The data set `geyser` in the R package `MASS` contains data on the eruptions of the Old Faithful geyser in Yellowstone National Park in the USA. It records two variables: the eruption duration and the time until the next eruption (both in minutes). We might be interested in finding out whether there is a relationship between these two variables, or whether the eruptions fall into groups in one or both of the variables. Here we look at the duration variable and use histograms to find its distribution. The following R code investigates two things. It first plots histograms with different bin sizes, figure 7.1. We see that increasing the number of bins increases the amount of detail, but taken to an extreme we would end up with just one or zero objects per bin, which doesn't tell us anything beyond the original data. This is another example of the bias-variance trade-off discussed in section 4.8. Broad bins have small variance because they average over a lot of points, but a large bias because they are not a very accurate representation of the density at all positions across of the bin. Narrow bins are the opposite: the bias is small, but the variance is large because small changes in bin size or position result in big changes in the histogram.

The second part of the code below plots histograms all with the same bin width, but with different bin centers, shifting them by a fifth of the bin centre each time. The result is shown in figure 7.2. We see how sensitive the results are to this shift. We are unlikely to have any information telling us which set of bin centres is the most appropriate.

Let us now see how we can generalize the idea of histograms.

R file: `histograms.R`

```
##### Investigate histograms
##### Modified from the example given by Venables & Ripley (2002)

library(MASS) # for truehist and geyser data
attach(geyser)
# Look at sensitivity to bin width
pdf("histograms1.pdf", 12, 8)
par(mfrow=c(2,3), mar=c(3.5,3.5,1.5,0.5), oma=c(0.5,0.5,0.5,0.5),
    mgp=c(2.2,0.8,0), cex=1.0)
for(h in c(2, 1, 1/2, 1/4, 1/8, 1/16)) {
  truehist(duration, h=h, x0=0.0, xlim=c(0,6), ymax=0.8, ylab="density",
```

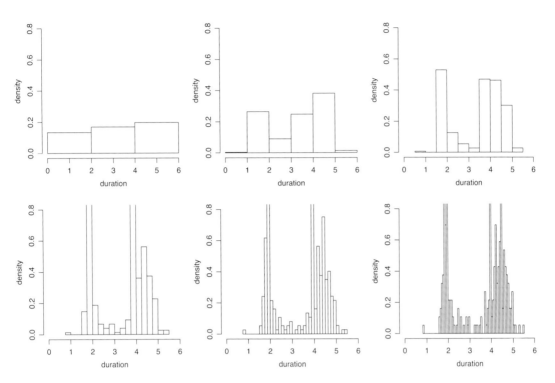

Fig. 7.1 Six histograms of the same data (geyser duration) with bin sizes ranging from 2 to $1/16$ in steps of factors of $1/2$. In the bottom row the highest peaks extend off the scale, the highest being for the narrowest bins (up to around 3.0 at a duration of about 4).

```
                   col="white")
}
dev.off()
# Look at sensitivity to placing of bin centres
pdf("histograms2.pdf", 12, 8)
par(mfrow=c(2,3), mar=c(3.5,3.5,1.5,0.5), oma=c(0.5,0.5,0.5,0.5),
    mgp=c(2.2,0.8,0), cex=1.0)
binwidth <- 0.5
for(x0 in binwidth*(0:5)/5) {
   truehist(duration, h=binwidth, x0=x0, xlim=c(0,6), ymax=0.8,
            ylab="density", col="white")
}
dev.off()
detach(geyser)
```

7.2.2 Kernel density estimation

A histogram models the density by assuming it to be constant over some finite range of the variable, namely the bin width. Every point is assigned to a single bin, and the density

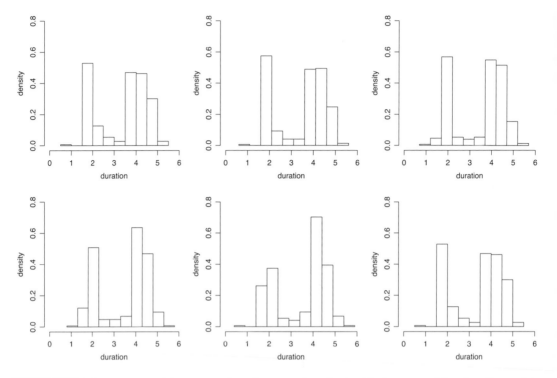

Fig. 7.2   Histograms of the same data (geyser duration) with the same bin size ($h = 1/2$), but bin centers offset by $1/5$ of the bin width each time.

is the average number of points in the bin. We could do something more sophisticated than assuming a constant density. This generalization leads us to the idea of *kernel density estimation*. In this context a kernel is a weighted function of data. The weights are usually a function of the distance from the point at which the kernel is being evaluated.

Given a one-dimensional data set $\{x_i\}$ of size N, we can estimate the density at an arbitrary point x as being

$$f(x) = \frac{1}{N\lambda} \sum_{i=1}^{N} K(u_i) \quad \text{where} \quad u_i = \frac{x - x_i}{\lambda}. \tag{7.17}$$

K is the *kernel*, which must be normalized ($\int K(u)du = 1$), and as it is being used for density estimation it must also be non-negative. The bandwidth λ characterizes the length scale over which the kernel operates. We may set this ourselves, or we can estimate it from the data (we'll see one way of doing this later in this section). It need not be fixed. $N\lambda$ is a normalization constant which ensures $f(x)$ is a density. We are free to choose any appropriate kernel. A simple choice is the rectangular kernel

$$K(u) = \begin{cases} 1 & \text{if } |u| < 1/2 \\ 0 & \text{otherwise.} \end{cases} \tag{7.18}$$

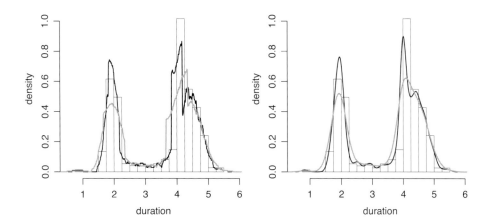

Fig. 7.3   Kernel density estimation (using `density` in R) on the geyser duration data set using a rectangular kernel (left) and a Gaussian kernel (right) with bandwidth parameter 0.1 (in black) and 0.2 (in grey). A histogram of the same data with bin width 0.25 is overplotted in both cases for comparison.

This is similar to the histogram in the sense that the density at point x is determined just by the number of points that fall within the bin, i.e. within $\pm\lambda/2$ of x. But unlike the histogram this kernel can be computed at any point x, and not just at a pre-defined set of bin centres separated by the bin width. We can think of moving a box of width λ along the x-axis and computing the density continuously. This is known as a *moving average*. Whereas the densities in each histogram bin are independent of their neighbouring bins (they share no data), here they are not. The resulting density estimate is shown for two different choices of λ in the left panel of figure 7.3.

As with the histogram, we see that the rectangular kernel does not give a very smooth variation of the density. There are other kernels which generally give better estimates of the density. Here I mention just two.

The first is a *Gaussian kernel*. We put down a Gaussian at x (i.e. with mean x) and make a weighted sum of all the data $\{x_i\}$, where the weights are given by the value of the Gaussian function. The standard deviation of the Gaussian is the bandwidth λ. The kernel is

$$K(u) \; = \; \frac{1}{\sqrt{2\pi}} \exp\left(-\frac{1}{2}u^2\right) \tag{7.19}$$

with the density at any point x again given by equation 7.17. An example is shown in the right panel of figure 7.3. It tends to produce smoother density estimates than the rectangular kernel, for a given bandwidth.

Another commonly used kernel is the *k-nearest neighbours kernel*, whereby k is a fixed positive integer. At a specified point x we identify the k nearest neighbours among the $\{x_i\}$ and compute the length $\Delta(x)$ that they span: that is, $\Delta(x)$ is the distance between the most distant neighbour above x and most distant one below x. The kernel size therefore varies with x, in order to include a fixed number of neighbours. The kernel is just $K(u) = k$, and it follows from equation 7.17 that the density function is

$$f(x) = \frac{k}{N\Delta(x)} \,. \tag{7.20}$$

The smoothness of the density curve produced by a kernel density estimate is determined by the size of the kernel, or by the number of nearest neighbours. The larger the kernel (or the smaller the number of nearest neighbours) the smoother the curve will be. This may smooth out "real" features. If the kernel is too small (or we have too few neighbours), the variations in density may just be artefacts (noise) from having too few data in the density estimate at each point.

Note an important difference between the two types of kernel. The rectangular and Gaussian kernels have a fixed bandwidth, and so include a variable number of points that depends on the local density of the data. The nearest neighbour kernel instead includes a fixed number of points, but has a variable bandwidth. If we have large spans of the variable where there are no data, the latter kernel will always give a (non-zero) density estimate, but this may be inaccurate because it relies on distant data points. In regions of high density, the nearest neighbour kernel may be more precise than a fixed-width kernel, because the former will shrink and so will follow density variations more closely.

I will introduce a couple of other kernels in section 12.4 when we look at using them for regression.

The following R code produces the plots in figure 7.3. The density estimation is performed by the function density. The parameter n in this function is the number of points at which the density is estimated. These points are spread uniformly over the range of the data (although the range can be specified using from and to). Provided n is sufficiently high, its exact choice will have no impact on the appearance of the density estimate. I encourage you to experiment with changing the kernel, specified by kernel, and the bandwidth, specified by bw. Note that the bw parameter may not give a size for the bandwidth which you expect. This is because the functions internal to density may scale this by an amount that depends on the choice of kernel. Read the documentation to find out more.

R file: kde.R

```
##### Investigate kernel density estimation in 1D
##### Modified from the example given by Venables & Ripley (2002)

library(MASS) # for truehist and geyser data
attach(geyser)
pdf("kde.pdf", 8,4)
par(mfrow=c(1,2), mar=c(3.5,3.0,0.5,0.5), oma=c(0.5,0.5,0.5,0.5),
    mgp=c(2.2,0.8,0), cex=1.0)
truehist(duration, h=0.25, xlim=c(0.5, 6), ymax=1.1, ylab="density",
         col="white", lwd=0.5)
lines(density(duration, kernel="rectangular", bw=0.1, n=2^10), lwd=1.5)
lines(density(duration, kernel="rectangular", bw=0.2, n=2^10), lwd=2,
      col="grey60")
truehist(duration, h=0.25, xlim=c(0.5, 6), ymax=1.1, ylab="density",
         col="white", lwd=0.5)
```

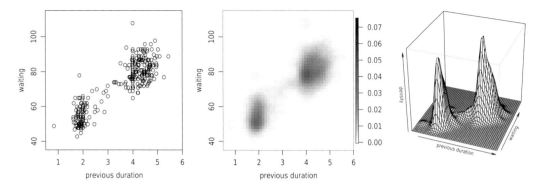

Fig. 7.4 Two-dimensional kernel density estimation with a Gaussian kernel applied to the geyser data. Left: original data. Centre and right: kernel density estimation using a bandwidth of 0.75 for "previous duration" and 5 for "waiting". The central panel shows the kernel density estimation as a density map, the right panel as a three-dimensional perspective mesh.

```
lines(density(duration, kernel="gaussian",   bw=0.1, n=2^10), lwd=1.5)
lines(density(duration, kernel="gaussian",   bw=0.2, n=2^10), lwd=2,
      col="grey60")
dev.off()
detach(geyser)
```

You may be interested in looking at what the function density actually produces. If you do

```
attach(geyser)
mydense <- density(duration, kernel="gaussian", bw=0.2, n=2^10)
```

then attributes(mydense) lists all the things computed and stored by density. This produces the following.

```
$names
[1] "x"  "y"  "bw"  "n"  "call"  "data.name"  "has.na"
$class
[1] "density"
```

The attribute x is the vector of n points at which the density was estimated, to give the values y. You can retrieve these values with mydense$x. Note that density estimation does not produce an analytic function, but rather evaluations of the estimator at a pre-defined set of points. The function lines in the above code just connects the x and y values in order to plot an apparently continuous function.

The idea of kernel density estimation extends to higher dimensions. The following code is an example in two dimensions applied to the geyser data. It produces figure 7.4.

R file: kde2d.R

```
##### Investigate kernel density estimation in 2D
##### Modified from the example given by Venables & Ripley (2002)
```

```
library(fields) # for image.plot
library(MASS)
library(RColorBrewer)
mypalette <- colorRampPalette(brewer.pal(9, "Greys"), space="rgb",
                   interpolate="linear", bias=2)
mycols <- mypalette(64)
geyser2 <- data.frame(as.data.frame(geyser)[-1, ],
                   pduration=geyser$duration[-299])
attach(geyser2)
pdf("kde2d.pdf", 12, 4)
par(mfrow=c(1,3), mar=c(3.5,3.5,0.5,1), oma=0.5*c(1,1,1,1),
    mgp=c(2.2,0.8,0), cex=1.0)
plot(pduration, waiting, xlim=c(0.5, 6), ylim=c(35, 115), xaxs="i",
    yaxs="i", xlab="previous duration", ylab="waiting")
f1 <- kde2d(pduration, waiting, n=50, h=c(0.75, 10),
         lims=c(0.5, 6, 35, 115))
image.plot(f1, zlim=c(0, 0.075), xlim=c(0.5, 6), ylim=c(35, 115), xaxs="i",
         yaxs="i", xlab="previous duration", ylab="waiting", col=mycols)
persp(f1, phi=30, theta=20, d=5, xlim=c(0.5, 6), ylim=c(35, 115), xaxs="i",
    yaxs="i", xlab="previous duration", ylab="waiting", zlab="density")
dev.off()
detach(geyser2)
```

Good kernel density estimation requires that we adopt a suitable bandwidth. There are various ways to achieve this. Experimenting with various values and inspecting the results can be quite effective. Judgement will anyway be required, no matter what theoretical scheme is used. A widely used scheme is to compute the *mean integrated squared error* (also called the L^2 *risk function*), defined as

$$\text{MISE} = E\left[\int (\hat{f}(x; \lambda) - f(x))^2 \, dx\right] \tag{7.21}$$

where $\hat{f}(x; \lambda)$ is our kernel density estimate of the true, unknown function $f(x)$. The expectation value in the above equation is taken with respect to the sample of data, i.e. it represents what we would get if we had infinite data. Minimizing the MISE with respect to λ gives us a good choice of bandwidth. The problem is that the expression involves the very function we do not know, $f(x)$. But we can work around this by first expanding the square to write it as

$$\text{MISE} = E\left[\int \hat{f}^2(x; \lambda) \, dx\right] - 2E\left[\int \hat{f}(x; \lambda) f(x) \, dx\right] + E\left[\int f^2(x) \, dx\right]. \tag{7.22}$$

The first term we can estimate directly from the data and our kernel density estimator. The last term is independent of λ, so can be ignored. The middle term may be estimated by *leave-one-out cross-validation*: given a set of N data points, we select one point x_i, then use all the other points to estimate the density (with our estimator) at that one point. Label this estimated density as $\hat{f}_{-i}(x_i; \lambda)$. Essentially we are using all the other data points to approximate $f(x)$ for the sake of computing $\hat{f}_{-i}(x_i, \lambda)$. We then repeat this for all N points and average. To find the optimal bandwidth we therefore minimize the following metric with respect to λ

$$\int \hat{f}^2(x;\lambda)\, dx - \frac{2}{N} \sum_{i=1}^{N} \hat{f}_{-i}(x_i;\lambda). \tag{7.23}$$

This is sometimes called the *estimated risk*.

The R function density implements various ways of estimating the bandwidth of a kernel from the data. Details on these can be obtained with help("bw.nrd"). An example is the function bw.nrd. It can be used to find the bandwidth by typing bw.nrd(geyser$duration) which gives the value 0.389. We can instead do this directly within the call to density by setting bw="nrd".

We will use kernels for density estimation in chapter 9, and we will encounter kernels again in section 12.4 for doing local regression.

8   Monte Carlo methods for inference

Monte Carlo methods are essential for sampling distributions and for estimating integrals, both of which lie at the heart of probabilistic inference. I start this chapter with a discussion of why sampling is non-trivial, and will then summarize where we use integration in inference and how we can estimate such integrals. We shall then look at the basic concepts of Monte Carlo sampling, and learn about a widely used Markov Chain Monte Carlo method: the Metropolis–Hastings algorithm.

8.1   Why we need efficient sampling

To recap section 3.3, the central equation in Bayesian inference tells us that the posterior PDF over the model parameters (θ) of a model (M) given the data (D) is the product of the likelihood and prior divided by the evidence

$$P(\theta|D, M) = \frac{P(D|\theta, M)P(\theta|M)}{P(D|M)}. \tag{8.1}$$

As the denominator is independent of the model parameters, we don't need it if we just want to find the shape of the posterior. In that case we can work with the unnormalized posterior,

$$P^*(\theta|D, M) = P(D|\theta, M)P(\theta|M). \tag{8.2}$$

With just one parameter, the posterior is a one-dimensional PDF. If appropriate, we can choose a conjugate prior for the given likelihood, in which case the posterior has the same functional form as the prior (see section 5.4). With the coin tossing example from section 5.1 the likelihood was binomial and the prior and posterior were beta distributions. If the posterior is not a standard distribution we could still evaluate it on a dense grid, as we did in sections 3.5 and 5.1 in one dimension and in section 6.3 in two dimensions. We also used these grid evaluations to integrate the distribution and thus to find the properly normalized posterior.

Evaluating on a grid is not as straightforward as it first seems, for a number of reasons. We must ensure that the grid covers the full range of the posterior. If the distribution has infinite support, we could use a grid only if we cover all the high probability density regions. Yet if the expression for the posterior is complicated, it will not be easy to identify these regions. And even if we can, there may still be a lot of probability contained in extended, low density tails. These cannot be neglected if we want to calculate any quantity involving the integral of the posterior, such as the mean or median. The posterior could also be highly peaked, in which case we need to use a very fine sampling around this region. The posterior could even be multimodal, so just sampling around the first maximum found will be insufficient. We could attempt to work out the optimal sampling range and density by trial and error: visualize the posterior and study whether quantities like the mean or mode are stable with changes in the sampling. But this is not a realistic option if we are processing a large number of data sets.

For problems with three or more parameters the situation is even more complicated. All of the above issues apply, but now the posterior exists in a higher dimensional parameter space, where more complex surfaces are possible that are impossible to visualize completely. It is also much more time consuming to determine the posterior by evaluating it on a regular, but now multi-dimensional, parameter grid. This is because of the curse of dimensionality. If we need N grid points in order to sample one parameter with sufficient density, then for two parameters we will need N^2 grid points. For J-dimensions we will need N^J grid points. Even with $J = 5$ and $N = 100$ (quite modest), we already need 10^{10} grid points, an infeasible number of calculations of the likelihood (which is normally more time-consuming to compute than the prior). Moreover, the likelihood and/or prior (and therefore the posterior) is likely to be vanishingly small at the overwhelming majority of these points (increasingly so in high dimensions), so most of these computations would be pointless anyway.

We could avoid the inefficiency of a grid by sampling from the posterior PDF directly. That is, we would like to be able to draw samples from an arbitrary PDF – let's call it $g(\theta)$ – in such a way that the frequency distribution of the samples is equal to $g(\theta)$ in the limit of a large number of draws. Let's assume that we can evaluate $g(\theta)$ up to some multiplicative constant; that is, we do not require $g(\theta)$ to be normalized. How can we sample from it? This is not self-evident, because in order to sample efficiently we would need to draw preferentially from regions of high relative probability. But it is not obvious how we could locate these regions without explicitly evaluating the function everywhere, which is no better than evaluating on a grid.

It turns out that there are efficient ways to sample an arbitrary probability distribution using *Monte Carlo* methods. We can then use a set of samples $\{\theta\}$ to represent that distribution. This is the case even when $g(\theta)$ is unnormalized: rescaling the distribution by a constant factor will not change the relative frequency with which samples are drawn. Thus once we have such a set of samples we can approximate essentially any property of $g(\theta)$ – such as the mean, variance, and confidence intervals – directly from the set of samples. This applies also to some transformation of the parameter, $f(\theta)$. So if we wanted to calculate the variance of a function f of the samples, we just compute the variance of the transformed samples $\{f(\theta)\}$.

It is also straightforward to use such samples to approximate integrals. Integration is almost endemic in Bayesian inference, so before looking into Monte Carlo sampling methods in section 8.4, let us examine where we need integration (section 8.2) and how we can approximate it using Monte Carlo samples (section 8.3).

8.2 Uses of integration in Bayesian inference

There are at least four different tasks in Bayesian inference that require integration. The first three we have met before.

8.2.1 Marginal parameter distributions

The first task is marginalization, which we already met in section 6.1 and used in section 6.2. If we have a two-dimensional posterior PDF over the parameters (θ_1, θ_2) then we can determine the posterior PDF over just one parameter – that is, regardless of the value of the other parameter – by integration

$$P(\theta_1 \,|\, D, M) = \int P(\theta_1, \theta_2 \,|\, D, M) \, d\theta_2. \tag{8.3}$$

8.2.2 Expectation values (parameter estimation)

The expectation value of θ is defined as

$$E[\theta] = \int \theta \, P(\theta \,|\, D, M) \, d\theta. \tag{8.4}$$

This requires $P(\theta \,|\, D, M)$ to be the normalized posterior. If we only have the unnormalized posterior (equation 8.2), then we write this as

$$E[\theta] = \frac{1}{\int P^*(\theta \,|\, D, M) \, d\theta} \int \theta \, P^*(\theta \,|\, D, M) \, d\theta. \tag{8.5}$$

8.2.3 Model comparison (marginal likelihood)

The denominator in equation 8.1 can be thought of as a normalization constant, because it is the integral of the numerator over all θ

$$P(D \,|\, M) = \int P(D \,|\, \theta, M) P(\theta \,|\, M) \, d\theta = \int P^*(\theta \,|\, D, M) \, d\theta. \tag{8.6}$$

But as we shall see in chapter 11, this also plays a central role in model comparison where it is often called the marginal likelihood, or evidence.

8.2.4 Prediction

Suppose we have the data set $D = \{y\}$ obtained at fixed $\{x\}$, and have determined the posterior PDF over the parameters. This model might be a simple straight line, for example. Given a new point x_p, what is the model prediction of y (call it y_p)? The maximum likelihood approach is to find the "best" parameters of the model, to use these in the model equation to predict y, and finally to attempt to propagate the errors in some way (see section 4.1). But this would not take into account the uncertainty in the parameters, which is reflected by the (ignored) finite width of the posterior.

The Bayesian approach is instead to find the posterior PDF over y_p. Specifically, we want to find $P(y_p | x_p, D, M)$, which is the *posterior predictive distribution* at the specified point x_p given all of the original data D. The model parameters θ do not appear in this expression because in a prediction problem we are not interested in them. The parameters are just a means to an end here. We can remove parameters by marginalizing over them. Dropping M for brevity (it is implicit everywhere), we can therefore write

$$
\begin{aligned}
P(y_p | x_p, D) &= \int P(y_p, \theta | x_p, D)\, d\theta \\
&= \int P(y_p | x_p, \theta, D)\, P(\theta | x_p, D)\, d\theta \\
&= \int \underbrace{P(y_p | x_p, \theta)}_{\text{likelihood}}\, \underbrace{P(\theta | D)}_{\text{posterior}}\, d\theta.
\end{aligned}
\tag{8.7}
$$

In going from the second line to the third line I have exploited the fact that I am allowed to remove variables to the right of the "|" symbol when the variables to the left are *conditionally independent* of them. $P(y_p | x_p, \theta, D)$ is independent of D once we specify the model parameters, because for determining the PDF over y_p, D contains no additional information beyond what we have in the parameters. The parameters, together with x_p, are sufficient to determine y_p. In a similar way, the x_p disappears from the other term, $P(\theta | x_p, D)$, because our knowledge of the model parameters cannot depend on where we later chose to make a prediction. The net result is the third line, which contains two quantities we know: the likelihood and the posterior. This idea of conditional independence is useful when manipulating probability equations because it may lead us to realise that apparently unknown quantities are equivalent to things we already know.

Equation 8.7 tells us that the posterior predictive distribution at point y_p is the posterior-weighted average of the likelihood of y_p. Another way of looking at this is to think that for each θ we get a predictive distribution over y_p (the likelihood). We then average all of these distributions with a weighting factor (the posterior) which tells us how well that θ is supported by the data.

We will see an example of this prediction in action in section 9.1.3.

8.3 Monte Carlo integration

A simple solution to integrating a function is to evaluate it over a dense, regular grid, as we did in chapters 5 and 6. If the N values in this grid are represented by the set $\{\theta_i\}$, then

$$
\int f(\theta)\, d\theta \simeq \sum_{i=1}^{N} f(\theta_i)\, \delta\theta = \frac{\Delta\Theta}{N} \sum_{i=1}^{N} f(\theta_i)
\tag{8.8}
$$

for some function $f(\theta)$, in which $\Delta\Theta$ is the total range of the grid of θ values, and $\delta\theta = \Delta\Theta/N$ is the spacing between them. This method of approximation is called the rectangle method, because we approximate the integral using a sequence of rectangles (figure 3.2). Equation 8.8 is also valid for a irregular grid (e.g. when $\{\theta_i\}$ is drawn from a uniform distribution) in which case $\delta\theta$ is the mean spacing between the grid points.

If θ is a vector of parameters, then the integral in equation 8.8 is multi-dimensional and we think of $\delta\theta$ as being a small N-dimensional hypervolume centered on the sample, and $\Delta\theta$ as being the total hypervolume of the grid. However, as explained in section 8.1, this numerical estimate of the integral suffers from the curse of dimensionality: to keep the uncertainty constant, we would need to increase the number of samples exponentially with the number of dimensions.

We can integrate more efficiently in higher dimensions once we have managed to sample explicitly from a PDF. The Monte Carlo approximation of the integration of some function $f(\theta)$ over the PDF $g(\theta)$ is

$$\frac{\int g(\theta)f(\theta)\,d\theta}{\int g(\theta)\,d\theta} = \langle f(\theta)\rangle \simeq \frac{1}{N}\sum_{i=1}^{N} f(\theta_i) \tag{8.9}$$

where the samples $\{\theta_i\}$ have been drawn from $g(\theta)$, which we assume to be unnormalized. If it were normalized then the denominator on the left would be unity. The symbols $\langle\,\rangle$ denote an expectation value, in this case over $g(\theta)$. The set of samples we get from $g(\theta)$ by Monte Carlo sampling is the same whether or not it is normalized. So if we want to integrate $f(\theta)$ over a normalized PDF $Zg(\theta)$, for some unknown normalization constant Z, then we can safely sample from the unnormalized PDF $g(\theta)$. This is because that constant doesn't make any difference to the relative frequency with which samples are drawn.

Although $g(\theta)$ does not have to be normalized, it does have to be normalizable (proper) so that we can sample from it. The uniform distribution, for example, is only proper if we constrain its range. If $g(\theta) = 1$ is the (unnormalized) uniform distribution over the range $\Delta\Theta$, then $\int g(\theta)\,d\theta = \Delta\Theta$, and equation 8.9 reduces to equation 8.8. So in this particular case the normalization constant of the sampling distribution, $\Delta\Theta$, is still present in the Monte Carlo approximation of the integral.

How precise is the Monte Carlo estimate of the integral? Recalling the discussion in section 2.4, provided the Monte Carlo samples are independent of each other, the uncertainty in $\langle f(\theta)\rangle$ is given by the standard deviation in the mean, which is

$$\sigma(\langle f(\theta)\rangle) = \frac{\sigma_f}{\sqrt{N}} \quad \text{where}$$

$$\sigma_f = \sqrt{\frac{1}{N-1}\sum_i (f(\theta_i) - \langle f(\theta)\rangle)^2} \tag{8.10}$$

is the sample standard deviation in $f(\theta)$. As σ_f fluctuates around some constant value, the uncertainty in $\langle f(\theta)\rangle$ varies as $N^{-1/2}$: the uncertainty in the Monte Carlo estimate of the integral decreases with the square root of the sample size. This is independent of the dimensionality of the integral, which is precisely the advantage that Monte Carlo methods are intended to bring. This is in contrast to the rectangle method, for which the uncertainty varies as $N^{-1/J}$, where J is the number of dimensions (due to the curse of dimensionality). Thus the Monte Carlo method will generally achieve much smaller errors – or equivalently, will require far fewer function evaluations to achieve a given uncertainty – than grid methods.[1] However, this is only an estimate of the error and not an error bound. The actual performance depends on the specific algorithm.

We can use Monte Carlo integration for all four of the tasks listed in section 8.2.

(1) Suppose we have drawn a set of samples $\{\theta_1, \theta_2\}$ from a two-dimensional distribution $P(\theta_1, \theta_2 \,|\, D, M)$. Each sample is a two-element vector (θ_1, θ_2). Because the set of samples is representative of the full distribution, the set of values $\{\theta_1\}$ is representative of $P(\theta_1 \,|\, D, M)$ (equation 8.3). Thus we marginalize over θ_2 simply by ignoring the values of θ_2.

For the other three cases we use equation 8.9, in which we draw samples from $g(\theta)$ and compute the average of $f(\theta)$ at them.

(2) With $g(\theta)$ equal to the posterior and $f(\theta) = \theta$, we get the *expectation value* of θ (equation 8.4). This tells us that the expectation value of θ with respect to $g(\theta)$ is just the average of samples drawn from $g(\theta)$ (even if it's unnormalized).[2]
(3) With $g(\theta)$ equal to the prior and $f(\theta)$ equal to the likelihood, we get the *marginal likelihood* or evidence (equation 8.6). That is, by drawing samples from the prior and then calculating the likelihood at these, the average of these likelihoods is the evidence. We shall use this in section 11.3.
(4) With $g(\theta)$ equal to the posterior and $f(\theta)$ equal to the likelihood at a new point (x_p, y_p), we obtain the value of the *posterior predictive distribution* at this one point (equation 8.7). That is, we average the likelihood at this new point over the samples drawn from the posterior. We will see in section 9.1.3 how to apply this in practice.

8.4 Monte Carlo sampling

In its most general sense a Monte Carlo method is a means of selecting samples at random to solve what would otherwise be a hard computational problem. Here I use the term to mean sampling from an arbitrary PDF $g(\theta)$, using random numbers drawn from a simpler distribution, i.e. one we can easily draw from.

Having samples drawn *from* $g(\theta)$ is quite different from defining an arbitrary set of values of θ (perhaps a regular grid) at which we then evaluate $g(\theta)$. The power of Monte Carlo sampling is that the samples represent the distribution. Anything we would want to

---

[1] Note that in one dimension the rectangle method has an error that scales as N^{-1}, so may be more efficient than a Monte Carlo method. The trapezium rule in one dimension even has an error that scales as N^{-2}. But these are only scaling relations. Whether either of these is *actually* more efficient than Monte Carlo depends on the specific problem.
[2] In chapters 5 and 6 we *did* need the normalization constant to calculate the expectation values. That was because we did not have samples *drawn from* the posterior (normalized or not). We instead had evaluations on a regular grid.

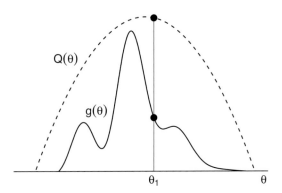

Fig. 8.1 Illustration of rejection sampling of the distribution $g(\theta)$ (solid curved line) using the proposal distribution $Q(\theta)$ (dashed line). We draw a sample θ_1 from $Q(\theta)$. We then draw a random number u from the uniform distribution $\mathcal{U}(0, Q(\theta_1))$. If $u \leq g(\theta_1)$, then θ_1 is accepted to our set of samples. Otherwise it is rejected. We repeat this for a large number of draws from $Q(\theta)$.

do with or compute from the distribution we can approximate using the samples. The mean of the distribution is approximated by the mean of the samples, for example. When we want to plot the distribution, we do not need to know the values of the posterior probability density (normalized or not). We simply take the samples we have drawn from it, and use them to construct a histogram or other density estimate (as done in section 7.2).

We look now at methods of sampling $g(\theta)$. In the following we assume that we can evaluate $g(\theta)$, but we do not require that $g(\theta)$ be normalized: we are free to scale it by any constant.

8.4.1 Rejection sampling

The idea behind rejection sampling is to define a simpler function that we *can* draw samples from, the *proposal distribution*. (We discussed ways to draw random numbers in section 1.8.) Let $Q(\theta)$ be the proposal distribution and define it such that $Q(\theta) \geq g(\theta)$ for all θ. We now draw samples from $Q(\theta)$, but only accept each sample to our set with a probability $g(\theta)/Q(\theta)$; otherwise the sample is rejected. This is illustrated in figure 8.1. The set of samples retained is equivalent to having been drawn from $g(\theta)$.

Unfortunately this method is often inefficient, because in practice we have to use a $Q(\theta)$ that is much larger than $g(\theta)$ at most θ in order for the former to be simple enough to draw from easily. This results in a large number of rejections, so we need a large number of draws and function evaluations. This problem generally gets more acute the higher the dimensionality of $g(\theta)$.

8.4.2 Importance sampling

Importance sampling is a method of calculating expectation values, i.e. the integral in equation 8.9, rather than drawing samples from $g(\theta)$.

Here we again make use of a proposal distribution $Q(\theta)$ that can easily be sampled from (and again this need not be normalized). Having drawn a set of samples $\{\theta_n\}$ from $Q(\theta)$, the idea is to correct for the fact that we have not drawn from $g(\theta)$ by reweighting. Equation 8.9 can be written[3]

$$\langle f(\theta) \rangle = \frac{1}{\int w(\theta)Q(\theta)\, d\theta} \int w(\theta)Q(\theta)f(\theta)\, d\theta \qquad (8.11a)$$

$$\simeq \frac{1}{\sum_i w(\theta_i)} \sum_i w(\theta_i)f(\theta_i) \qquad (8.11b)$$

where $w(\theta) = g(\theta)/Q(\theta)$ is the *importance weight*, and the sample $\{\theta_i\}$ has now been drawn from $Q(\theta)$. $w(\theta_i)$ is the ratio of the target density to the proposal density at θ_i. This looks like an easy solution, but in order to be easy to draw from, $Q(\theta)$ may be so different from $g(\theta)$ that we still need a very large number of samples.

8.5 Markov Chain Monte Carlo

The main drawback of the above Monte Carlo methods is that they waste a lot of time drawing samples in regions where $g(\theta)$ is low. This happens because we take random draws from the proposal distribution. There is nothing that makes us sample preferentially in regions where $g(\theta)$ is high. We want to do this, but in such a way that the resulting set of samples is still representative of $g(\theta)$. It turns out that we can achieve this if we relax the constraint of drawing samples independently. We cannot draw in an arbitrary fashion, but we can draw using any process that provides samples in the same proportions as $g(\theta)$.

The principle of a *Markov Chain Monte Carlo* (MCMC) method is to set up a random walk over the parameter space which preferentially explores regions of high probability density. The random walk is performed using a *Markov chain*. This is a random process in which the probability of moving from state θ_t to θ_{t+1} is defined by a transition probability $Q(\theta_{t+1}|\theta_t)$ that depends only on the current state θ_t, and not on any previous ones. This is sometimes referred to as a memoryless process.

We can use this process to select samples from a distribution $g(\theta)$; the sequence of samples is called the *chain*. The key requirement of the Markov chain is that it asymptotically reaches a stationary distribution that is equal to $g(\theta)$. This requirement would not be met if some part of the parameter space over which $g(\theta)$ is defined were not reachable from another part of the parameter space, for example. The requirement can be met if the chain satisfies a condition known as *detailed balance* (which turns out to be a sufficient but not necessary condition). This essentially means that the chain is reversible: if we pick a point

---

[3] Use equation 8.9 to re-write both the numerator and denominator of equation 8.11a. A term $(1/N) \int Q(\theta)\, d\theta$ cancels to leave equation 8.11b.

and transitioned to another, then it is just as likely that we would pick point θ_a and transition to point θ_b as it is that we would pick point θ_b and transition to point θ_a. Once this has been achieved the chain is in some kind of equilibrium state and the samples we obtain are representative of $g(\theta)$. More information on the required properties of the chains can be found in books on random processes or Monte Carlo methods (e.g. MacKay, 2003). The main point is that not any Markov chain can be used.

There are many different MCMC algorithms including Gibbs sampling, slice sampling, parallel tempering, Hamiltonian Monte Carlo, nested sampling, and the affine-invariant ensemble sampler (Goodman & Weare, 2010; Foreman-Mackey et al., 2013). Here I outline a simple yet widely-used MCMC method, the Metropolis–Hastings algorithm.

8.5.1 Metropolis–Hastings algorithm

As with rejection sampling, the Metropolis–Hastings algorithm also uses a proposal distribution $Q(s|\theta)$. This is a distribution from which we can easily draw a candidate sample s for the next point in the chain, θ_{t+1}, given the current parameter value θ_t. We initialize the chain at some value. The algorithm then iterates the following two tasks.

(1) Draw at random a candidate sample s from the proposal distribution $Q(s|\theta_t)$. This distribution could be (and often is) a multivariate Gaussian, in which case the mean is the θ_t, and the covariance matrix specifies the typical size of steps in the chain in each dimension of θ. Generally we want a proposal distribution that gives a higher probability of picking nearby points than far away ones.

(2) Decide whether or not to accept the candidate sample based on the value of the *Metropolis ratio*

$$\rho = \frac{g(s)}{g(\theta_t)} \frac{Q(\theta_t|s)}{Q(s|\theta_t)}. \tag{8.12}$$

If $\rho \geq 1$ then we accept the candidate and set $\theta_{t+1} = s$. If $\rho < 1$ we only accept it with a probability of ρ (i.e. we draw a number from a uniform distribution $\mathcal{U}(0,1)$ and compare it to ρ). If we don't accept the candidate then we set $\theta_{t+1} = \theta_t$, i.e. we repeat the existing sample in the chain.

The algorithm iterates between these two tasks and is stopped after some number of iterations. The number required to get a good sampling depends on the problem, but 10^4 to 10^6 iterations is typical.

If we use a symmetric proposal distribution (such as the multivariate Gaussian), then we don't need the term $Q(\theta_t|s)/Q(s|\theta_t)$ in the definition of ρ because it is always 1. The algorithm is then often referred to as the Metropolis algorithm (as it's what we had before Hastings came along and generalized it). I provide an R implementation of the Metropolis algorithm using a multivariate Gaussian proposal distribution in section 8.6.2.

Why does the Metropolis algorithm work? The decision rule in task 2 ensures that the walk will always go uphill – towards higher probability densities – if it can. But downhill moves are also allowed. This is essential, otherwise the algorithm will only move toward

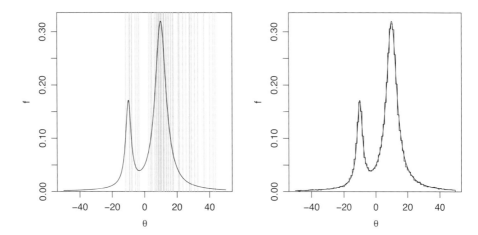

Fig. 8.2 A demonstration of the Metropolis algorithm. The smooth black curve (identical in both panels) is the function we wish to sample from. The left panel overplots 100 individual samples (grey vertical lines). The right panel overplots a histogram built from 10^5 samples.

and then remain at some local maximum, which is not what we want. By also going down-hill we can explore the entire distribution. The larger a proposed downhill move, the less likely it is to be accepted. This ensures that regions of higher probability density are pref-erentially sampled. But given enough iterations even large downhill moves will be made, although we will tend not to stay there, because when near the bottom of a hill most direc-tions lead up. The remarkable thing about this algorithm (not proven here) is that it will, eventually, produce a set of samples that is representative of the entire distribution.

Depending on where in the distribution the chain is initialized, the initial samples may not be representative of it. These should not be retained. The discarded initial samples are called the *burn-in*. With a good initialization the burn-in may only be a few percent of the chain.

A demonstration of the Metropolis algorithm in one dimension is shown in figure 8.2. The function we wish to sample is the smooth black curve. The proposal distribution is a Gaussian with standard deviation $\sigma = 10$. The algorithm is initialized at $\theta = -5$. A burn-in is not used. The left panel shows the first 100 samples (some may be repeated). The right panel show a histogram density estimate of 10^5 samples. R code for running this demonstration in real time, whereby you can see the individual candidate samples being proposed and then accepted or rejected, is provided in section 8.6.1.

8.5.2 Analysing the chains

While the above algorithm works in principle, it may not produce a representative set of samples in practice. This is true for any MCMC method, so we should always inspect the resulting chains to see whether they have the right properties.

There are many ideas concerning what to use as the covariance matrix of the proposal distribution, how long the burn-in period should be, how many iterations we expect to need before convergence, etc. For many proposal distributions the Metropolis algorithm does have the essential property that its stationary distribution is the desired one, and can be reached within a sufficiently large number of iterations. The difficulty in practice is not only that a very large number of samples may be required, but also that there is no simple way of knowing how many samples are sufficient.

One of the simplest ways to check whether the chain has reached some kind of steady state is to rerun the sampling several times, each starting at a different point. All chains should converge to roughly the same region of parameter space. The degree of convergence can be measured by comparing the variance between the chains to the variance within each of the chains. Various metrics exist for quantifying this, such as the one from Gelman & Rubin (1992).

The samples in the chain show some degree of correlation, as the sampling is a Markov chain, not totally random. We can quantify this using the autocorrelation function (ACF). This is the correlation coefficient (equation 1.67) of the chain with an offset version of itself, for different values of this offset, known as the *lag*. Thus the autocorrelation function for a chain of length N at lag h follows from the definition of the sample covariance (equation 1.66) and is

$$\text{ACF}(h) = \frac{\frac{1}{N-h}\sum_{t=1}^{N-h}(\theta_t - \bar{\theta})(\theta_{t+h} - \bar{\theta})}{\frac{1}{N-1}\sum_{t=1}^{N}(\theta_t - \bar{\theta})^2} \tag{8.13}$$

where θ_{t+h} is the chain offset by h steps. $\text{ACF}(h)$ therefore measures how closely the chain is correlated with itself h steps later. It is a rather noisy estimator, so we often need quite long chains for it to be representative. The characteristic length of the chain over which there is a significant correlation is the *autocorrelation length*, defined as

$$\lambda = 1 + 2\sum_{h=1}^{\infty}\text{ACF}(h) \tag{8.14}$$

where in practice the sum has to be truncated to a lower value (at the most N). We can then define the *effective sample size* – the effective number of independent samples in the chain – as N/λ.

Autocorrelation can be a problem because it may make us think the chain has converged to its stationary distribution when in fact it has not. A positive correlation will also lead to an underestimate of the variance of the distribution, if it is computed as the variance of all the samples. This in turn will underestimate the estimated error (equation 8.10 in $\langle f(\theta)\rangle$). The mean and mode will not be affected by the autocorrelations, however.

The degree of correlation we get from MCMC sampling depends on the typical step size, i.e. on the covariance matrix of the proposal distribution. If the steps are large then the chain quickly loses memory of where it has been, so the autocorrelation function decays to small values after a small lag. This suggests we are covering a wide range of the function, which is good, although such large steps may also fail to sample $g(\theta)$ finely enough, which is bad. Conversely, if the steps are small then the ACF drops off slowly because we are sampling a lot around the same place. This is inefficient, and if the chain is too short the resulting PDF will not be representative of $g(\theta)$. We can get some idea of how adventurous the search is by recording the time-variable acceptance rate, the fraction of proposed samples that have been accepted in some time window. This is done by the code in section 8.6.2. We should aim for moderate acceptance rates, say 0.3 to 0.7.

One way to reduce the correlation in a chain after it has been produced is by *thinning*. This means to retain only every Kth sample in the chain ($K > 1$), thereby yielding a less correlated sequence of samples. The drawback of this is that we discard a lot of samples, or rather we must run the algorithm for K times as many iterations to achieve the same number of (retained) samples.

Figure 8.3 plots the evolution of the MCMC chains (left column) and the correspond-ing autocorrelation function (right column) when sampling the function shown in figure 8.2. The proposal distribution is a Gaussian with standard deviation σ. The top row is for $\sigma = 10$ and no thinning ($K = 1$). The chain has a long correlation length of $\lambda = 203$, and although there are 10^5 samples, there are only 493 effective samples. This nonetheless gives a distribution very close to the true one (right panel of figure 8.2). If we now thin this by a factor of $K = 20$ (the second row), the autocorrelation length is significantly reduced, to 13. The chain plot looks more or less the same, because of the strong correla-tions when not thinning. Even removing 19 out of every 20 samples does not change the overall features of the chain. A histogram plot using the 5000 remaining samples is slightly noisier than with the full set of 10^5 samples, but looks very similar. Despite the thinning, it turns out that the variance has hardly changed (in this particular case). The bottom two rows show the results for a step size ten times smaller, $\sigma = 1$. Without any thinning the correlation length is large. Thinning significantly improves this, although there is still quite some autocorrelation apparent in the chain.

One big drawback of the Metropolis algorithm is that it uses a fixed step size, the mag-nitude of which can be hard to determine in advance, yet its value may have a big impact on the efficiency of the sampling. Metropolis is not always the best choice, in particular when there are many dimensions. Kass *et al.* (1998) give some practical advice on using MCMC.

8.5.3 Summarizing the distribution from the sample

We already discussed in section 5.5 the various metrics that can be used to summarize a distribution. Many of these summaries are simple to compute once we have a set of samples $\{\theta\}$ drawn from the distribution. The moments – mean, variance, etc. – can be computed directly from equation 1.42. The quantiles can be computed by first sorting the samples in ascending order, to be $\theta_1, \ldots, \theta_t, \ldots, \theta_N$. The quantile at probability p is approximately $\theta(t = pN)$, whereby we have to think about how to deal with repeated values and how to interpolate. This is done by the R function `quantile`, which has different options for dealing with those issues. If we have multiple parameters, then we could work out the quantiles on the one-dimensional marginalized distributions. The highest density interval (HDI) can be computed from a set of samples using the R function `HPDinterval` in the package `coda`.

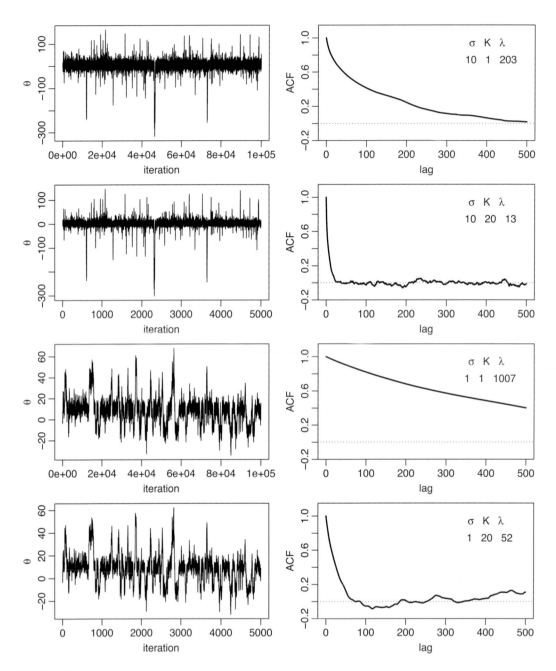

Fig. 8.3 MCMC chains (left column) and the corresponding autocorrelation function (right column) for a Metropolis sampling of the function shown in figure 8.2. In all cases the proposal distribution was a zero mean Gaussian initialized at $\theta = -5$. The four rows correspond to two different values of σ (10 and 1) and two different degrees of the thinning, namely $K = 1$ (no thinning) and $K = 20$. All chains were run for $N = 10^5$ samples before (in the case of $K = 20$) being thinned. λ is the computed correlation length.

8.5.4 Parameter transformations

Sometimes it is more efficient to sample over a transformed parameter. For example, if the parameter θ is strictly positive, then it is more appropriate to sample over (i.e. use a proposal distribution in) $\ln \theta$, as this ensures that proposed values of θ cannot be negative.[4] However, this corresponds to drawing samples from $P(\ln \theta)$, not $P(\theta)$, so this is what we must use in equation 8.12. For this particular example

$$P(\theta)\, d\theta \;=\; P(\ln \theta)\, d(\ln \theta) \quad \Rightarrow \quad P(\ln \theta) \;=\; \theta P(\theta). \tag{8.15}$$

Thus when we use a symmetric proposal distribution $(Q(\theta_t\,|\,s)/Q(s\,|\,\theta_t) = 1$ in equation 8.12) the Metropolis ratio becomes

$$\rho \;=\; \frac{sP(s)}{\theta_t P(\theta_t)}. \tag{8.16}$$

The base of the logarithm in the transformation is unimportant as it corresponds to a constant factor that cancels in the ratio. For a general transformation, the relevant quantity for such transformations is the Jacobian determinant (section 1.9). If we transform from parameters $(\theta_1, \theta_2, \ldots, \theta_J)$ to parameters $(\phi_1, \phi_2, \ldots, \phi_J)$, then we form the Jacobian determinant of the original parameters with respect to the transformed parameters, which is

$$\mathcal{J} \;=\; \left| \frac{\partial(\theta_1, \theta_2, \ldots, \theta_J)}{\partial(\phi_1, \phi_2, \ldots, \phi_J)} \right|. \tag{8.17}$$

The Metropolis ratio is then

$$\rho \;=\; \frac{P(s)}{P(\theta_t)} \frac{\mathcal{J}_s}{\mathcal{J}_{\theta_t}} \tag{8.18}$$

where the subscripts on the Jacobian determinants indicate the samples they are evaluated at.

8.6 R code

8.6.1 Demonstration of the Metropolis algorithm in one dimension

You can use the following code to sample the test function `testfunc`, or any other function you choose to provide. I recommend you go through the code in chunks rather then sourcing the entire file. With `plotev` set to TRUE the code will introduce a delay of `delay` seconds between each iteration. (Actually it's just the plotting that is decelerated; all the samples are calculated in advance by `metrop`.) At each iteration the code plots the current value of the chain in blue, and then the proposed candidate value in red and also prints

---

[4] We could instead just let the function we want to sample return zero in forbidden regions, as this would prevent such proposals ever being accepted. But this could get quite inefficient if we are near a boundary.

this candidate value to the console. If the proposal is accepted it turns green. If it is not accepted it stays red and the current value turns green, because it is accepted again. The value to be used as the next current value is then printed to the console. The next iteration starts by turning the current point blue. The final plot can either overplot the individual samples or a histogram (density estimate) of the set of samples. The former is only suitable if Nsamp is less than a few hundred. I had to fiddle around a bit to get the histogram to appear right (see the comments in the code). I suggest you experiment with different initial values, thetaInt, and different step sizes, sampleSig.

You can calculate the effective sample size N_{eff} using the function effectiveSize in the package coda. This actually uses a different formula from that given in section 8.5.2. The syntax is as follows.

```
effectiveSize(as.mcmc(chain$funcSamp[,3]))
```

I then calculate the correlation length as $\lambda = N/N_{\text{eff}}$, where N is the number of samples.

R file: mcmc_demo.R

```
##### MCMC demonstration in 1D

plotev     <- TRUE  # do we want to plot the evolution?
delay      <-  1    # time delay in seconds between steps in evolution
thetaInit <- -5     # MCMC initialization
sampleSig <-   10   # MCMC step size
Nsamp      <-  1e5  # number of MCMC samples

source("metropolis.R") # provides metrop()

# Test function
testfunc <- function(theta) {
  return(dcauchy(theta, -10, 2) + 4*dcauchy(theta, 10, 4))
}

# Interface to test function for metrop()
testfunc.metrop <- function(theta) {
  return(c(0,log10(testfunc(theta))))
}

# Establish range, and plot test function
x <- seq(from=-50, to=50, length.out=1e4)
y <- testfunc(x)
ymax <- 1.05*max(y)
# Compute normalization function: used later to put on same as histogram
Zfunc <- sum(y)*diff(range(x))/(length(x))
par(mfrow=c(1,1), mar=c(3.0,3.0,0.5,0.5), oma=c(0.5,0.5,0.5,0.5),
    mgp=c(2.2,0.8,0), cex=1.0)
plot(x, y, type="l", yaxs="i", lwd=2, ylim=c(0, ymax),
     xlab=expression(theta), ylab="f")

# Run MCMC
set.seed(120)
chain <- metrop(func=testfunc.metrop, thetaInit=thetaInit, Nburnin=0,
                Nsamp=Nsamp, sampleCov=sampleSig^2, verbose=Inf, demo=TRUE)
```

```
# Plot evolution of chain to the screen
if(plotev) {
  par(mfrow=c(1,1), mar=c(3.0,3.0,0.5,0.5), oma=c(0.5,0.5,0.5,0.5),
      mgp=c(2.2,0.8,0), cex=1.0)
  for(i in 1:nrow(chain$thetaPropAll)) {
    plot(x, y, type="l", yaxs="i", lwd=2, ylim=c(0, ymax),
         xlab=expression(theta), ylab="f")
    if(i==1) {
      segments(thetaInit, 0, thetaInit, 1, col="blue", lwd=2)
      Sys.sleep(delay)
    } else {
      segments(chain$funcSamp[1:(i-1),3], 0, chain$funcSamp[1:(i-1),3],
               ymax, col="green")
      segments(chain$funcSamp[i-1,3], 0, chain$funcSamp[i-1,3], ymax,
               col="blue", lwd=2)
    }
    lines(x, y, lwd=2) # replot to bring to front
    segments(chain$thetaPropAll[i,1], 0, chain$thetaPropAll[i,1], ymax,
             col="red", lwd=2)
    cat(formatC(i, digits=3), "Proposal: ", formatC(chain$thetaPropAll[i,1],
                                            digits=2, width=6, format="f"))
    Sys.sleep(delay)
    segments(chain$funcSamp[i,3], 0, chain$funcSamp[i,3], ymax,
             col="green", lwd=2)
    cat("  =>  Current: ", formatC(chain$funcSamp[i,3], digits=2, width=6,
                                   format="f"), append=TRUE)
    if(chain$funcSamp[i,3]==chain$thetaPropAll[i,1]) {
      cat(" ACCEPTED")
    }
    cat("\n")
    Sys.sleep(delay)
  }
}


# Plot function, samples, and histogram density estimate of samples. The
# histogram is rescaled to have the same normalization as the function.
# Samples outside the range of x are not included in the histogram.
# lines(type="s") plots the vertical steps at the x values given, so we
# must supply it with the histogram breaks, not midpoints. There is one
# more break than counts in the histogram, and each element of hist$counts
# is associated with the lower break, so we must provide an additional
# count=0 to complete the histogram.
pdf("mcmc_demo.pdf", 4, 4)
par(mfrow=c(1,1), mar=c(3.2,3.2,0.5,0.5), oma=c(0.5,0.5,0.5,0.5),
    mgp=c(2.2,0.8,0), cex=1.0)
plot(x, y, type="n", yaxs="i", ylim=c(0, 1.05*max(y)),
     xlab=expression(theta), ylab="f")
# Uncomment the following if you want to plot every sample
#segments(chain$funcSamp[,3], 0, chain$funcSamp[,3], 1, col="green")
lines(x, y, lwd=1) # plot function
# Build and plot histogram of samples
sel <- which(chain$funcSamp[,3]>=min(x) & chain$funcSamp[,3]<=max(x))
hist <- hist(chain$funcSamp[sel,3], breaks=seq(from=min(x), to=max(x),
                                   length.out=100), plot=FALSE)
Zhist <- sum(hist$counts)*diff(range(hist$breaks))/(length(hist$counts))
lines(hist$breaks, c(hist$counts*Zfunc/Zhist,0), type="s", lwd=1)
```

```
dev.off()

# Plot ACF
acor <- acf(chain$funcSamp[,3], lag.max=1e3, plot=FALSE)
plot(acor$lag, acor$acf, xlab="lag", ylab="ACF", type="l", lwd=2)
abline(h=0, lty=3, lwd=1)
```

8.6.2 Implementation of the Metropolis algorithm

The function metrop in the file metropolis.R is the Metropolis algorithm. It is mostly explained by the inline documentation, but four things should be highlighted. First, it uses a multivariate Gaussian proposal distribution. This is symmetric about its mean, so the factor $Q(\theta_t|s)/Q(s|\theta_t)$ is always 1 in the definition of the Metropolis ratio (equation 8.12). Second, the algorithm works with the logarithm of the Metropolis ratio. This is because it also uses the logarithm of the density of the function being sampled (for dynamic range reasons), so we save computation time and retain this dynamic range if we also make the selection in the logarithm. Third, the R function that metrop calls, func, must return a two-element vector, the sum of which is the logarithm of the density of the function being sampled. I designed it this way because in Bayesian applications I use metrop to sample an unnormalized posterior, which is the product of a prior and a likelihood. I then define func to return the logarithm of the prior and the logarithm of the likelihood, the sum of which is the logarithm of the unnormalized posterior. It is useful to have these available separately for the sake of subsequent analyses.[5] It is also useful when using other techniques that use the individual likelihoods. One example is the cross-validation likelihood technique, to be discussed in section 11.6.1. If we didn't return the likelihood values we would have to recalculate them. If you want to use metrop to sample a distribution that does not separate into the product of a prior and a likelihood, just define func so that one of the two elements it returns is zero. This is what I did in the function testfunc.metrop in the demonstration code listed in the previous section. Finally, the initial values of the parameters, thetaInit, passed into metrop must produce finites value of the function, and not zero, NA, NaN, etc., otherwise the code will throw an error. Ensure you use a sensible initialization.

R file: metropolis.R

```
##### The Metropolis algorithm

library(mvtnorm) # for rmvnorm

# Metropolis (MCMC) algorithm to sample from function func.
# The first argument of func must be a real vector of parameters,
# the initial values of which are provided by the real vector thetaInit.
# func() returns a two-element vector, the logPrior and logLike
# (log base 10), the sum of which is taken to be the log of the density
# function (i.e. unnormalized posterior). If you don't have this separation,
# just set func to return one of them as zero. The MCMC sampling PDF is the
# multivariate Gaussian with fixed covariance, sampleCov. A total of
# Nburnin+Nsamp samples are drawn, of which the last Nsamp are kept. As the
```

---

[5] Some implementations of MCMC algorithms don't return the function values at all.

```
# sampling PDF is symmetric, the Hasting factor cancels, leaving the basic
# Metropolis algorithm. Diagnostics are printed very verbose^th sample:
# sample number, acceptance rate so far.
# ... is used to pass data, prior parameters etc. to func().
# If demo=FALSE (default), then
# return a Nsamp * (2+Ntheta) matrix (no names), where the columns are
# 1:  log10 prior PDF
# 2:  log10 likelihood
# 3+: Ntheta parameters
# (The order of the parameters in thetaInit and sampleCov must match.)
# If demo=TRUE, return the above (funcSamp) as well as thetaPropAll, a
# Nsamp * Ntheta matrix of proposed steps, as a two element named list.
metrop <- function(func, thetaInit, Nburnin, Nsamp, sampleCov, verbose,
                   demo=FALSE, ...) {

  Ntheta    <- length(thetaInit)
  thetaCur  <- thetaInit
  funcCur   <- func(thetaInit, ...) # log10
  funcSamp  <- matrix(data=NA, nrow=Nsamp, ncol=2+Ntheta)
  # funcSamp will be filled and returned
  nAccept   <- 0
  acceptRate <- 0
  if(demo) {
    thetaPropAll <- matrix(data=NA, nrow=Nsamp, ncol=Ntheta)
  }

  for(n in 1:(Nburnin+Nsamp)) {

    # Metropolis algorithm. No Hastings factor for symmetric proposal
    if(is.null(dim(sampleCov))) { # theta and sampleCov are scalars
      thetaProp <- rnorm(n=1, mean=thetaCur, sd=sqrt(sampleCov))
    } else {
      thetaProp <- rmvnorm(n=1, mean=thetaCur, sigma=sampleCov,
                           method="eigen")
    }
    funcProp  <- func(thetaProp, ...)
    logMR <- sum(funcProp) - sum(funcCur) # log10 of the Metropolis ratio
    #cat(n, thetaCur, funcCur, ":", thetaProp, funcProp, "\n")
    if(logMR>=0 || logMR>log10(runif(1, min=0, max=1))) {
      thetaCur   <- thetaProp
      funcCur    <- funcProp
      nAccept    <- nAccept + 1
      acceptRate <- nAccept/n
    }
    if(n>Nburnin) {
      funcSamp[n-Nburnin,1:2] <- funcCur
      funcSamp[n-Nburnin,3:(2+Ntheta)] <- thetaCur
      if(demo) {
        thetaPropAll[n-Nburnin,1:Ntheta] <- thetaProp
      }
    }

    # Diagnostics
    if( is.finite(verbose) && (n%%verbose==0 || n==Nburnin+Nsamp) ) {
      s1 <- noquote(formatC(n,       format="d", digits=5, flag=""))
      s2 <- noquote(formatC(Nburnin, format="g", digits=5, flag=""))
```

```
      s3 <- noquote(formatC(Nsamp,      format="g", digits=5, flag=""))
      s4 <- noquote(formatC(acceptRate, format="f", digits=4, width=7,
                           flag=""))
      cat(s1, "of", s2, "+", s3, s4, "\n")
    }

  }

  if(demo) {
    return(list(funcSamp=funcSamp, thetaPropAll=thetaPropAll))
  } else {
    return(funcSamp)
  }

}
```

9 Parameter estimation: Markov Chain Monte Carlo

In chapters 5 and 6 we found the posterior probability density function for one or two parameter problems either through an analytic approach or by computing the posterior on a grid. Here we will use the Metropolis algorithm described in chapter 8 to sample posteriors with more than two parameters. By way of example we will look at fitting curves (both linear and quadratic) $y = f(x) + \epsilon$, whereby we also infer the noise level in the data. We will see how we can use a mixture model to take account of outliers. I will finally show that it is conceptually straightforward in the Bayesian approach to accommodate uncertainties in both the x and y variables.

9.1 Fitting a straight line with unknown noise

In chapter 4 we looked at fitting a straight line by minimizing the sum of squares of the residuals of the data about that line. This is equivalent, for a Gaussian likelihood, to both maximum likelihood and to finding the maximum of the posterior with a uniform prior. A simple first-order propagation of errors (valid for small errors) allowed us to estimate uncertainties in the parameters and in the predictions.

If we don't just want to find the maximum, if we have more prior information, or if the errors are not small, we will want to properly characterise the distribution over the parameters by finding the posterior PDF. This will tell us more about the solutions than just a local maximum of the likelihood. We use the Monte Carlo methods discussed in the previous chapter to sample the PDF.

Suppose we have a two-dimensional set of N points $\{x_i, y_i\}$. The model M predicts the values of y as being

$$y = f(x) + \epsilon \quad \text{where}$$
$$f(x) = b_0 + b_1 x \tag{9.1}$$

which is a straight line with parameters b_0 (intercept) and b_1 (gradient). $f(x)$ is the generative model: it gives the noise-free predictions of the data given the parameters. The residuals $\epsilon = y - f(x)$ are modelled as a zero-mean Gaussian random variable with standard deviation σ, i.e. $\epsilon \sim \mathcal{N}(0, \sigma)$. This is the *noise model*. Assuming the $\{x\}$ are noise free, this tells us that the likelihood is

$$P(y_i | x_i, \theta, M) = \frac{1}{\sigma\sqrt{2\pi}} \exp\left[-\frac{[y_i - f(x_i; b_0, b_1)]^2}{2\sigma^2} \right] \tag{9.2}$$

where $\theta = (b_0, b_1, \sigma)$ are the parameters of the model. It may come as a surprise that we will try to infer the uncertainty in the data points from the data. Yet σ is as a model parameter just like the others. Although the x values are supplied with the data, we assume them to be fixed: they are not described by a measurement model. Thus the data are $D = \{y_i\}$. Assuming that the various y measurements are independent, the log likelihood for all N data points is

$$\ln P(\{y_i\} \mid \{x_i\}, \theta, M) = \sum_{i=1}^{N} \ln P(y_i \mid x_i, \theta, M). \qquad (9.3)$$

From now on I will drop the conditioning both on $\{x_i\}$ for the sake of brevity, and on M because we only consider a single model. In general none of the three parameters is known in advance, so we want to infer their posterior PDF from the data, which is given as usual by

$$P(\theta \mid D) \propto P(D \mid \theta) P(\theta). \qquad (9.4)$$

As we will be sampling from the posterior we do not need to compute the normalization constant.

Given a set of data, the procedure to compute the posterior is as follows:

(1) define the prior PDF over the parameters. I will use plausible yet convenient priors, and I will make use of a variable transformation;
(2) define the covariance matrix of the proposal distribution. I will use a diagonal, multi-variate Gaussian;
(3) define the starting point (initialization) of the MCMC;
(4) define the number of burn-in iterations and the number of sampling iterations.

Once we have run the MCMC we perform the following analyses:

(5) thin the chains (see section 8.5.2);
(6) plot the chains and the one-dimensional marginal posterior PDFs over the parameters. I do the latter via kernel density estimation (see section 7.2.2);
(7) plot the two-dimensional posterior distributions of all three pairs of parameters, sim-ply by plotting the samples (we could do two-dimensional kernel density estimation instead). I do this to look for correlations between the parameters;
(8) calculate the maximum a posteriori (MAP; see section 4.4.4) values of the model pa-rameters from the MCMC chains, calculate and plot the resulting model, and compare to the original data;
(9) calculate the predictive posterior distribution over y at a new data point (following the approach described in section 8.2.4).

Note that because we have samples drawn from the posterior, we don't need the actual values of the posterior density in order to plot the posteriors. We likewise don't have to do any integration to get the one-dimensional marginal distributions (see point number 1 at the end of section 8.3).

I should point out that, with a suitable choice of priors, this problem has an analytic solution for the posterior. But in the general case we need to sample the posterior.

9.1.1 Priors

I adopt the following priors on the three model parameters.

- Intercept, b_0: $P(b_0) = \mathcal{N}(\mu, s)$, a Gaussian with mean μ and standard deviation s. One can estimate the two parameters of this prior by inspection of the general properties of the data.[1] This might be difficult in practice if the data are far from $x = 0$, in which case it would be better to centre the data. We will do this in section 9.1.6, but for now we use the data as they come.
- Gradient, b_1: We can write the gradient as $b_1 = \tan \alpha$, where α (in radians) is the angle between the horizontal and the model line. I assume that we have no prior knowledge of the slope of the line (not even its sign). This means we should use a uniform distribution in α, $P(\alpha) = 1/2\pi$ (or $1/\pi$, as a line is invariant under a rotation of π radians; but as we don't need the normalization constant this consideration is irrelevant). In contrast, a uniform distribution over the gradient would assign much less probability to lines with small angles than to lines with large ones.[2] It is preferable to use α as a parameter in an MCMC algorithm. With finite characteristic step sizes, a sampling over b_1 can never move the line from large positive to large negative gradients, because it cannot step over $b_1 = \pm\infty$.
- Standard deviation, σ: We argued in section 5.3.1 that, in the absence of any other information, a scale parameter such as the standard deviation of a Gaussian should be assigned a Jeffreys prior, $P(\sigma) \propto \log \sigma$. This also prevents σ from becoming negative. This is an improper prior, but this is often not a problem for determining the unnormalized posterior.

Given these priors, my model parameters are now $(b_0, \alpha, \log \sigma)$. These are the parameters that the Monte Carlo algorithm will sample over. The prior distributions are likewise defined over the parameters, as Gaussian, uniform, and uniform respectively. I therefore do not need to include the Jacobian mentioned in section 8.5.4 into the Metropolis algorithm. That would only be needed if the prior was specified as a density function over a different function of the parameters than that which is used by the sampling algorithm.

9.1.2 Sampling the posterior

The R code in section 9.1.4 below performs all of the above-mentioned tasks using some simulated data. I draw ten data points at random between $x = 0$ and $x = 10$, evaluate the straight line model (equation 9.1) with $b_0 = 0$ and $b_1 = 1$, then add zero mean Gaussian noise with standard deviation $\sigma = 1$ to produce the y values. The data are plotted in figure 9.1. From inspection of the data I assign mean zero and a generous standard deviation of two to the Gaussian prior on the intercept b_0. The priors on α and $\log \sigma$ have no parameters. The priors are set in the file linearmodel_functions.R. I initialize the MCMC using

---

[1] The parameters of a prior are sometimes called *hyperparameters* to distinguish them from the parameters of the model.

[2] You can easily show, using the method in section 1.9.1, that a uniform distribution over the angle α corresponds to a Cauchy distribution over the slope b_1.

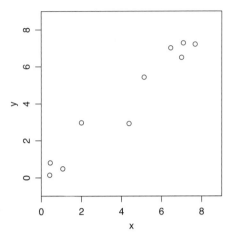

Fig. 9.1 Data used for the straight line fitting. They have been drawn at fixed x values from a straight line with $b_0 = 0$ and $b_1 = 1$ (equation 9.1), to which zero mean Gaussian noise with standard deviation 1 has been added.

values of the parameters that are intentionally far from being a good model – $b_0, \alpha, \log \sigma$ equal to $2, \pi/8, \log 3$ respectively – so that we can see how well/quickly the MCMC sampler finds regions of high probability density. The step sizes I set to $0.1, 0.02, 0.1$ for the three parameters (respectively), which intuitively seem to be small compared to the likely precision with which we can determine the parameters given these data.

We can now use MCMC to sample the posterior. Remember that it is sufficient if we use the unnormalized posterior – the product of likelihood and prior – in the MCMC, because we will draw samples in the same relative frequency whether it is normalized or not (see section 8.1). However, the posterior must be normaliz*able*. That is, it must be a proper distribution. If the posterior were improper we could still run an MCMC and we would still get a bunch of samples. But no finite number of samples would be representative of an entire improper distribution. If the prior is improper then in practice the likelihood will often ensure that the posterior is a proper distribution, but this is not guaranteed.

In contrast to the posterior, it is essential that the likelihood be normalized (not just normalizable). This is because it is a PDF over the data, so its normalization constant will generally be a function of the parameters we are sampling.

I run the MCMC for 50 000 iterations without burn-in. The average acceptance rate is around 0.56. I apply a thinning factor of 25, i.e. I retain only every twentyfifth sample in order to reduce the autocorrelation in the chain (see section 8.5.2). All results and plots that follow are based on the remaining 2000 samples. These are stored in the matrix postSamp[,], for which the five columns are the log prior, log likelihood, b_0, α, and $\log \sigma$ respectively, with one row per iteration. The chains are shown in the left column of figure 9.2. We see how the sampler quickly moves to a new region of parameter space in the first few iterations. We should strictly remove these from the posterior estimates (i.e. use a burn-in). After this the search looks reasonably stable, and the chains look reasonable.

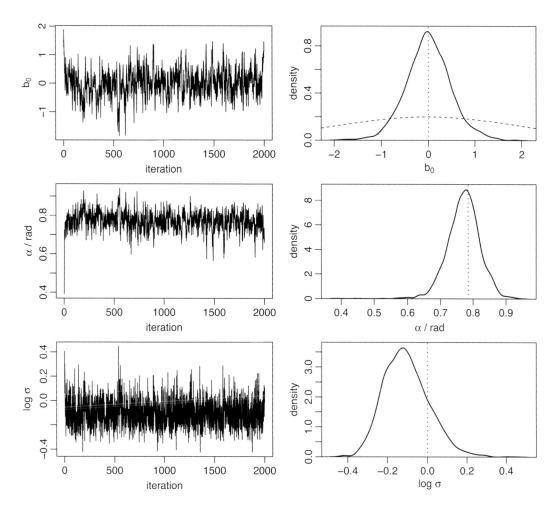

MCMC chains (left columns) and resulting marginal posterior PDFs (right columns) for the straight line fitting problem. These one-dimensional posteriors have been computed by a kernel density estimate of the samples. The vertical dotted lines indicate the true parameters. These are only for reference: the parameters that best fit the noisy data are not necessarily the same as those that would best fit noise-free data. The curved dashed line in the panel for b_0 shows the prior distribution. The other priors are uniform.

The joint posterior distribution is the three-dimensional distribution over the MCMC samples. The one-dimensional marginalized distributions are obtained by making a density estimation of the samples for each parameter. These are shown in the right column of figure 9.2. Using the set of samples we can compute various statistics, for example the maximum or mean of the posterior as a single best estimate. In the present example the maximum is at $(b_0, \alpha, \log \sigma) = (0.036, 0.77, -0.19)$. This is the global maximum of the joint three-

Table 9.1 Summary of the covariance of the posterior PDF for the line-fitting problem. The leading diagonal gives the standard deviation of the three parameters. The off-diagonal elements give the correlation coefficients.

	b_0	α	$\log \sigma$
b_0	0.48		
α	-0.83	0.050	
$\log \sigma$	0.038	-0.073	0.11

dimensional posterior. This is not necessarily equal to the maxima of the corresponding one-dimensional marginalized distributions (although in this case they are very close). The mean of the posterior is $(b_0, \alpha, \log \sigma) = (0.0042, 0.77, -0.11)$. If we want to find the mean of the posterior over the original model parameters – (b_0, b_1, σ) – then we must transform the individual samples first and then compute the statistic (and not vice versa). This is because the mean is generally not invariant under a parameter transformation (see section 4.4.4). As the set of samples is representative of the distribution, transforming them essentially transforms the distribution. To find the mean of the transformed samples we do

```
mean(tan(postSamp[,4])) # transform alpha to b_1
mean(10^(postSamp[,5])) # transform log10(sigma) to sigma
```

which gives $(b_0, b_1, \sigma) = (0.0042, 0.98, 0.81)$. As a measure of the precision of these estimates we can calculate their standard deviations as well as the correlations between the parameters. These are shown in table 9.1. We can also inspect the correlations by plotting the parameter samples in two dimensions against each other. This is shown in figure 9.3 and confirms the values for the correlation coefficients in table 9.1. The strong negative correlation between b_0 and α arises because if we increase the slope of the line we need to decrease the intercept in order to retain a similar quality of fit for these data.

Having sampled the posterior we could now summarize it with, for example, the mean or maximum values, and see what this model looks like in the data space. This is shown in figure 9.4 using the maximum.[3] In this case the model at the mean of the parameters is almost identical, as is the least squares fit. This is not surprising, because I adopted broad priors on the parameters and the data are relatively informative. Note, however, that the least squares fit does not infer σ.

9.1.3 Posterior predictive distribution

Once we have decided on the "best" values for the model parameters (e.g. the maximum of the posterior), we could use these in the model to predict the value of y at any speci-

---

[3] Peek ahead to the right panel of figure 9.8 if you want to see a set of models drawn from the posterior, as opposed to just this single model at the mode.

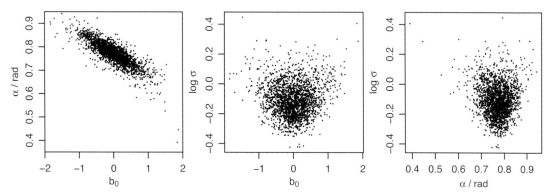

Fig. 9.3 The MCMC samples from the posterior PDF of the straight line model for each parameter, plotted pairwise to show the correlations.

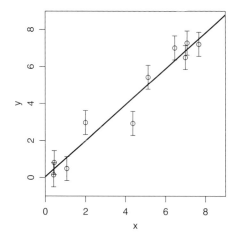

Fig. 9.4 The open circles are the data from figure 9.1. The straight line and error bars show the model obtained with the parameters set to the maximum of the posterior PDF (MAP), which has values $b_0 = 0.036$, $b_1 = 0.97$, $\sigma = 0.65$. The standard deviations in these and the correlations between them are listed in table 9.1.

fied x, call it x_p. We might further characterise the uncertainty in this prediction using the likelihood (equation 9.2). But the problem with this approach is that it gives a prediction and uncertainty estimate conditioned on a *single* set of parameters, thereby ignoring the uncertainty in the parameters reflected by the finite width of the posterior. How can we accommodate this? We saw in section 8.2.4 that the rules of probability lead us to incorporate uncertainties in parameters by marginalizing over them. This gave us the posterior predictive distribution

$$P(y_p \,|\, x_p, D) = \int \underbrace{P(y_p \,|\, x_p, \theta)}_{\text{likelihood}} \underbrace{P(\theta \,|\, D)}_{\text{posterior}} \, d\theta. \qquad (9.5)$$

This distribution is computed in the code below in two ways. The "direct" approach is to explicitly evaluate $P(y_p|x_p, D)$ over a grid $\{y_p\}$, which is called ycand in the code. At a fixed value of y_p we take our set of N_s posterior samples $\{\theta_l\}$ (obtained by MCMC), calculate the likelihood at each of these, and then average these likelihoods, i.e.

$$P(y_p|x_p, D) \simeq \frac{1}{N_s} \sum_{l=1}^{N_s} P(y_p|x_p, \theta_l) \tag{9.6}$$

which was point number 4 at the end of section 8.3. Each of these likelihood calculations requires an evaluation of the generative model, as this gives the mean of the likelihood (see equation 9.2). We repeat this for each value in $\{y_p\}$ and plot these probability densities (called ycandPDF in the code) vs y_p. Equation 9.6 makes it clear that the posterior predictive distribution is a posterior-weighted average of the predictions (the likelihood) made at each θ. While this approach is accurate, it is slow, because it involves many likelihood evaluations for each value of y_p.

An alternative, "indirect" approach is to sample the joint distribution $P(y_p, \theta|x_p, D)$, and then to marginalize over θ. This helps because we can factorize the joint distribution as

$$P(y_p, \theta|x_p, D) = P(y_p|x_p, \theta)P(\theta|D). \tag{9.7}$$

whereby variables could be removed because of conditional independence (as explained immediately after equation 8.7). Each of the two PDFs on the right side can be represented by samples drawn from them. The second term is the posterior PDF; we already obtained the set of samples $\{\theta_l\}$ from this with the MCMC. The first term is the likelihood. We now sample this once (obtain a single value of y_p) for each value in $\{\theta_l\}$. As the likelihood is a univariate Gaussian (equation 9.2), it may be sampled using a standard function (rnorm in R). Its mean is the evaluation of the straight line at $(b_0, b_1)_l$, and its standard deviation is σ_l. Doing this for all N_s posterior samples gives a set of N_s predictions $\{y_p\}$ (called likeSamp in the code). The implementation in R is shorter than this explanation:

```
likeSamp <- rnorm(n=length(modPred), mean=modPred, sd=10^postSamp[,5])
```

where modPred (of length N_s) is the evaluations of the straight line at the posterior samples. We now have samples of θ and y_p. We marginalize their joint distribution simply by ignoring the θ, to give the required distribution $P(y_p|x_p, D)$. We then use density estimation to compute and plot their distribution (likeDen in the code).

In both the direct and indirect approaches we are effectively making predictions of y_p using all possible models (lines), and then weighting them by their posterior probability density computed from the data.

The left panel of figure 9.5 shows the posterior predictive distribution for $x_p = 6$ computed by both methods. We can take the maximum of $P(y_p|x_p, D)$ as the estimate of y_p, and use the 15.9% and 84.1% quantiles of this distribution as a measure of the (asymmetric) precision. These would correspond to the mode plus/minus 1 standard deviation if the posterior predictive distribution were Gaussian. Doing this we get $y_p = 5.86^{+0.84}_{-0.84}$ computed by the direct method (and $y_p = 5.80^{+0.92}_{-0.80}$ by the indirect method). This is shown in relation to the data in the right panel of figure 9.5. This value of x_p lies within the range

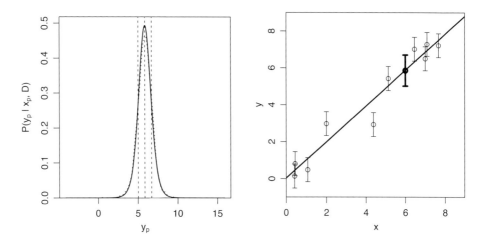

Fig. 9.5 The posterior predictive distribution $P(y_p | x_p, D)$ for $x_p = 6$ (left). The solid line is for the direct method, and the (barely distinguishable) dotted line is for the indirect method. Vertical dashed lines indicate the maximum as well as the 15.9% and 84.1% quantiles (from the direct method). The right plot shows this maximum as a thick point and these two quantiles as a thick error bar, plotted over the data and MAP model prediction as in figure 9.4.

of the original data. How does this method fare when we predict much further from the data (extrapolation)? An example is given in figure 9.6 for $x_p = 25$. Our predicted values are now $y_p = 24.4^{+2.1}_{-2.0}$ (and $y_p = 24.1^{+2.5}_{-1.7}$ with the indirect method). The posterior is necessarily wider in this case, because a given uncertainty in the gradient is amplified to a larger uncertainty in y_p the further the prediction is made from the data (think of the line as a "lever arm" pivoting about its intercept). Recall that the standard deviation of a prediction arising from the least squares fit in the case of known σ – derived in section 4.1.2 – showed a similar behaviour.

9.1.4 R code for fitting a straight line

The analysis described above is all done with the R script `linearmodel_posterior.R`, with explanations provided as comments in the code. The code looks long, but a reasonable chunk of it is actually concerned with plotting and analysing the results. The code initially sources two other files. The first file, `linearmodel_functions.R`, defines functions that compute the (logarithm of the) prior, likelihood, and posterior. This is listed below. The second file is the Metropolis algorithm listed in section 8.6.2. The rest of the code is then executed.

To better appreciate how MCMC works in this example, I recommend you experiment with the code, and change in particular:

- the initialization of the parameters;
- the parameter step sizes;

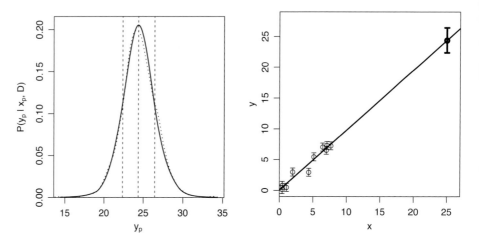

Fig. 9.6   As figure 9.5, but for a prediction at $x_p = 25$.

- the number of iterations and the length of the burn-in;
- the value of the standard deviation of the prior on b_0;
- the prior on b_0 from a Gaussian to an improper uniform prior. Do this by setting b0Prior to unity in the function logprior.linearmodel;
- the amount of data. Try both more data points (e.g. 100) and fewer, including – before you read section 9.1.5 – just three, two, and one data points.

R file: linearmodel_posterior.R

```
##### Bayesian inference of a 3-parameter linear model to 2D data

library(gplots) # for plotCI
source("metropolis.R")
source("linearmodel_functions.R") # provides logpost.linearmodel

########## Define true model and simulate experimental data from it

set.seed(50)
Ndat <- 10
x <- sort(runif(Ndat, 0, 10))
sigTrue <- 1
modMat <- c(0,1) # 1 x P vector: coefficients, b_p, of sum_{p=0} b_p*x^p
y <- cbind(1,x) %*% as.matrix(modMat) + rnorm(Ndat, 0, sigTrue)
# Dimensions in the above: [Ndat x 1] = [Ndat x P] %*% [P x 1] + [Ndat]
# cbind does the logical thing when combining a scalar and vector,
# then do vector addition
y <- drop(y) # converts into a vector
pdf("linearmodel_data.pdf", width=4, height=4)
par(mfrow=c(1,1), mar=c(3.5,3.5,0.5,0.5), oma=0.1*c(1,1,1,1),
    mgp=c(2.0,0.8,0), cex=1.0)
plot(x, y, xlim=c(0,9), ylim=c(-1,9), xaxs="i", yaxs="i")
#abline(a=modMat[1], b=modMat[2], col="red") # true model
dev.off()
```

```
# True parameters, transformed to conform with model to be used below
thetaTrue <- c(modMat[1], atan(modMat[2]), log10(sigTrue))
obsdata <- data.frame(cbind(x,y)) # columns must be named "x" and "y"
rm(x,y)

### Define model and infer the posterior PDF over its parameters

# Model to infer: linear regression with Gaussian noise
# Parameters: intercept b_0, gradient b_1; Gaussian noise sigma, ysig.
# MCMC works on: theta=c(b_0, alpha=tan(b_1), log10(ysig)), a 1x3 vector.
# Prior PDFs:
# b_0:          N(mean=m, sd=s); m,s estimated from global properties of data
# alpha:        Uniform (0 to 2pi)
# log10(ysig):  Uniform (improper)

# Define covariance matrix of MCMC sampling PDF:
# sigma=c(b_0, alpha, log10(ysig))
sampleCov <- diag(c(0.1, 0.02, 0.1)^2)
# Set starting point
thetaInit <- c(2, pi/8, log10(3))
# Run the MCMC to get samples from the posterior PDF
set.seed(150)
allSamp <- metrop(func=logpost.linearmodel, thetaInit=thetaInit, Nburnin=0,
                  Nsamp=5e4, sampleCov=sampleCov, verbose=1e3,
                  obsdata=obsdata)
# 10^(allSamp[,1]+allSamp[,2]) is the unnormalized posterior at each sample
thinSel  <- seq(from=1, to=nrow(allSamp), by=25) # thin by factor 25
postSamp <- allSamp[thinSel,]

# Plot MCMC chains and use density estimation to plot 1D posterior PDFs.
# We don't need to do any explicit marginalization to get the 1D PDFs.
pdf("linearmodel_mcmc.pdf", width=8, height=7)
par(mfrow=c(3,2), mar=c(3.0,3.5,0.5,0.5), oma=0.5*c(1,1,1,1),
    mgp=c(1.8,0.6,0), cex=0.9)
parname <- c(expression(b[0]), expression(paste(alpha, " / rad")),
             expression(paste(log, " ", sigma)))
for(j in 3:5) { # columns of postSamp
  plot(1:nrow(postSamp), postSamp[,j], type="l",
       xlab="iteration", ylab=parname[j-2])
  postDen <- density(postSamp[,j], n=2^10)
  plot(postDen$x, postDen$y, type="l", lwd=1.5, yaxs="i",
       ylim=1.05*c(0,max(postDen$y)), xlab=parname[j-2], ylab="density")
  abline(v=thetaTrue[j-2], lwd=1.5, lty=3)
  if(j==3) { # overplot prior
    b0Val <- seq(from=-8, to=8, by=0.01)
    lines(b0Val, dnorm(b0Val, mean=0, sd=2), lty=2)
  }
}
dev.off()

# Plot gradient and intercept of samples in 2D.
# Fix range for b_0 vs alpha to enable comparison to centered data case
pdf("linearmodel_parameter_correlations.pdf", width=6, height=2)
par(mfrow=c(1,3), mgp=c(2.0,0.8,0), mar=c(3.0,3.0,0.5,0.5),
    oma=0.1*c(1,1,1,1))
plot(postSamp[,3], postSamp[,4], xlab=parname[1], ylab=parname[2],
```

```
      pch=".", xlim=c(-2,2), ylim=c(0.35,0.95), xaxs="i", yaxs="i")
plot(postSamp[,3], postSamp[,5], xlab=parname[1], ylab=parname[3], pch=".")
plot(postSamp[,4], postSamp[,5], xlab=parname[2], ylab=parname[3], pch=".")
dev.off()

# Find MAP and mean solutions.
# MAP is not the peak in each 1D PDF, but the peak of the 3D PDF.
# Mean is easy, as samples were drawn from the (unnormalized) posterior.
posMAP     <- which.max(postSamp[,1]+postSamp[,2])
thetaMAP   <- postSamp[posMAP, 3:5]
thetaMean <- apply(postSamp[,3:5], 2, mean) # Monte Carlo integration
cov(postSamp[, 3:5]) # covariance
cor(postSamp[, 3:5]) # correlation

# Overplot these solutions with original data
pdf("linearmodel_fits.pdf", width=4, height=4)
par(mfrow=c(1,1), mar=c(3.5,3.5,0.5,0.5), oma=0.1*c(1,1,1,1),
    mgp=c(2.0,0.8,0), cex=1.0)
plotCI(obsdata$x, obsdata$y, xlim=c(0,9), ylim=c(-1,9), xaxs="i", yaxs="i",
       xlab="x", ylab="y", uiw=10^thetaMAP[3], gap=0)
abline(a=thetaMAP[1], b=tan(thetaMAP[2]), lw=2)     # MAP  model
#abline(a=modMat[1],   b=modMat[2], col="red", lw=2) # true model
# Compare this with the result from ML estimation from lm()
#abline(lm(obsdata$y ~ obsdata$x), col="black", lty=2)
dev.off()

### Make prediction: determine PDF(ycand | xnew, obsdata)

# Model and likelihood used here must be consistent with logpost.linearmodel

# Example 1
xnew <- 6
xlim <- c( 0,9) # xlim and ylim for plotting only
ylim <- c(-1,9)
# Example 2
#xnew <- 25
#xlim <- c( 0,27)
#ylim <- c(-1,29)

# Evaluate generative model at posterior samples (from MCMC).
# Dimensions in matrix multiplication: [Nsamp x 1] = [Nsamp x P] %*% [P x 1]
modPred <- cbind(postSamp[,3], tan(postSamp[,4])) %*% t(cbind(1,xnew))

# Direct method
# ycand must span full range of likelihood and posterior
dy      <- 0.01
ymid   <- thetaMAP[1] + xnew*tan(thetaMAP[2]) # to center choice of ycand
ycand <- seq(ymid-10, ymid+10, dy) # uniform grid of y with step size dy
ycandPDF <- vector(mode="numeric", length=length(ycand))
for(k in 1:length(ycand)) {
  like <- dnorm(ycand[k], mean=modPred, sd=10^postSamp[,5]) # [Nsamp x 1]
  ycandPDF[k] <- mean(like) # integration by rectangle rule. Gives a scalar
}
# Note that ycandPDF[k] is normalized, i.e. sum(dy*ycandPDF)=1.
# Find peak and approximate confidence intervals at 1sigma on either side
peak.ind   <- which.max(ycandPDF)
```

```
lower.ind <- max( which(cumsum(dy*ycandPDF) < pnorm(-1)) )
upper.ind <- min( which(cumsum(dy*ycandPDF) > pnorm(+1)) )
yPredDirect <- ycand[c(peak.ind, lower.ind, upper.ind)]

# Indirect method. likeSamp is [Nsamp x 1]
likeSamp <- rnorm(n=length(modPred), mean=modPred, sd=10^postSamp[,5])
likeDen  <- density(likeSamp, n=2^10)
# Find peak and confidence intervals
yPredIndirect <- c(likeDen$x[which.max(likeDen$y)], quantile(likeSamp,
                   probs=c(pnorm(-1), pnorm(+1)), names=FALSE))

# Plot the predictive posterior distribution
pdf("linearmodel_prediction6_PDF.pdf", width=4, height=4)
par(mfrow=c(1,1), mar=c(3.0,3.5,0.5,0.5), oma=0.5*c(1,1,1,1),
    mgp=c(2.2,0.8,0), cex=1.0)
plot(ycand, ycandPDF, type="l", lwd=1.5, yaxs="i",
     ylim=1.05*c(0,max(ycandPDF)), xlab=expression(y[p]),
     ylab=expression(paste("P(", y[p], " | ", x[p], ", D)")))
abline(v=yPredDirect, lty=2)
# overplot result from the indirect method
lines(likeDen$x, likeDen$y, type="l", lty=3, lwd=1.5)
dev.off()

# Compare predictions between the two methods
rbind(yPredDirect, yPredIndirect)

# Overplot direct prediction with original data and the MAP model
pdf("linearmodel_prediction6_ondata.pdf", width=4, height=4)
par(mfrow=c(1,1), mar=c(3.5,3.5,0.5,0.5), oma=0.1*c(1,1,1,1),
    mgp=c(2.0,0.8,0), cex=1.0)
plotCI(obsdata$x, obsdata$y, xlim=xlim, ylim=ylim, xaxs="i", yaxs="i",
       uiw=10^thetaMAP[3], gap=0, xlab="x", ylab="y")
abline(a=thetaMAP[1], b=tan(thetaMAP[2]), lwd=2) # MAP   model
plotCI(xnew, ycand[peak.ind], li=ycand[lower.ind], ui=ycand[upper.ind],
  gap=0, add=TRUE, lwd=3)
dev.off()
```

R file: `linearmodel_functions.R`

```
##### Functions to evaluate the prior, likelihood, and posterior
##### for the linear model, plus to sample from the prior

# theta is vector of parameters
# obsdata is 2 column dataframe with names [x,y]
# The priors are hard-wired into the functions

# Return c(log10(prior), log10(likelihood)) (each generally unnormalized)
# of the linear model
logpost.linearmodel <- function(theta, obsdata) {
  logprior <- logprior.linearmodel(theta)
  if(is.finite(logprior)) { # only evaluate model if parameters are sensible
    return( c(logprior, loglike.linearmodel(theta, obsdata)) )
  } else {
    return( c(-Inf, -Inf) )
  }
}
```

```
# Return log10(likelihood) (a scalar) for parameters theta and obsdata
# dnorm(..., log=TRUE) returns log base e, so multiply by 1/ln(10) = 0.434
# to get log base 10
loglike.linearmodel <- function(theta, obsdata) {
  # convert alpha to b_1 and log10(ysig) to ysig
  theta[2] <- tan(theta[2])
  theta[3] <- 10^theta[3]
  modPred <- drop( theta[1:2] %*% t(cbind(1,obsdata$x)) )
  # Dimensions in mixed vector/matrix products: [Ndat] = [P] %*% [P x Ndat]
  logLike <- (1/log(10))*sum( dnorm(modPred - obsdata$y, mean=0,
                                   sd=theta[3], log=TRUE) )
  return(logLike)
}


# Return log10(unnormalized prior) (a scalar)
logprior.linearmodel <- function(theta) {
  b0Prior      <- dnorm(theta[1], mean=0, sd=2)
  alphaPrior   <- 1
  logysigPrior <- 1
  logPrior <- sum( log10(b0Prior), log10(alphaPrior), log10(logysigPrior) )
  return(logPrior)
}
```

9.1.5 Discussion

Why is the inferred straight line sometimes quite different from the true straight line?
Don't be mislead by the fact that with noisy data the inferred model is not identical to the
true model. We can only work with the noisy data; we never have access to the true model.
If we are "unlucky" that the data drawn are far from the true model, then we will infer a
different straight line. For this reason I intentionally did not plot the true model in figure
9.1 and similar figures (but you can do so by uncommenting a line in the code). When
we know the true model, we should find that the inferred model (e.g. maximum of the
posterior) converges to the true one as we use more data.

How can we infer the errors bars as well as the line?
Because the data are noisy, they will not normally lie on a straight line. So it should be
clear that if we guessed a value of σ that was small compared to the true (unknown) value,
then the likelihood of the data for any given line would be small (as there would be a large
spread in the data compared to our σ, putting us right down in the tails of the likelihood
for most data points; see equation 9.2). Increasing our guess for σ slightly would increase
the likelihood. You might think that an ever larger σ would always increase the likelihood
for a given data point, because the Gaussian becomes wider. But don't forget that the
Gaussian likelihood is normalized, so as it gets wider it also also gets lower (see figure
1.5). Mathematically, the exponential part of the Gaussian gets larger, but the $1/\sigma$ term in
front gets smaller: as $\sigma \to \infty$, the likelihood tends to zero (for finite data). Thus for given
data and a given line (set by the other two parameters), there must be an intermediate
value of σ that maximizes the likelihood.

What happens if we have much more data?
As we saw in chapter 5, in particular figure 5.6, the more informative the data – which often corresponds to having more data – the narrower the likelihood becomes as a function of the parameters. The prior remains unchanged. The posterior thus becomes increasingly determined by the data and less dependent on the choice of prior.

How are the results affected by the choice of priors?
You can easily investigate this by editing the function `logprior.linearmodel`, and I encourage you to do so. In the set-up above I have used uniform priors on α and $\log \sigma$, which are reasonably uninformative, but I used a Gaussian prior on b_0. If we instead use a uniform improper prior on b_0, the resulting posteriors (with ten data points) are hardly any different: this Gaussian prior was already quite uninformative compared to these data. I have intentionally taken an abstract example here without a scientific context, so that we can concentrate on the mechanics of posterior inference using MCMC. In practice we would have some information that would influence our choice of priors. For example, we invariably have some idea of the scale of the measurement uncertainties. We could use this information to set the standard deviation of the prior on σ. I adopted here a prior that is uniform over α. Quite often we would have background information which tells us a priori that the slope must be constrained to a narrower region. For example, if y were the distance travelled by an electron in time x, then we know that the slope – the speed – cannot be larger than the speed of light. We could scale the range of standard distributions (such as the beta) to impose a variable prior with hard constraints, for example. Note that the uniform prior over α is not invariant with respect to a change of the measurement units of x and y. This may not be such a problem, however, as physical constraints on its values would be expressed in the same units.

What happens if we reduce the size of the data set to two points, or even just one?
Even with fewer data points than parameters, the likelihood function is still defined, so we still get a posterior PDF. The posterior is likely to become a lot wider, because with less data, the less well we can determine the parameters. But we still have a distribution from which we can estimate the mode of the parameters (and the mean if it is still a normalizable distribution). If you don't believe me, try it.[4] But do look carefully at the MCMC chains: they are likely to be poor – not in a steady state – when we have few data. If that is the case then they are not representative of the posterior. To remedy this you may need a lot more samples, a longer burn-in, more thinning, and/or a different covariance matrix for the proposal distribution.

But how, logically thinking, can we estimate three parameters at all given only one or two data points? Surely the solution is somehow "underdetermined"?
No, it's not. You're not only using the data in your inference. You also have a prior. The prior on α is uniform, but the prior on the intercept (b_0) is not. We used a Gaussian with

---

[4] The R code will work with just one data point. This is not immediately obvious, however, because when a vector or matrix only has one element, then R (sometimes) automatically converts it into a scalar, which may then be an incompatible data type for some calculation (e.g. a matrix multiplication).

mean zero and standard deviation 2. You can think of this as acting as a constraint on the model. Specifically, the prior on the intercept is like an additional fuzzy data point (y value) at $x = 0$. So when combined with a single data point, this will tell you something about the gradient. If we increased the width of the intercept prior, then the posterior over this parameter (and typically that over the gradient) would become broader. More fundamentally, the posterior represents all the information we have on the parameters. Even with little data the posterior constrains the parameters more than the prior. Note that the maximum likelihood (least squares) solution, as produced by lm, is not defined when there are fewer data points than parameters. The Bayesian solution, in contrast, always exists.

What if we have no data?
In that case the likelihood function is not defined (it doesn't even exist). The posterior is then just equal to the prior. Indeed, that's exactly what the prior is: our knowledge of the PDF over the parameters in the absence of data.[5]

9.1.6 Centring the data

In the above example I used a Gaussian prior $\mathcal{N}(\mu, s)$ on the intercept parameter b_0. I extrapolated the data by eye to the vertical axis in figure 9.1 in order to chose suitable values of μ and s. Extrapolating like this is generally cumbersome, especially if the data lie far from the origin. It is more convenient to first subtract the mean from the data, i.e. transform them to $x' = x - \bar{x}$ and $y' = y - \bar{y}$. This is called *centring* the data. We can then set $\mu = 0$ and s to a generous estimate of the possible range of the data at $y' = 0$. Estimating priors from the general properties of the data in this way is known as *empirical Bayes*.

As centring is just a linear shift of the data, it does not affect the priors we adopt for α and $\log \sigma$. If is simple to modify the R code in section 9.1.4 to deal with centred data:

- Redefine obsdata

```
xMean <- mean(x)
yMean <- mean(y)
obsdata <- data.frame(cbind(x=x-xMean,y=y-yMean))
```

- Set the prior on the intercept for the model of the centred data. In fact, $\mu = 0$ and $s = 2$ are still reasonable values, so you do not have to change anything.
- If you want to plot the true model remember that it is defined with an intercept b_0 in the original data space. In the centred data space the intercept is $b_0 - \bar{y} + b_1\bar{x}$.

Without making any further changes the code can be run again to produce the posterior samples. You will find that the chains and posteriors are almost identical to what we had before. But there is one important difference. Now that the data are centred, the gradient

[5] The R code will not work properly if you set data to be NULL. This is because the code does not accommodate the likelihood being undefined (although that would be easy to fix). The code actually does return a value, but it is not the value you want.

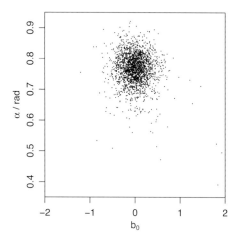

Fig. 9.7 The posterior MCMC samples for b_0 and α for the straight line model applied to the centred data. We see that the correlation between the parameters has been reduced by the centring.

and intercept are much less correlated. This can be seen from the plot of the samples in figure 9.7. Compare this to the case when we did not centre the data (the left panel of figure 9.3). Using centred data the correlation decreases in magnitude from -0.83 to -0.11. We saw when doing least squares fitting in section 4.1.1 that the uncertainties in the parameters also become decorrelated when $\bar{x} = 0$ (equation 4.17).

Figure 9.4 showed how one particular posterior model, the one corresponding to the maximum of the posterior PDF, appears in the data space. It is instructive to plot a sample of posterior models. This is shown in the right panel of figure 9.8. Each grey line corresponds to one set of parameters drawn from the posterior (although it does not show the σ parameter). The variation in intercepts and gradients reflects the finite width of the posterior PDF. We can compare this to a sample of models drawn from the prior, show in the left panel. The prior is uniform in α and has a small range of b_0, so the lines cover all position angles and pivot near to the origin. We see that these prior models cover a much broader range of the data space than the posterior models.[6] This again illustrates how the process of inference combines the prior with the likelihood to form the posterior: we update our prior knowledge by using the data. Here the data are informative, so our knowledge of the model parameters is improved.

The prior models we used with the original data were actually quite inappropriate because they were also lines of all angles pivoting around the origin, whereas the data were uncentred. We nonetheless managed to get good results (equally tight posteriors) because the data were informative, so the likelihood dominated the prior to make the posterior.

---

[6] Had I used an improper uniform prior on b_0 I would not have been able to draw from the prior for b_0 so I could not have made this plot.

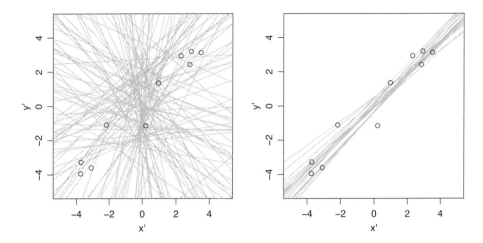

Fig. 9.8 Models drawn from the prior (left) and from the posterior (right) for the straight line model. The open circles show the centred data.

9.2 Fitting a quadratic curve with unknown noise

We can repeat the same procedure as in the previous section, but now for a quadratic generative model

$$y = f(x) + \epsilon \quad \text{where} \tag{9.8}$$
$$f(x) = b_0 + b_1 x + b_2 x^2 \tag{9.9}$$

and again $\epsilon \sim \mathcal{N}(0, \sigma)$. The R script `quadraticmodel_posterior.R` and the functions file `quadraticmodel_functions.R` (both available online only) are similar to the straight line model in the previous section, but now with a quadratic term added. I again use a Gaussian prior on b_0, transform b_1 to $\alpha = \arctan(b_1)$ and use a uniform prior on this, and transform σ to $\log \sigma$ and use an improper uniform prior on this too. To b_2 I apply a Gaussian prior.

In the code I draw 20 data points at random between $x = 0$ and $x = 10$, evaluate the quadratic model (equation 9.9) with $b_0 = 25$, $b_1 = -10$, and $b_2 = 1$, and add zero mean Gaussian noise with standard deviation $\sigma = 2$ to produce the y values. These simulated data are plotted in figure 9.9. For the Gaussian priors I use $\mathcal{N}(0, 10)$ for b_0 and $\mathcal{N}(0, 5)$ for b_2. These are quite broad and are adopted purely for illustration purposes. The priors on α and $\log \sigma$ are again uniform and have no parameters. The priors are set in `quadraticmodel_functions.R`.

For the four parameters b_0, α, b_2, and $\log \sigma$ I use step sizes (Gaussian standard deviations) in the MCMC of 0.1, 0.01, 0.01, and 0.01 respectively. In principle I can initialize the sampling anywhere, but it could then take an extremely large number of steps to locate the high density regions of the posterior. The more parameters we have, the more acute this problem becomes. It is far better to take a good guess at the values of the parameters. Here I

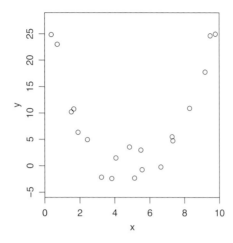

Fig. 9.9 Data used for the quadratic model fitting. They have been drawn from a quadratic curve at fixed x values with $b_0 = 25$, $b_1 = -10$, $b_2 = 1$ (equation 9.9), to which zero mean Gaussian noise with standard deviation 2 has been added.

do one better by finding the maximum likelihood (least squares) solution of the parameters b_0, α, and b_2 (using lm) and use these as the initial condition for the Markov chain. For the initial value of σ I use the root mean square of the residuals about this fit.

To achieve good chains we nonetheless need more iterations than in the straight line problem, on account of the higher complexity of the model (we have one more parameter). After a burn-in of 20 000 iterations, I sample for a further 200 000 iterations. To reduce the autocorrelation I use a thinning factor of 100. The resulting chains (for the 2000 samples retained) are shown in the left column of figure 9.10. We still see some evidence for autocorrelations. Yet the marginal posterior PDFs, show in the right column, are not very sensitive to this. Their shapes and summary statistics do not change much if we double or half the number of iterations or the degree of thinning.

If we take the parameter values at the maximum of the posterior, this gives a good model for the data: this is the curve and error bars shown figure 9.11. The mean of the parameter samples (not shown) gives an almost identical fit. We should be careful with the mean, however: it takes into account all samples, so it can be more affected if the sampling algorithm produces unrepresentative samples by getting stuck in an island of low probability.

Figure 9.12 shows the correlations between the samples. We see correlations between all parameters in the generative model, in particular a strong anticorrelation between α and b_2. This is not that surprising, because we can compensate for a small change in one of these parameters by a small change in the other, i.e. in order to retain more or less the same curve. There is barely any correlation between the inferred noise σ and the generative model parameters.

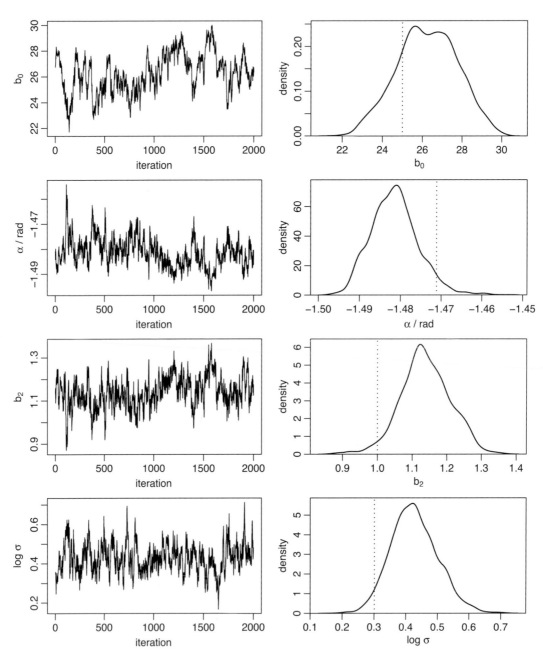

Fig. 9.10 MCMC chains (left columns) and resulting marginal posterior PDFs (right columns) for the quadratic model fitting problem. The one-dimensional posteriors have been computed by a kernel density estimate of the samples. The vertical dotted lines indicate the true parameters. The priors for b_0 and b_2 are Gaussian, with zero mean and standard deviation 10 and 5 respectively, so they are virtually flat over the range of the parameters plotted. The priors for the other two parameters are uniform.

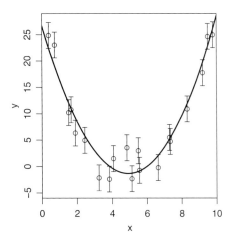

The open circles are the data from figure 9.9. The curve and error bars show the model obtained with the parameters set to the maximum of the posterior PDF (MAP), which has values $b_0 = 26.7$, $b_1 = -11.4$, $b_2 = 1.16$, $\sigma = 2.45$.

9.3 A mixture model: fitting a straight line with an outlier model

In the straight line and quadratic model examples in the previous sections, we assumed that the residuals $y - f(x)$ followed a zero mean Gaussian distribution (the likelihood) with common, but unknown, standard deviation σ. But what if the data contains outliers, as data tend to? These could distort the line fit and/or increase the estimated value of σ. One could attempt to remove or manually down-weight outliers, but such approaches tend to be ad hoc and unstable. Could we instead deal with them consistently – probabilistically – in the modelling procedure?

Virtually by definition an outlier is a data point that we believe is not drawn from the adopted likelihood. The fact that it sticks out (rather than in, where we would not notice it) suggests it has come from a broader likelihood distribution. We could model outliers with a second Gaussian, but more convenient is a Cauchy distribution (section 1.4.7), because it has much broader tails. Its PDF for point (x, y) is

$$L_{\text{out}}(y) = \frac{1}{\pi w[1 + (\frac{y - f(x)}{w})^2]} \tag{9.10}$$

where w is the half-width at half-maximum, and we have set the mode equal to the model-predicted values (the outlier distribution is symmetric about this). We can then combine this with the main likelihood distribution

$$L_{\text{main}}(y) = \frac{1}{\sigma\sqrt{2\pi}} \exp\left[-\frac{[y - f(x)]^2}{2\sigma^2}\right] \tag{9.11}$$

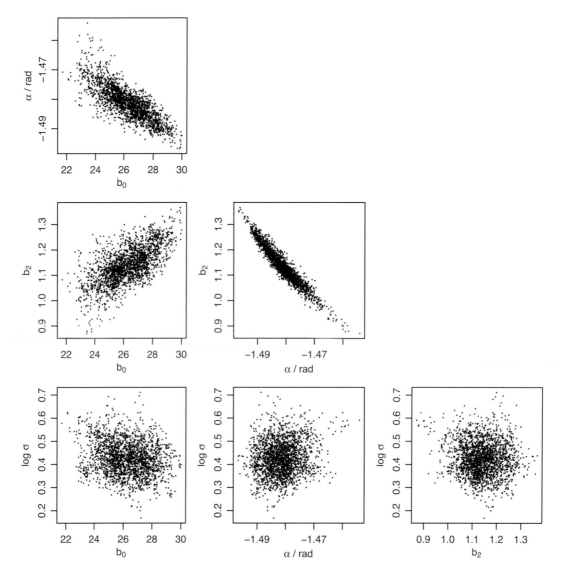

Fig. 9.12 The MCMC samples from the posterior PDF of the quadratic model, plotted in two dimensions to show the correlations between them.

as a *mixture model*

$$L_{\text{total}}(y) = (1 - p)L_{\text{main}}(y) + pL_{\text{out}}(y) \quad \text{where} \quad 0 \le p \le 1 \qquad (9.12)$$

where p is the *mixing factor*. The log likelihood for all data points is $\sum_i \log L_{\text{total}}(y_i)$. Our mixture model in principle has five parameters: b_0, α, σ, w, and p. With all other parameters fixed, we might think that a larger w would allow a better fit to outliers. But remember that the outlier distribution (equation 9.10) is normalized, so making it broader actually decreases the probability density at smaller values of $|y - f(x)|$. For a fixed w, the degree to which points are outliers can be controlled by the parameter p (to some extent – this will be discussed further below). I therefore fix $w = 1$ in the following.

We can now proceed with fitting a straight line as before (section 9.1), but with a different likelihood and an extra parameter, p. This needs a prior. As it is constrained to lie between 0 and 1, a natural choice is the beta distribution (section 1.4.3). Clearly a uniform prior is not appropriate: isn't the very nature of outliers that they are rare? In the code (see below) I set the two shape parameters of the beta distribution to be 1 and 20, which gives a strong peak at $p = 0$ and decreases monotonically to zero at $p = 1$ (overplotted later in figure 9.13 with the posterior). In a real application we should adjust this prior according to what we know about the frequency of outliers, and also investigate how sensitive the results are to the choice.

Having found the posterior PDFs, and having chosen a suitable estimator for the parameters, we can evaluate the probability that a particular point is an outlier by taking the ratio of the (unnormalized) probability that the point is an outlier to the (unnormalized) probability that the point is either an outlier or not

$$P_{\text{outlier}}(y) = \frac{pL_{\text{out}}(y)}{L_{\text{total}(y)}} = \frac{pL_{\text{out}}(y)}{(1 - p)L_{\text{main}}(y) + pL_{\text{out}}(y)}. \tag{9.13}$$

The R code for fitting this model, `linearmodel_outlier_posterior.R`, is available online. It is similar in structure to that used for fitting the line without outliers (section 9.1.4), except that now I define the posterior function at the beginning of the file. The new parameter is the mixing factor, p. It must lie within the range 0–1, yet the MCMC sampler – which uses a multivariate Gaussian proposal distribution – can propose values outside of this range. These are automatically assigned zero prior probability by the `dbeta` function, but we must manually set L_{total} to zero (equation 9.12). The posterior is then zero, so this proposal will always be rejected by the Metropolis algorithm.

9.3.1 Application and results

I apply the model to the same data as that used in section 9.1, but with one point replaced with a single outlier. The outlier is ad hoc; it is not drawn from the outlier model. The MCMC is run for 10^5 iterations (with an initial burn-in of 10^4) and the resulting chains are thinned by a factor of 50. The resulting chains and one-dimensional marginalized posteriors are shown in figure 9.13. The chains are not particularly good (I have made little effort to optimize them), but they suffice for this demonstration. Figure 9.14 shows the original data (open circles) as well as the model fit from the maximum of the posterior (the solid line as well as the error bars). For comparison, the dashed line shows the least squares fit to all the data, i.e. without using an outlier model. We see that this is biased upwards by the presence of the outlier. Figure 9.15 shows the posterior samples in two dimensions. The only substantial correlation is between b_0 and α, for the same reason as when we had no outlier (the data are not centred).

The outlier probabilities for the points are shown in table 9.2. We see that not only does the mixture model prevent the outlier from distorting our fit too much, it also identifies the

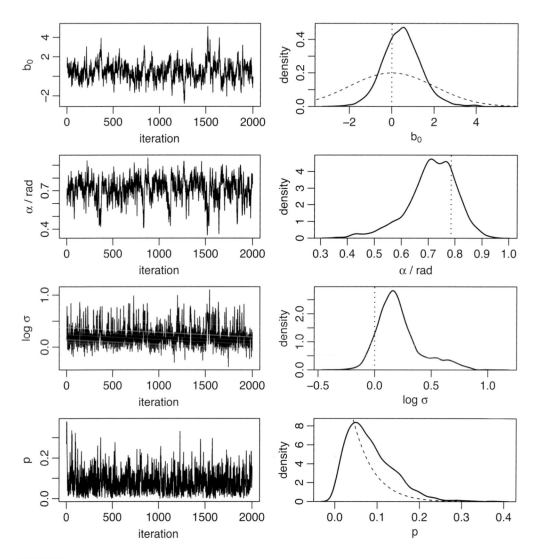

Fig. 9.13 MCMC chains (left columns) and resulting one-dimensional marginal posterior PDFs (right columns) for the straight line model with outlier model fitting problem. The one-dimensional posteriors have been computed by a kernel density estimate of the set of samples. The vertical dotted lines indicate the true parameters (there is no definition of the true value for the mixing coefficient p). The curved dashed lines in the panels for b_0 and p show the prior distributions for those parameters. The priors for the other two parameters are uniform.

outlier ($n = 5$). In this case identification would have been easy by eye, but this would be harder with a higher dimensional data set, a nonlinear problem, or for more marginal cases.

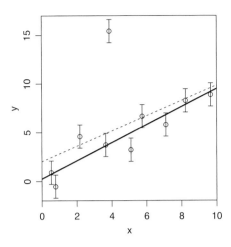

Straight line fitting with an outlier model. The open circles are the data. These
have been drawn from a straight line at fixed x values with $b_0 = 0$ and $b_1 = 1$
(equation 9.1). Zero mean Gaussian noise with standard deviation 1 has been
added to all of these points, except the point at $x = 3.88$, which was made into an
outlier by adding 10 to the function evaluation. The solid straight line and error
bars show the model obtained with the parameters set to the maximum of the
posterior PDF (MAP), which has values $b_0 = 0.25$, $b_1 = 0.93$, $\sigma = 1.19$, and
$p = 0.048$. The dashed line is the least squares (maximum likelihood) solution to
all the data, which assumes that all the data come from a Gaussian likelihood
with common standard deviation.

9.3.2 Discussion

As outliers are presumably rare, the mixing factor p must be small. We might want to think
of it as the average probability that a given point is an outlier *under this model*. Here I
have fixed the value of the width parameter in the Cauchy distribution to 1. This is not
robust, as the distribution is not scale independent: if we rescaled the data (e.g. measured
y in different units), then we would need to change this parameter. However, provided it is
on roughly the right scale, p will effectively act as the width parameter, because it scales
the whole Cauchy distribution up and down, thus changing the contribution to the total
likelihood. This has its limits, though, because p must be less than 1.

I do not claim that this outlier model is particularly good. I have used this simple method
primarily to demonstrate the concept of using a likelihood mixture model.

There are several ways one could change or extend this approach. One is to make w a
parameter that we infer along with the other four parameters. However, there is likely to be
a correlation ("degeneracy") between w and p, as they can play a similar role. This is not a
fundamental problem, but it might require many more MCMC samples to get good chains
and suitable posterior PDFs. Another improvement would be to assign a different p to each
data point (with w fixed again). This would not only give the model more flexibility, it

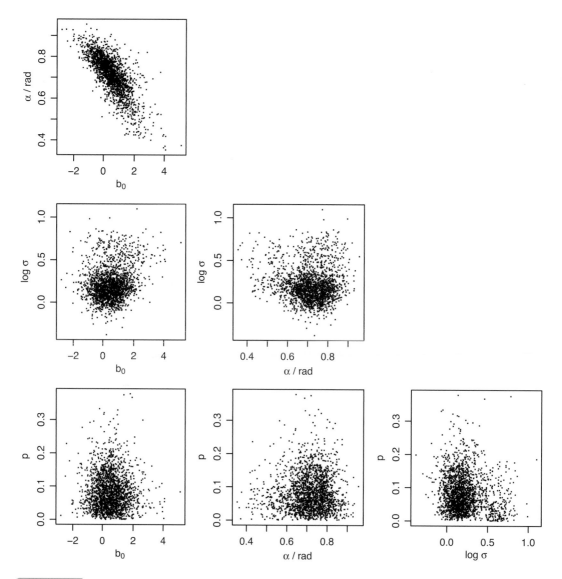

Fig. 9.15 The MCMC samples from the posterior PDF of the straight line model with an outlier model, plotted in two dimensions to show the correlations between them.

would allow us (but not force us) to assign a different prior to each point, which is useful if we are more suspicious of some points that others. This would increase the number of parameters, and in fact increase it to be more than the number of data points. This is not a problem for a Bayesian analysis per se. But such flexibility can present new difficulties, not least for the MCMC because it now has to sample a much higher dimensional space. This often requires many more samples and better tuning of the step sizes. (Metropolis is not well suited to sampling parameters spaces with many dimensions.)

Table 9.2 Outlier probabilities (equation 9.13) for the ten points shown in figure 9.14

i	x	y	P_{outlier}
1	0.5	0.9	0.045
2	0.8	−0.5	0.032
3	2.2	4.6	0.048
4	3.7	3.7	0.045
5	3.9	15.4	1.000
6	5.1	3.2	0.033
7	5.8	6.7	0.032
8	7.1	5.8	0.033
9	8.2	8.3	0.042
10	9.7	8.9	0.043

9.4 Fitting curves with arbitrary error bars on both axes

Sometimes we have uncertainties not only in the y values but also in the x values. We saw one way to deal with this in a least squares sense in section 4.7. The Bayesian approach is more general and more flexible. For example, it allows us to accommodate any noise model, not just a Gaussian one.

A general approach is to make a distinction between the (noisy) measured values of x and y and their true (noise-free) but unknown values, which I will denote x' and y' respectively. Thus

$$x = x' + \epsilon_x$$
$$y = y' + \epsilon_y$$

(9.14)

where ϵ_x and ϵ_y are random numbers which in general depend on both x' and y'. We then write the noise model, with parameters ϕ, for a single data point as $P(x, y \,|\, x', y', \phi)$. An example of this is a bivariate Gaussian distribution, perhaps with correlated errors in x and y, in which case ϕ is the covariance matrix. As before the generative model relates the noise-free quantities as $y' = f[x'; \theta]$, where θ are its parameters. This could be a polynomial model or a nonlinear model, for example. Whatever the noise and generative models, we can write the likelihood for a single data point as a marginalization over the true, unknown values

$$P(x, y \,|\, \theta, \phi) = \iint P(x, y \,|\, x', y', \phi) \, P(x', y' \,|\, \theta) \, dx' dy'.$$

(9.15)

The second term under the integral we can write as

$$P(x', y' \,|\, \theta) = P(y' \,|\, x', \theta) P(x' \,|\, \theta).$$

(9.16)

In many situations we may assume that, prior to making any measurements, all values of x' are equally probable over an arbitrarily large range of x' between x'_{lo} and x'_{hi}. In that case we can set $P(x'|\theta) = 1/\Delta x'$ inside this range, and zero outside the range, where $\Delta x' = x'_{hi} - x'_{lo}$. We will see in a moment that none of these values actually matter. If we further assume that the generative model is deterministic (we have only considered such models in this book), then

$$P(y'|x',\theta) = \delta(y' - f[x';\theta]) \tag{9.17}$$

where $\delta()$ is the delta function. Inserting this into equation 9.16 and that into equation 9.15, we see that the integration over y' is trivial on account of the delta function: the integral is only non-zero at $y' = f[x';\theta]$. Thus the likelihood for one data point is

$$P(x,y|\theta,\phi) = \frac{1}{\Delta x'} \int_{x'_{lo}}^{x'_{hi}} P(x,y|x',y'=f[x';\theta],\phi)\, dx'. \tag{9.18}$$

This likelihood is normalized,[7] so it must become arbitrarily small at extreme values of x. Thus provided $\Delta x'$ is large enough – and we can make it arbitrarily large – then the values of x'_{lo} and x'_{hi} are irrelevant, because the integrand in equation 9.18 will become arbitrarily small before we hit these limits. Furthermore, $\Delta x'$ will often be independent of θ, in which case it can be absorbed into the normalization constant for the posterior. We just have to ensure that the integration over x' extends out to where the likelihood becomes very small.

If we have a set of N data points $D = \{x_i, y_i\}$ that have been measured independently, then the likelihood of all the data is

$$P(D|\theta,\phi) = \prod_i P(x_i, y_i|\theta,\phi). \tag{9.19}$$

The unnormalized posterior is therefore

$$P^*(\theta,\phi|D) = P(\theta,\phi)\prod_i \int_{x'_{lo}}^{x'_{hi}} P(x_i, y_i|x',y'=f[x';\theta],\phi)\, dx' \tag{9.20}$$

where $P(\theta,\phi)$ is the prior.

We can now proceed as in the earlier sections: we decide on a prior then sample from the posterior using MCMC. The only difference now is that each evaluation of the posterior involves N one-dimensional integrations. In general these will need to be solved numerically, so the posterior may be computationally expensive (slow) to compute. Yet this approach allows us to fit arbitrary curves to data with arbitrary noise models on both x and y.

To illustrate this method, let us suppose we want to fit a straight line

$$y' = f[x'] = b_0 + b_1 x' \tag{9.21}$$

to some data $D = \{x,y\}$, for which we have known Gaussian uncertainties in x and y of σ_x and σ_y respectively. These are the same for each data point. Thus the noise model

---

[7] This is important, because in general the normalization constant of the likelihood depends on the parameters θ, and so cannot be ignored when we are interested in the posterior $P(\theta|D)$.

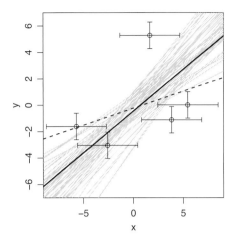

Fig. 9.16 Straight line fitting with uncertainties in both x and y. The open circles are the data, and the error bars show the (known) one-sigma uncertainties. The solid black line shows the straight line model corresponding to the maximum of the posterior. The grey lines show 50 models drawn from the posterior. The dashed line is the least squares fit ignoring the uncertainties in x.

parameters ϕ are fixed. The likelihood for one data point is therefore given by a bivariate Gaussian

$$P(x,y\,|\,x',y'=f[x';\theta]) \;=\; \frac{1}{2\pi|\Sigma|^{1/2}}\exp\left(-\frac{1}{2}[(x,y)-(x',y')]^{\mathsf{T}}\Sigma^{-1}[(x,y)-(x',y')]\right)$$

where

$$\Sigma = \begin{bmatrix} \sigma_x^2 & 0 \\ 0 & \sigma_y^2 \end{bmatrix}. \tag{9.22}$$

We could easily generalize this approach to have different known uncertainties in x and y for each point, and even a different non-zero correlation for each data point. Alternatively, we could let σ_x, σ_y, and the correlation between them be unknown parameters that we want to determine from the data. But for now we assume there is just a single σ_x and a single σ_y, and these are known. The parameters of this model are therefore (b_0, b_1).

I simulate some data by drawing five points from a uniform distribution between 0 and 10. These are the noise-free x values. The corresponding noise-free y values are computed from equation 9.21 with $b_0 = 0$ and $b_1 = 1$. I then add zero mean Gaussian noise with standard deviation $\sigma_x = 3$ to the noise-free x values to get the x data (and then I centre these; see section 9.1.6). I likewise add zero mean Gaussian noise with standard deviation $\sigma_y = 1$ to noise-free y values to get the y data (and also centre these). I do not use the noise-free data in the inference, of course.

The simulated data are shown in figure 9.16. I adopt a uniform prior on $\alpha = \arctan b_1$, and a Gaussian prior on b_0 with mean zero and standard deviation 5. I use the Metropolis algorithm to sample the posterior over the parameters (b_0, α), using a Gaussian proposal

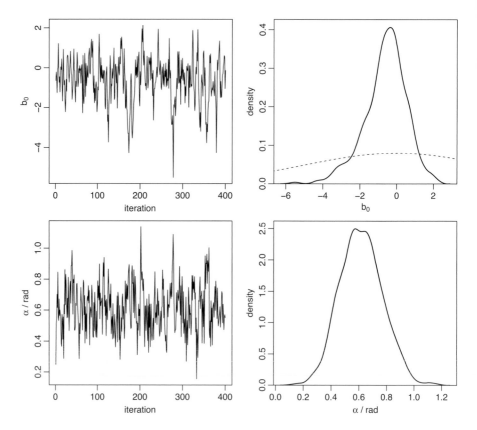

Fig. 9.17 MCMC chains (left columns) and resulting marginal posterior PDFs (right columns) for fitting a straight line with uncertainties in both x and y. These one-dimensional posteriors have been computed by a kernel density estimate of the samples. The curved dashed line in the panel for b_0 shows the prior distribution. The prior over α is uniform.

distribution with standard deviations (step sizes) of $(0.25, 0.05)$. I initialize Metropolis at the least squares solution for b_0 and α, sample for 10 000 iterations, and apply a thinning factor of 25. (Each iteration involves five numerical integrations.) This took about ten minutes to run on my laptop.

The resulting chains and marginal posteriors are shown in figure 9.17.[8] The posteriors are quite broad, i.e. the parameters are not precisely determined, but this is not surprising given that we have only five data points and large error bars. The straight line corresponding to the maximum of the posterior is shown as the thick solid black line in figure 9.16. The thin grey lines show 50 models drawn from the posterior to illustrate how the uncertainty represented by the finite width of the posterior transforms to the data space. The dashed

---

[8] Although the chains have been thinned, they still show some signs of correlations, even though the correlation lengths (post-thinning) are about 5 and 2 for b_0 and α respectively.

line shows the ordinary least squares solution (section 4.1). It is quite different, primarily because it ignores the uncertainties in x.

The online R file `linearmodel_xyerr_posterior.R` performs the above experiment. This also makes use of the file `linearmodel_xyerr_functions.R`. These are analogous to the two files of similar name listed in section 9.1.4. The one-dimensional integral in equation 9.18 is done numerically using the R function `integrate` (which uses the technique of adaptive quadrature). We can think of this as integrating a bivariate Gaussian along the line $y' = f[x'; \theta]$.

The approach described above up to and including equation 9.16 is completely general. I then assumed a uniform prior for $P(x'|\theta)$. This could of course be generalized. For example, if the data were a time series, such that x is time and y is the signal, then the phenomenon being measured might be one in which events are more likely to occur at some times than others, or maybe cannot occur in certain time ranges at all. The phenomenon itself might even be non-deterministic, in which case we could use a stochastic model for $f[x'; \theta]$. The principle of marginalizing over x' and y' to determine the likelihood for each data point (x, y) still applies. For those who want to know more, see Bailer-Jones (2012).

Frequentist hypothesis testing

In this chapter we look at frequentist hypothesis testing and the resulting concept of p values. I will introduce the Gaussian z test and the (Student's) t test, as well as the χ^2 distribution as a means of assessing goodness of fit. I conclude the chapter with a discussion of the problems with this approach to assessing models. Although this is a book on Bayesian inference, and I believe frequentist hypothesis testing should often be avoided, a basic understanding of this topic is necessary.

10.1 The principles of frequentist statistics and hypothesis testing

In previous chapters we learned how to infer the posterior probability distribution over parameters, $P(\theta \,|\, D, M)$, as well the posterior probability of the model, $P(M \,|\, D)$. With this probabilistic (Bayesian) method we take a direct approach: we infer the probabilities of things we don't know (parameters, models), given the things we do (data, priors). Data are noisy and samples are incomplete, so our inferences cannot be 100% certain. Our uncertainties in both the data and the parameters are quantified by probabilities.

The frequentist approach is different. It assumes that the model parameters are fixed, and it is only the data which are variable. Parameters do not have probability distributions: we can only derive fixed values of them from the data. The idea is to establish how the data might have varied in imaginary repetitions of the measurements, in order to assess our confidence in parameter values. As the name suggests, in the frequentist approach probabilities are defined in terms of frequencies of occurrences over these imagined sets of data.

In frequentist hypothesis testing a hypothesis (or model) is either true or false: it is not possible to attach a probability to it. To decide whether or not it is true, we first assume it is, then compute a statistic (from the data). On the basis of its value we either (a) reject the hypothesis, or (b) do not reject the hypothesis. You would be right to think that this is a rather indirect approach to assessing the plausibility of a model. From this perspective it is perhaps surprising that it was standard practice in science for much of the twentieth century, and was widely taught, often as the only approach for model assessment. Part of the reason for this may have been its apparent simplicity. In addition, the probabilistic alternative using the Bayesian evidence (chapter 11) often requires numerical integration that was often beyond the computational means of the time.

The general procedure for frequentist hypothesis testing is as follows. We do an experiment, gather the data, and then compute a statistic η which summarizes these data. This might be the ratio of the mean to the standard deviation, for example. The statistic is chosen to be one which, when assuming a particular hypothesis – call it H_0 – to be true, has a predictable, one-dimensional PDF $P(\eta \,|\, H_0)$. This is called the *sampling distribution*. If we imagine repeating the experiment a large number of times and plotting the distribution of this statistic over these imaginary data sets, the statistic would follow the distribution $P(\eta \,|\, H_0)$, assuming H_0 (which is a model for the data) is true. We then use $P(\eta \,|\, H_0)$ to calculate the probability that we could have obtained (in the imaginary repetitions) a value of the statistic which is *more extreme* than the one we actually observed. This probability is called the *p value*. If the p value is sufficiently small then we say the data are unlikely to have been so extreme, and so we reject the hypothesis. If the p value is not small then we cannot reject the hypothesis. Note that this is not the same as accepting the hypothesis. In frequentist hypothesis testing we can never accept a hypothesis. The best we can do is fail to reject it, and simultaneously reject all other rival hypotheses. What counts as "sufficiently small" is a matter of choice (and will be discussed later).

The reason for the curious procedure of imagining "data more extreme than those observed" is that the probability of obtaining exactly any specific value of the statistic is zero. This is because the probability *density* function is a continuous, finite function and so has zero probability at any point (an infinitesimally narrow range)

$$\text{probability} \;=\; P(\eta)\,\delta\eta \;\rightarrow\; 0 \quad \text{as} \quad \delta\eta \rightarrow 0. \tag{10.1}$$

We must instead consider this concept of "more extreme", namely

$$p \;=\; \int_{\eta}^{\infty} P(\eta')\,d\eta' \tag{10.2}$$

which is an actual probability, the p value (for a one-sided test: more on this later). If we imagine repeating the experiment many times, then the p value is the fraction of experiments in which the measured statistic would exceed the actual measured value of η. Note that the p value is *not* the probability that the hypothesis is true. This cannot be calculated, because in frequentist hypothesis testing hypotheses do not have probabilities.

Bayesian model comparison avoids reference to "more extreme data", because in comparing models it uses the *ratio* of two probability densities, which is well defined at a point. It therefore treats competing models symmetrically, so it can, within the set of models considered, accept as well as reject hypotheses (at some quantified probability level). Frequentist hypothesis testing instead compares the observed data (via the statistic) to hypothetical "more extreme" data sets.

We normally use frequentist hypothesis testing to try to reject a so-called *null hypothesis*. The null hypothesis is generally chosen to be something for which the statistic has a simple distribution (so in practice the statistic and the null hypothesis are chosen together). The null hypothesis is usually a model which shows "no interesting signal", making our attempt to reject the null hypothesis something like proof by contradiction.

If we manage to reject the null hypothesis, then this suggests that some alternative hypothesis may be true. But it does not tell us anything about what that alternative hypothesis is. The only time this is possible is when the alternative is the complement of the null, i.e. the two models are mutually exclusive and exhaustive.

In discussions of frequentist hypothesis testing you will sometimes encounter type I and type II errors, which are two distinct types of error one can make.[1]

Type I error: probability of rejecting the hypothesis when it is true.

Type II error: probability of not rejecting the hypothesis when it is false.

These errors trade off against each other. If we set our confidence threshold sufficiently high to greatly reduce the chance of making type II errors, then we will naturally increase the chance of making type I errors and vice versa. I shall not use these terms, however.

Because the concept of hypothesis testing relies on coming up with a distribution of the test statistic, there is a plethora of hypothesis tests designed for different types of data (t test, F test, KS test, signed rank test, U test, etc.). There are statistical recipe books dedicated to providing a suitable test for given ingredients. Here I will just look at a few widely-used examples which assume Gaussian-distributed data. They differ according to what properties of the parent distribution are known and what null hypothesis we are trying to test. After we have seen how hypothesis testing works, I will return at the end of this chapter to discuss some of the conceptual implications and problems of hypothesis testing.

10.2 One-sample hypothesis tests

One-sample hypothesis tests are used when we have a single sample of data and typically want to ask whether the mean of the sample differs from some specified value. There are two widely used tests, both of which assume Gaussian distributions for the data. They differ in what information we are given. The z test assumes the standard deviation is known. The t test assumes it is not.

10.2.1 Gaussian z test

A company has long produced batteries which, due to inevitable variability in production, have an energy storage capacity that can be described by a Gaussian distribution with a mean of 200 MJ and a standard deviation of 20 MJ. The company is now selling a battery which it claims has a higher mean capacity (but same standard deviation), based on new technology. One of these new batteries is measured to have a capacity of 230 MJ. Is this new technology really better?

To formulate this question statistically we ask "what is the probability that this capacity could be achieved by the old technology?". We need to find the probability that the old process would produce a battery with a capacity as high as 230 MJ. We therefore test the following null hypothesis:

---

[1] Sometimes these are referred to as false positive and false negative probabilities, but which is which depends on whether you consider rejecting a hypothesis as positive or negative.

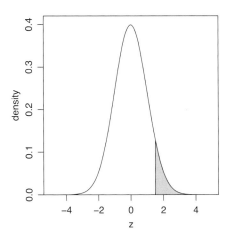

Fig. 10.1 The standardized Gaussian distribution as used in the Gaussian z test. The grey region indicates $z > 1.5$, which has a probability of 0.067.

H_0: the battery was drawn from a Gaussian with mean $\mu = 200$ MJ and standard deviation $\sigma = 20$ MJ.

As we're interested in a battery that has a capacity of $x = 230$ MJ *or even higher*, we carry out a so-called one-sided test. We want to find the probability that we would get a battery with as much capacity as the one observed, assuming the null hypothesis to be true. We do this by calculating the z value or z statistic

$$z = \frac{x - \mu}{\sigma}. \tag{10.3}$$

As μ and σ are fixed and x has a Gaussian distribution $\mathcal{N}(\mu, \sigma)$, the statistic z is the standardized Gaussian distribution $\mathcal{N}(0, 1)$, which has no free parameters. In the present example

$$z = \frac{230 - 200}{20} = 1.5. \tag{10.4}$$

The probability of getting this value of z or higher, the p value, is the area under the distribution with $z > 1.5$, shown graphically in figure 10.1. We compute the area by integration using the cumulative distribution function (equation 1.64), for which R provides functions for standard distributions. For the Gaussian distribution, this area is[2]

```
1 - pnorm(1.5) = 0.067
```

This is of course equivalent to considering the probability of drawing a value of 230 or higher from a $\mathcal{N}(200, 20)$ distribution:

```
1 - pnorm(230, mean=200, sd=20) = 0.067
```

We interpret this result to mean that there is a probability of 0.067 that we would get a battery with at least the measured capacity from the old technology (the null hypothesis).

[2] As the Gaussian is symmetric, `1 - pnorm(q) = pnorm(-q)`.

As this probability is quite small, it suggests that the old technology was not the one used, i.e. the null hypothesis may not be true, so we may want to reject it. We may then take this to imply that the company's claim is true. Remember, however, that by rejecting a null we do not automatically accept some alternative, even if it is the complement of the null.[3] Nonetheless, an even smaller p value would give you more confidence that the null hypothesis is not true.

Strictly speaking we should not be talking about the null hypothesis being true or not, because we are doing a one-sided test. We are only interested in whether the battery has a *larger* capacity than before. That is, we would not reject the null hypothesis if we measured a very small capacity (a large negative value of z), even though getting such extreme values is also unlikely under the null hypothesis. In the current one-sided test, the implicit *alternative hypothesis H_1* is therefore

H_1: the battery was drawn from a Gaussian distribution with mean $\mu > 200$ MJ.

This is only the complement of H_0 for a one-sided test. A two-sided test would be to test whether the capacity is significantly higher or lower than 200 MJ (we'll do this later). For such a two-sided test we might reject H_0 on the grounds that z is large and negative, but this would obviously not lead us to think H_1 can be favoured.

Some practitioners decide to reject hypotheses at a fixed confidence level. If we reject at the 90% confidence level, this means we reject the null hypothesis if $p < 0.1$. So in the above example we would. If we choose the 95% confidence level we need $p < 0.05$ to reject, which is not met here. At what level one chooses to reject a hypothesis is a matter of choice and convention, and varies between disciplines. Often $p < 0.01$ or smaller is used. The p value is sometimes reported in terms of the equivalent number of sigma for a Gaussian distribution. We just saw that $p = 0.067$ corresponds to 1.5σ for a one-sided test. Note that 2, 3, 4, and 5 times σ correspond to p values of 2.3×10^{-2}, 1.3×10^{-3}, 3.2×10^{-5}, and 2.9×10^{-7} respectively, for a one-sided test. A level of 5σ or higher is often used in physics. The computed value of p should always be reported in addition to a reject/no-reject decision, so others can make their own decision.

We return to the example. Instead of one battery we are now given $N = 4$ batteries and measure their mean capacity to be $\bar{x} = 230$ MJ (same value as before). What do we now say about the company's claim?

As we have more batteries the given mean capacity must be more secure, i.e. it must have a standard deviation less than 20 MJ. Given what we learned in section 2.3 from the central limit theorem about repeated measurements, we see that the standard deviation in $\bar{x}$ is a factor of $\sqrt{N}$ smaller than the single battery standard deviation (which is given to us; we don't have to infer it from the data). So the z value is now

$$z = \frac{\bar{x} - \mu}{\sigma/\sqrt{N}} = \frac{230 - 200}{20/\sqrt{4}} = 3 \tag{10.5}$$

and we can once again calculate the p value – the probability that a sample of four batteries drawn from the original distribution would have a capacity of 230 MJ or more – which is

```
1 - pnorm(3) = 0.0013
```

---
[3] Would you accept a hypothesis that you haven't actually tested?

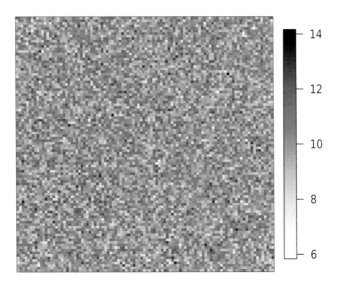

14

12

10

8

6

Fig. 10.2 An image with 100×100 pixels with a flux level in each pixel generated from a standardized Gaussian distribution $\mathcal{N}(10, 1)$.

As this is smaller than the value of 0.067 from before, we can now be more confident about rejecting the null hypothesis.

To use this test it's actually not necessary for the parent distribution of the data $\{x\}$ to be Gaussian. Given the definition of z, it is sufficient if the sample mean $\bar{x}$ has a Gaussian distribution. Provided the central limit theorem applies to the parent distribution, then in the limit of large N its sample mean will have a Gaussian distribution.

Example: detecting a signal in an image

Suppose we want to look for a signal (a single bright pixel) in a digital image in the presence of a noisy but otherwise uniform background illumination. An example is shown in figure 10.2. In the absence of a signal, the flux in the pixels will vary around the mean background level due to noise. Given a model for this background+noise – here a Gaussian with mean $\mu = 10$ and standard deviation $\sigma = 1$ – we want to set a threshold above which it is unlikely that the noise would give rise to such a large flux. A 5σ detection, for example, means that the probability of the noise producing a flux of 15 (or brighter) in any one pixel is $p = 2.9 \times 10^{-7}$ (pnorm(-5)). In an image with $N = 100 \times 100$ pixels, the probability that at one or more of the pixels would be this bright due to noise is $1 - (1 - p)^N = 0.003$. So if we did measure a pixel flux at 5σ in an image of this size, the standard response is to say that this is unlikely to be a result of noise (or more correctly, of *this* noise model). We may then either reject such a measurement by claiming it to be an outlier – an implicit admission that our noise model is inferior[4] – or claim that it is due to something of interest. One scientist's noise is another scientist's signal!

The following code makes figure 10.2. It looks better if you plot it with a diverging colour scale, which you can achieve by replacing Greys with BrBG in the definition of mypalette.

R file: noisy_image.R

```
##### Plot a Gaussian noise image

library(fields) # for image.plot
library(RColorBrewer) # for colorRampPalette
mypalette <- colorRampPalette(brewer.pal(9, "Greys"), space="rgb",
                      interpolate="linear", bias=1)
mycols <- mypalette(64)
pdf("noisy_image.pdf", 5, 4)
par(mfrow=c(1,1), mgp=c(2.0,0.8,0), mar=c(1,1,1,1), oma=0.1*c(1,1,1,1))
set.seed(100)
x <- 1:1e2
y <- 1:1e2
z <- matrix(data=rnorm(length(x)*length(y))+10, nrow=length(y),
            ncol=length(x))
image.plot(z=z, x=x, y=y, xaxt="n", xlab="", yaxt="n", ylab="", nlevel=1024,
           zlim=c(-4.1,4.1)+10, col=mycols)
dev.off()
```

10.2.2 Student's t test

One-sided test

Consider a subtle but important variation of the battery problem variation on page 208.

We now measure the capacity of four batteries (which were all produced by the same technology) to have a mean of $\bar{x} = 230\,\text{MJ}$ and a standard deviation of $\hat{\sigma} = 16\,\text{MJ}$. How likely is it that they come from a technology which produces batteries with a mean capacity of $\mu = 200\,\text{MJ}$?

This appears to be the same problem: we take the same null and alternative hypotheses and also want a one-sided test. But now we have estimated the standard deviation from the sample itself, by calculating the sample standard deviation, $\hat{\sigma}$ (equation 2.12). Before we just took the manufacturer's quoted value, whereas now we aren't given one. We can still write down something which looks like the z value, but it is now called the t value or t statistic

$$t = \frac{\bar{x} - \mu}{\hat{s}} \tag{10.6}$$

because $\hat{s}$ is now an estimate *obtained from the data* of the standard deviation in $\bar{x}$. This estimate is the standard error in the mean, $\hat{s} = \hat{\sigma}/\sqrt{N}$ (equation 2.13). Because there is

---

[4] A Gaussian noise model is unlikely to apply to this kind of problem in reality. There are invariably additional noise sources which occur only rarely but can have a large impact. Examples are hot pixels in the detector, scattered light, and interference in the read-out electronics. In practice we would want to use a noise model with heavier tails, or a mixture model to handle large, occasional noise contributions (cf. section 9.3).

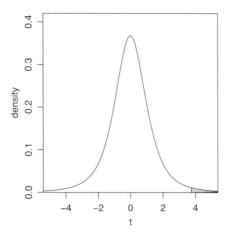

Fig. 10.3 The t distribution with three degrees of freedom. The grey region indicates $t > 3.75$, which has a probability of 0.017.

uncertainty in the value of $\hat{s}$ itself, the significance of a given deviation between $\bar{x}$ and μ is now less than when s was known. The degree of uncertainty in $\hat{s}$ depends on the degrees of freedom (see section 4.3). Whereas the z value in equations 10.3 and 10.5 followed a standardized Gaussian distribution, which has no free parameters, it can be shown that t follows a Student's t distribution. This has one free parameter, the degrees of freedom ν. This is equal to $N - 1$ here, since one degree of freedom is "used up" to estimate $\hat{s}$.

We encountered the Student's t distribution in section 6.2: it is the marginal posterior PDF for the mean of a Gaussian distribution given N data points when the standard deviation is unknown (and assigned a Jeffreys prior). Its form is given in equation 6.12 and shown in figure 6.2. It is similar in shape to the standardized Gaussian but has more probability in the tails; this reflects the extra uncertainty arising from having to estimate the standard deviation from the data. As the number of degrees of freedom tends toward infinity the distribution becomes a Gaussian. Equivalently, the standard deviation of the data would then be estimated exactly, so the conditions for the z test are fulfilled. Note, therefore, that if we have enough data we can just use the Gaussian z test rather than the t test.

Let me be more precise about the definition of the t statistic in equation 10.6. The quantity $\hat{s}$ is actually the standard deviation of $\bar{x} - \mu$. But if μ is given (a constant) then its standard deviation is zero,[5] so $\hat{s}$ is the standard deviation of $\bar{x}$. This is the case when we test the hypothesis that the data were drawn from a Gaussian with some specified mean μ. The Student's t distribution is then the distribution of the sample mean relative to the specified mean, scaled by the standard error in the sample mean.

Let us return to the battery question. $N = 4$, so the relevant t statistic is

$$t = \frac{\bar{x} - \mu}{\hat{\sigma}/\sqrt{N}} = \frac{230 - 200}{16/\sqrt{4}} = 3.75 \qquad (10.7)$$

---

[5] In section 10.3.2 we will compare two samples, in which case μ is not constant and the denominator includes the standard deviation in both terms in the numerator.

and this follows a t distribution with $N - 1 = 3$ degrees of freedom. This is plotted in figure 10.3. The probability of getting a t value this large or larger – the p value – is $P(t \geq 3.75 \,|\, H_0)$. This is given by integrating the t distribution. This is done in R with the function pt, which is analogous to pnorm for the Gaussian distribution. The area under the curve is

```
1 - pt(3.75, df=3) = 0.017
```

Thus the probability that we would get a mean battery capacity as large as we did (or larger), given these four samples, is quite small ($<2\%$) when the null hypothesis is true. So we can reject the null hypothesis at a 98% confidence level (but not at a 99% level). Compare this with $1 - \text{pnorm}(3.75) = 8.8 \times 10^{-5}$, which is much smaller. That is, if we *knew* that $s = 16$ MJ, rather than having to measure it from the data, we could be much more confident about rejecting the null hypothesis, because the same measured deviation from the expected mean is then more significant.

Two-sided test

Remaining with the same data, we now ask not whether the mean of the sample of four batteries has a larger capacity than that claimed by the old technology, but whether this mean is *different*, i.e. larger *or* smaller. Thus we test the null hypothesis
$H_0 : \mu = 200$ MJ
for which the alternative hypothesis is now
$H_1 : \mu \neq 200$ MJ
i.e. we use a two-sided test. The alternative hypothesis is now the complement of the null hypothesis.

The test statistic is the same as before, but we now calculate the probability that t is outside of the interval -3.75 to $+3.75$ (assuming the null hypothesis to be true). As the t distribution is symmetric, the probability is

```
2*(1-pt(3.75, df=3)) = 0.033
```

This is the probability that the data are consistent with the null hypothesis (more strictly: the probability of getting the measured capacity or something more extreme, assuming the null hypothesis is true). The probability with the two-sided test is twice as high as with the one-sided test, because we ask whether the mean differs from the specified value, rather than just whether it is larger. So for given data, the result of a two-sided test will be a larger p value. It is a weaker test.

10.2.3 General t testing in R and frequentist confidence intervals

The above is all you need for doing t tests with a single sample. Yet R provides a more convenient interface via the function t.test, which works directly on the original set of data. Suppose we have the following data set on batteries

```
x <- c(177, 194, 209, 228, 229, 235, 241, 244, 244, 287)
```

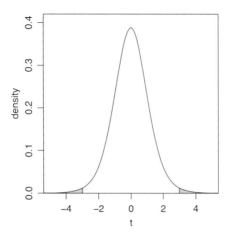

Fig. 10.4 The t distribution with nine degrees of freedom. The grey regions are those with $t < -2.99$ and $t > 2.99$, which together have a probability of 0.015.

We want to know whether their mean deviates from 200 (a two-sided test). We do

```
t.test(x, mu=200)
```

which produces

```
        One Sample t-test

data:   x
t = 2.9947, df = 9, p-value = 0.01508
alternative hypothesis: true mean is not equal to 200
95 percent confidence interval:
 207.0451 250.5549
sample estimates:
mean of x
    228.8
```

The output is largely self-explanatory. The t value is

$$t = \frac{\bar{x} - 200}{\hat{\sigma}/\sqrt{N}} \tag{10.8}$$

which you can check with R

```
(mean(x)-200)/(sd(x)/sqrt(length(x)))
```

The number of degrees of freedom ν is $N - 1 = 9$. The p value is `2*pt(-2.9947, df=9)`, and is the probability of getting a t value of this magnitude $|t|$, or more, assuming the null hypothesis to be true. The fact that it's quite small (see figure 10.4) suggests that the sample mean is significantly different from the mean specified by the null hypothesis.

The frequentist 95% confidence interval (quoted in the output) means that if we repeated this experiment many times, 95% of these confidence intervals (which would differ each time) would contain the true mean.[6] This interval is symmetric about the estimated mean because we're doing a two-sided test. We can calculate the interval using the quantile function for the t distribution $Q(p; \nu)$, which is the inverse of the cumulative distribution function (see section 1.5). $Q(p; \nu)$ gives the value of t below which the distribution contains probability p. In R it can be computed by the function qt(p,df), where df $= \nu$. By setting $p = 0.025$ we get the lower bound on the t distribution corresponding to the lower end of the 95% confidence interval. To scale this into a deviation from $\bar{x}$, it follows from equation 10.6 that we must multiply $Q(p; \nu)$ by $\hat{\sigma}/\sqrt{N}$. Thus the lower bound on the 95% confidence interval relative to $\bar{x}$ is

```
qt(p=0.025,df=9)*sd(x)/sqrt(length(x))
```

which gives -21.8. As the t distribution is symmetric (so qt(1-p,df) = -qt(p,df)), the upper bound on the 95% confidence limit (at $p = 0.975$) is at $+21.8$ relative to $\bar{x}$. Thus the estimate of the mean and its 95% confidence interval is

$$\bar{x} \pm Q(p = 0.975, \nu = N - 1)\,\frac{\hat{\sigma}}{\sqrt{N}} \tag{10.9}$$

which gives 228.8 ± 21.8, which is the range $(207.0, 250.6)$. This is the output from t.test above.

The default for t.test is a two-sided test. To do a one-sided test, we specify

```
t.test(x, mu=200, alternative="greater")
```

```
    One Sample t-test

data:  x
t = 2.9947, df = 9, p-value = 0.007542
alternative hypothesis: true mean is greater than 200
95 percent confidence interval:
 211.1712      Inf
sample estimates:
mean of x
    228.8
```

The parameter alternative in the function call defines the direction of the one-sided test. The p value here is the probability that we get a value of t equal to or greater than the one measured when the null hypothesis is true. This defines the alternative hypothesis: the mean is more than 200. As we get a sample with mean 228.8 and a standard error in this mean of 9.6 (sd(x)/sqrt(length(x))), then it's more likely that these data are consistent with the alternative hypothesis.

We can also do the one-sided test the other way. The p value is now the probability that we get a value of t equal to or less than the one measured when the null hypothesis (which

---

[6] This does *not* mean that there is a 95% chance that the true mean falls within the specific confidence interval measured. That is given by the Bayesian confidence interval, sometimes also called the credible interval (section 5.5). The distinction may appear subtle, but it is both substantial and important. A frequentist analysis makes statements about possible data. In this paradigm unobserved parameters do not have uncertainty (a distribution), so the frequentist cannot say anything about the probability that the true value lies within a specified range.

says the mean is 200) is true. This is of course quite high. The new alternative hypothesis is the complement of the previous one, so the p value is one minus what we got before

```
t.test(x, mu=200, alternative="less")
```

```
One Sample t-test

    data:  x
    t = 2.9947, df = 9, p-value = 0.9925
    alternative hypothesis: true mean is less than 200
    95 percent confidence interval:
        -Inf 246.4288
    sample estimates:
    mean of x
        228.8
```

to within the rounding error, at least.

Note that one of the confidence limit bounds in the one-sided test will always be infinite, because by construction we allow arbitrarily extreme values in one direction.

10.2.4 Summary: z test vs t test

In both tests we have a set of N measurements and compute their mean to be $\bar{x}$. We want to compare this to a mean μ specified by the null hypothesis.

We use the z test when we additionally know that the true standard deviation of each measurement is σ. Thus the standard deviation in the mean is $\sigma/\sqrt{N}$, and the resulting z statistic, $(\bar{x} - \mu)/(\sigma/\sqrt{N})$ has a standardized Gaussian distribution.

In the t test we do not know the standard deviation. We therefore estimate it from the data to be $\hat{\sigma}$, so the standard deviation in the mean is $\hat{\sigma}/\sqrt{N}$. The resulting t statistic $(\bar{x} - \mu)/(\hat{\sigma}/\sqrt{N})$ has a Student's t distribution with $N - 1$ degrees of freedom.

10.3 Two-sample hypothesis tests

10.3.1 Gaussian z test

In the previous section we compared statistics estimated from a sample of data to fixed values. Sometimes we want to compare two samples, for example to find out whether two samples have the same mean, or whether two samples with the same mean have different variances. In each case we must identify what knowledge we have a priori and what statistic we can and want to determine from the data. We can then establish the distribution of the statistic and identify the appropriate test to apply.

Let's take an example. We measure the boiling points of two liquids to be 127 ± 5 K and 141 ± 3 K. Are they significantly different?

The null hypothesis is that they are the same. We interpret the first measurement as saying that the boiling temperature is a Gaussian random variable with mean $\bar{x}_1$ and standard deviation σ_1, and similarly for the second liquid. Our task is to assess whether $\overline{x}_1 - \overline{x}_2$ is significantly different from zero. The difference of two Gaussian variables is another Gaussian variable with mean $\overline{x}_1 - \overline{x}_2$ and variance $\sigma_d^2 = \sigma_1^2 + \sigma_2^2$. The standard deviations are given, and not estimated from the data, so the z test is appropriate, with statistic

$$z = \frac{141 - 127}{\sqrt{3^2 + 5^2}} = \frac{14}{5.83} = 2.40. \tag{10.10}$$

The boiling points differ by $2.4\sigma_d$, which corresponds to the following p value (two-sided test).

```
2*(1-pnorm(2.4)) = 0.016
```

There is (just) a 1.6% chance that we would observe these two boiling points if they came from the same distribution. The complementary alternative hypothesis is that the boiling points differ, for which there is reasonable support. This conclusion assumes that the errors are indeed Gaussian. If they are not, then the z test does not apply.

10.3.2 Student's t test

We now turn to the more common case where we have to estimate the mean and standard deviation of each distribution from the data themselves. Suppose we have two samples, $\{x_1\}$ and $\{x_2\}$, with N_1 and N_2 members each. We first estimate the means of the two samples, $\overline{x}_1$ and $\overline{x}_2$, as well as their (sample) standard deviations, $\hat{\sigma}_1$ and $\hat{\sigma}_2$. In analogy to what we did earlier, the test statistic is a standardized difference between the two means

$$t = \frac{\overline{x}_1 - \overline{x}_2}{\hat{\sigma}_c \sqrt{\frac{1}{N_1} + \frac{1}{N_2}}} \tag{10.11}$$

where $\hat{\sigma}_c$ is the standard deviation of the combined samples. This is given by a weighted mean of the individual sample variances

$$\hat{\sigma}_c^2 = \frac{\nu_1 \hat{\sigma}_1^2 + \nu_2 \hat{\sigma}_2^2}{\nu_1 + \nu_2}$$
$$= \frac{(N_1 - 1)\hat{\sigma}_1^2 + (N_2 - 1)\hat{\sigma}_2^2}{N_1 + N_2 - 2} \tag{10.12}$$

where $\nu_1 = N_1 - 1$ and $\nu_2 = N_2 - 1$ are the number of degrees of freedom in the samples (each minus one is there because we are estimating the standard deviations from the data). This t statistic is appropriate if the samples have been drawn from two parent populations which have a common (but unknown) standard deviation, which we are estimating with $\hat{\sigma}_c$. (We don't assume the means are the same, of course). The denominator in equation 10.11 can be thought of as a generalization of the standard error in the mean. The above t statistic has a t distribution with $N_1 + N_2 - 2$ degrees of freedom. The -2 is there because we are calculating two standard deviations from the data.

If the two samples do not have a common standard deviation (or rather, if we relax this assumption), then the statistic is different

$$t = \frac{\overline{x}_1 - \overline{x}_2}{\sqrt{\frac{\hat{\sigma}_1^2}{N_1} + \frac{\hat{\sigma}_2^2}{N_2}}}.$$ (10.13)

This does not have a t distribution, but it can be approximated by one with a number of degrees of freedom given by

$$\nu = \frac{\left(\frac{\hat{\sigma}_1^2}{N_1} + \frac{\hat{\sigma}_2^2}{N_2}\right)^2}{\frac{(\hat{\sigma}_1^2/N_1)^2}{N_1-1} + \frac{(\hat{\sigma}_2^2/N_2)^2}{N_2-1}}$$ (10.14)

(you'll need to consult advanced classical texts for a proof of this). The statistical test for two samples using this statistic known as the *Welch two-sample t test*.

Needless to say, all of this can be done directly in R using t.test.

10.4 Hypothesis testing in linear modelling

We now return to chapter 4, and in particular to section 4.1.5, to see how the t test can be used in linear regression to assess the quality of a model fit.

We saw from equation 4.10 that the least squares estimate of the gradient of the line $\hat{b}$ (I now use the "hat" symbol to indicate it is an estimate) could be written as the weighted sum of several random variables, namely the $\{y_i\}$. The same can be shown for the estimate of the intercept, $\hat{a}$. If we assume – as we did – that each y_i has a Gaussian distribution with a common standard deviation, then the variable

$$t = \frac{\hat{b}}{\sigma_{\hat{b}}}$$ (10.15)

has a Student's t distribution with $N - 2$ degrees of freedom, where $\sigma_{\hat{b}}^2 = \text{Var}(\hat{b})$ (given by equation 4.15). In fact, the definition of the t statistic (equation 10.6) is $t = (\hat{b} - b_{\text{hyp}})/\sigma_{\hat{b}}$, where b_{hyp} is the value of the gradient under the hypothesis we want to test. So the expression in equation 10.15 only has a t distribution when $b_{\text{hyp}} = 0$. Thus by applying a t test to the statistic in equation 10.15, we are testing the null hypothesis that the data have zero gradient. This may be used to assess how important the parameter is for the model. This test has $N - 2$ degrees of freedom because we have "used up" two degrees of freedom in estimating a and b.

The lm function in R does this t test for all the parameters in the model. Specifically, it tests the null hypothesis that the particular parameter is zero, with all the other parameters held fixed at their fitted values. It prints the value of t and the resulting p value (see the code outputs in section 4.1.5).

In principle a small p value means the parameter is significant, i.e. we need it in the model to fit the data well. But this is only the case if we keep the other parameters fixed at their fitted values. Look back at the regression problem with nonlinear functions of the data in section 4.6. In the quadratic fit all the parameters were highly significant. This suggests that adding a cubic term would improve the fit. The cubic fit is indeed better – in terms of a lower residual standard error – but we see from the code output that the significance of the quadratic term has now dropped (larger p value). But this does not necessarily imply the quadratic term is not required. (Indeed it cannot mean this, because the cubic term has an even lower significance than the quadratic one.) The reason is that the t test on the quadratic parameter is applied by keeping the cubic parameter fixed at its fitted value. So we cannot fit a multiparameter model and identify the significance of each parameter using just the results of the t tests on this model. To find out if the quadratic or cubic term is better, we need to fit separate models. Specifically, we need to compare the results of model2, $y = b_0 + b_1 x + b_2 x^2$, with $y = b_0 + b_1 x + b_3 x^3$, which produces

```
lm(formula = y ~ x + I(x^3))
Residuals:
     Min      1Q   Median      3Q      Max
-2.15365 -0.75737 -0.00452  0.85481  2.28427
Coefficients:
                Estimate Std. Error t value Pr(>|t|)
(Intercept) 19.4897920  1.0350118  18.831 2.82e-10 ***
x           -1.8180150  0.2130960  -8.531 1.93e-06 ***
I(x^3)       0.0040257  0.0008941   4.503 0.000723 ***
Residual standard error: 1.355 on 12 degrees of freedom
Multiple R-squared:  0.9265,Adjusted R-squared:  0.9142
```

The quadratic fit is marginally better in terms of a lower residual standard error and a higher significance (smaller p value) on its nonlinear term. So on balance we need a quadratic term more than a cubic one. Having both (model3) improves the fit further, but only marginally.

10.5 Goodness of fit and the χ^2 distribution

When we fit a model to data, we don't usually expect (or want) a perfect fit, because data are noisy. But we do want to get an idea of how good the fit is. One measure of this is the χ^2 statistic. It is defined as

$$\chi^2 = \sum_{i=1}^{N} z_i^2 = \sum_{i=1}^{N} \left(\frac{y_i - f_i}{\sigma_i} \right)^2 \tag{10.16}$$

(which also defines z_i), where $\{y_i\}$ are the observed data values, $\{f_i\}$ are the corresponding values predicted by the model, and $\{\sigma_i\}$ are the standard deviations in the observed values, i.e. the uncertainties in the measurements. Sometimes these are known independently of the data. For example, when weighing objects the precision of the scales is often known in advance. We saw in section 4.1.4 how χ^2 could then be used as an error function to fit linear models, and we saw in section 4.4.2 that this is equivalent to maximizing the likelihood for known Gaussian uncertainties. In other situations the $\{\sigma_i\}$ must be estimated from the data (and we saw in section 9.1 that this can be done by computing the posterior).

The difference $y_i - f_i$ is the residual. If the residuals are typically of the same size as the

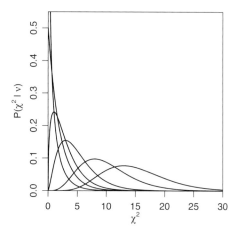

Fig. 10.5 The χ^2 distribution for 1, 2, 3, 5, 10, and 15 degrees of freedom. The higher the number of degrees of freedom, the further to the right is the curve.

expected error – suggesting a reasonable fit to the data – then $z_i \sim 1$ and χ^2 will be about N, the number of data points. If the residuals are much larger than the expected errors, then this suggests a poorer fit, so χ^2 will be much larger than N. If χ^2 is much smaller than N, this means the residuals are smaller than the expected errors. In that case either the fit is "too good" – the model is too finely tuned to the exact data and so is fitting the noise, something known as *overfitting* – or the errors have been overestimated. Clearly there will be a lot of scatter in the calculated value of χ^2 when N is small, so how much χ^2 can tolerably deviate from N depends on both N and the noise model.

The χ^2 statistic is useful for assessing the quality of fit for a given model. But it may not be used to find the best model among a set of models. I will explain why in chapter 11.

10.5.1 The χ^2 distribution

In equation 10.16, χ^2 can be calculated for any data with any distribution. However, if each z_i is an independent, standard Gaussian random variable – $\mathcal{N}(0, 1)$ – then χ^2 follows the χ^2 distribution with integer parameter $\nu = N$ (we shall prove this in section 10.5.3). Its density function is

$$P(\chi^2; \nu) = \frac{2^{-\frac{\nu}{2}}}{\Gamma(\nu/2)} (\chi^2)^{\frac{\nu}{2}-1} \exp(-\chi^2/2) \quad \text{where} \quad \nu \geq 1. \tag{10.17}$$

where $\Gamma()$ is the gamma function (equation 1.62). Examples of the distribution for different ν – the degrees of freedom – are shown in figure 10.5. The distribution has mean ν and variance 2ν. For $\nu = 1$, $P(\chi^2) \to \infty$ as $\chi^2 \to 0$. For $\nu = 2$ the mode is at $\chi^2 = 0$ and the distribution is just a negative exponential. For larger values of ν the distribution has a mode at larger values of χ^2 and becomes increasingly symmetric; from inspection of equation 10.17 we see that it in fact becomes increasingly like a Gaussian. The χ^2 distribution is a special case of the gamma distribution (section 1.4.6) with scale $\theta = 2$ and shape $k = \nu/2$.

Note that the χ^2 distribution is a distribution in χ^2, not χ. The distribution for χ is obtained by a change of variables, $P(\chi^2)d\chi^2 = P(\chi)d\chi$, to give $P(\chi) = 2\chi P(\chi^2)$, which is also parametrized by ν. For $\nu = 3$ the χ distribution is the Maxwell–Boltzmann distribution, which is the probability density function used to describe the speed $v = (v_x^2 + v_y^2 + v_z^2)^{1/2}$ of particles in a gas in three-dimensional space: $P(v) \propto v^2 \exp(-av^2/2)$ for some constant a. This follows because the velocity distributions of particles in each spatial dimension are assumed to be independent zero mean Gaussians (with a common standard deviation). In two dimensions ($\nu = 2$) this is the Rayleigh distribution, $P(\chi) \propto \chi \exp(-\chi^2/2)$.

10.5.2 Using χ^2 as a measure of goodness of fit

Suppose we have a linear model for some data which we believe is the true model, and the measurements differ from this only due to zero mean Gaussian noise. For point i the model predicts the value f_i and the expected standard deviation is σ_i. In that case each term $z_i = (y_i - f_i)/\sigma_i$ has a standard Gaussian distribution, and if the N measurements are independent, χ^2 will follow the χ^2 distribution with N degrees of freedom. This is the case when the model has been decided upon independently of these data. If we have instead fit the model to the data, then the residuals in this fit are no longer independent, so the degrees of freedom is reduced. If our (linear) model has p parameters, then χ^2 follows a χ^2 distribution with $\nu = N - p$ degrees of freedom. This last statement involves three important conditions.

(1) The residual are assumed to follow a Gaussian distribution. This is often not achieved in practice, for example due to outliers.
(2) The σ_i values are assumed to be known: they have not been estimated from the data. Estimating them would effectively reduce the number of degrees of freedom further, but the metric χ^2 would no longer have a χ^2 distribution.
(3) The model is linear in the parameters (see section 4.2). If it is not, then the concept of p degrees of freedom being "used up" by estimating p parameters does not apply (section 4.3).

Let's now use the χ^2 distribution to calculate a goodness of fit. We fitted a straight line to some data in 4.1.5. Suppose that the uncertainty in all the y-values is 2. We compute χ^2 as follows.

```
resid <- y - predict(model1)
err    <- 2
chi2   <- sum((resid/err)^2)
df     <- length(x)-2   # lose 2 dof due to the straight line fit
pchisq(chi2, df)
```

This gives $\chi^2 = 16.7$ on eight degrees of freedom. The last statement in the code above returns the probability that χ^2 has the specified value or less. This probability is 0.967. Let the null hypothesis H_0 be the statement that the model is the true one, i.e. the predictions agree with the measured data to within the expected errors. When H_0 is true we expect low χ^2 values (around 1 per degree of freedom). The probability that we would get a χ^2 as high as the one observed (or even higher), the p value, is $1-\texttt{pchisq(chi2,df)} = 0.033$. This suggests that it is unlikely we would observe such a large value of χ^2 for this number of degrees of freedom: either the model is a poor fit to the data, or we've underestimated the expected errors, or the data are erroneous (e.g. there are outliers). Note that even if we can rule out the last two alternatives, a low p value still does not "prove" the model to be wrong. All we can do is find an alternative (but still plausible) model which explains the data better (which for a frequentist can only be achieved by failing to reject it).

You will sometimes see reference to the *reduced chi-squared*. This is simply χ^2 divided by its expectation value (the number of degrees of freedom)

$$\chi_r^2 = \frac{\chi^2}{\nu} = \frac{1}{\nu}\sum_{i=1}^{N}\left(\frac{y_i - f_i}{\sigma_i}\right)^2. \tag{10.18}$$

You may encounter the rule of thumb "a fit is good if the reduced χ^2 is about 1". Although the mean of χ^2/ν for ν degrees of freedom is 1, the shape of the distribution still depends on ν.

We should not forget that χ^2 is only a statement about how likely the data are assuming the model to be true. It tells you nothing directly about how likely the model is given the data. The reason is that the probability of any particular data set – even from the true model – is generally very low (recall the discussion in section 5.2). Thus χ^2 can be large and the p value small even when the null hypothesis is true. Think of N cards drawn (with replacement) at random from a pack of cards. The probability of the data set is $1/52^N$ no matter which cards we draw. If some combinations of cards *seem* less likely (e.g. four aces), then this is only because we attach more meaning to these than to other combinations. We will discuss this further in section 11.1.

10.5.3 Derivation of the χ^2 distribution

I have not yet proven that the χ^2 distribution is the PDF of the sum of the squares of ν independent Gaussian variables $\{z_i\}$ each with zero mean and unit variance. This is quite straightforward.

Think of the variables as defining a point in a ν-dimensional space with coordinates $(z_1, z_2, \ldots, z_\nu)$. The PDF of each variable is proportional to $\exp(-z_i^2/2)$ so their joint PDF is their product

$$P(\{z_i\}) \propto \prod_i \exp\left(-z_i^2/2\right) = \exp\left(-\sum_i z_i^2/2\right) = \exp(-\chi^2/2). \tag{10.19}$$

In this ν-dimensional space a hypersphere with radius χ is a surface of constant probability. The probability that a point lies in a hypershell between χ and $\chi + d\chi$ is the above probability density (which is per unit volume) multiplied by the volume of this hypershell. The hypershell has surface area proportional to $\chi^{\nu-1}$ and thickness $d\chi$, so its volume is proportional to $\chi^{\nu-1}d\chi$. Thus

$$P(\chi)d\chi \propto \chi^{\nu-1}\exp(-\chi^2/2)d\chi. \tag{10.20}$$

To get $P(\chi^2)$ we do a transformation of variables

$$P(\chi^2) = P(\chi)\frac{d\chi}{d\chi^2}$$

$$\propto \frac{1}{\chi}\chi^{\nu-1}\exp(-\chi^2/2)$$

$$\propto (\chi^2)^{\frac{\nu}{2}-1}\exp(-\chi^2/2) \tag{10.21}$$

which is the χ^2 distribution (equation 10.17) to within a normalization constant.

A useful property of the χ^2 distribution follows from this derivation. If two variables x_1 and x_2 are described by χ^2 distributions with ν_1 and ν_2 degrees of freedom, then their sum, $x_1 + x_2$, must follow a χ^2 distribution with $\nu_1 + \nu_2$ degrees of freedom.

10.6 Issues with frequentist hypothesis testing

I dedicated this chapter to frequentist hypothesis testing because it is widely used, not because I think it is a sound way to test models. It has some significant problems, some of which I have already alluded to. I will now summarize these, before discussing some more fundamental issues on model inference.

(1) Frequentist hypothesis testing requires us to summarize the data via a test statistic and then find a convenient probability distribution for the statistic. This often involves significant approximations in practice.

(2) The approach hinges on calculating the probability that the hypothesis could produce a value of the statistic which is more extreme than the observed value of the statistic (this probability is the p value). It is not clear why the concept of imaginary "more extreme" data should be relevant to our inference.

(3) Although not a fault with the approach in itself, p values are frequently interpreted as being the probability that the hypothesis is true. This is wrong. The p value tells you nothing about the probability of the hypothesis; this concept is absent in the frequentist approach.

(4) A hypothesis is rejected on the basis that the data obtained were unlikely. This is odd, because we *did* get these data. By rejecting hypotheses in this way we ignore the possibility that the data may be equally unlikely under other plausible hypotheses.

(5) Rejecting the null hypothesis on the basis of a small p value does not imply that another (plausible) hypothesis exists which could explain the data better. In particular, rejecting the null does not imply we can accept some particular alternative. We should surely only accept a hypothesis if it has been explicitly tested. Yet in the frequentist paradigm we can never accept hypotheses, just fail to reject them (at some confidence level).

(6) The p values are sensitive to the size of a data set. The more data we have, the less probable any particular data set – even under the true model – and so the smaller the p value (this was seen in section 5.2). This makes it increasingly likely that we will reject a given model (null hypothesis) as we get more data. We could remedy this by raising

the rejection threshold as we get more data. But with complex problems in particular we cannot know how to set this threshold, and in practice fixed confidence levels (e.g. 95%, 99%, 3σ, 5σ) are used.

(7) As will be demonstrated in section 11.8, p values can depend on irrelevant information, which can nonetheless make the difference between rejecting and not rejecting a hypothesis.

(8) The relevance of a p value depends on what other tests have been done. Suppose we wanted to test whether there was a significant correlation between the annual per capita deaths by leukaemia, and each of a number of metrics, such as income, mobile phone use, number of doctors, etc. For each individual correlation comparison we might set $p = 0.05$ as the significance level for declaring a significant correlation. But if we did 20 such tests then we would expect to get one[7] significant correlation just by chance – indeed, by definition of the significance level. Yet if we reported just this one positive result it would appear to have statistical significance. This problem is known as *multiple testing*, more informally as *p hacking*. One can try to compensate for it by adjusting the significance level according to the number of tests made (e.g. the Bonferroni correction), or by looking at the fraction of false positive results among all tests made (e.g. with the false discovery rate).

Having highlighted some specific problems, I now turn to a more fundamental issue concerning frequentist hypothesis testing.

A popular impression of science is that it works by falsification. The idea is that a scientist specifies a hypothesis which makes a clear prediction. If an observation or measurement is made which contradicts this prediction, the hypothesis is falsified. A much used analogy is that of the black swan. The hypothesis is that all swans are white; the observation of one black swan falsifies this hypothesis. Null hypothesis testing attempts to put rejection of the hypothesis on a numerical basis by using the p value to specify a rejection confidence.

It's a nice story, but science does not work like this in practice. First, we rarely observe directly what the hypothesis states. In reality we observe something (e.g. the motions of planets) in order to infer the truth (or perhaps just the utility) of the hypothesis (e.g. "the Sun and planets orbit the Earth"). Second, a hypothesis rarely makes an explicit binary prediction. We must instead set up a model which generally predicts the size of an effect, not simply its presence or absence. Third, measurements are noisy (or are a sample from a population), so we cannot say with certainty whether a prediction is correct; we just get a measure of proximity between data and model predictions. The black swan is a poor caricature of science, and by sticking to it we would have to ask what we mean by "black" and indeed what we mean by "swan". What do we conclude if we observe a grey swan, or a bird that looks much like a swan but has small genetic differences? Testing models is not black and white (excuse the pun). Fourth, a model is only ever an approximation to reality. All models are therefore false at some level, so what does it even mean to falsify a model? If we required models to be true, we would end up rejecting them all. Yet science is full of approximate models which are very useful in practice: the model of gases

---

[7] The probability of getting at least one significant correlation by chance from 20 tests is $1 - (1 - 0.05)^{20} = 0.64$.

as non-interacting particles with zero volume; classical electromagnetism; Newton's law of gravity. All of these are false models of reality, yet they are very accurate in certain domains.

In the real world we cannot falsify a hypothesis or model any more than we "truthify" it. We can only determine a degree (probability) of model accuracy, of which "true" and "false" are the two extremes. Our model predictions will never agree with measurements exactly. We must therefore ask *to what extent* measured data support a specified model. As we have seen in earlier chapters, this is not an absolute measure: even under the true model, the observed data can be very unlikely (e.g. section 5.2). Thus not only are we unable to reject a model in an absolute sense, we cannot even do it probabilistically in any meaningful sense. All we can do is ask which of the available models explains the data best. That is, we must *compare* models. How to do this is the subject of the next chapter.

You may now be wondering why, given all these problems, we ever bother to do frequentist hypothesis testing. As we already saw in section 3.2, Bayesian model comparison is more direct, because it can give you $P(M|D)$. Actually, it can only give probabilities of the models if we can specify the complete set of models. This is often not possible, although we can still calculate the posterior odds ratio for any two models. But what if we only have one model? Bayesian model comparison cannot then tell you anything. This is, in some senses, logical: if we don't believe there is an alternative model, then the data *must* have come from the one and only model. We may nonetheless have reasons[8] to believe this model is suboptimal, even if we don't yet have a concrete proposal for a better alternative. Frequentist hypothesis testing may be able to help here, because a sufficiently small p value could indicate that the data are too unlikely, thereby triggering the search for a new model. Indeed, such tests seem quite natural in some circumstances, as in the example of looking for a signal on top of a noisy background (page 209). Given the pitfalls of p values, and in particular the fact that even data drawn from the true model (in a simulated set up) can result in arbitrarily small p values, it is difficult to know how low the p value should be before we initiate the search for a new model. Yet it does seem that, in practice, scientists are sometimes motivated to search for new models by low p values. Of course, once we have established a new model, we cannot rely on p values to establish its "truth".

The Bayesian approach to model assessment, discussed in more detail in the next chapter, avoids many of the pitfalls of frequentist hypothesis testing, although it is not without its own limitations, as we shall see. Frequentist hypothesis testing is useful – and used – because it is easy to perform, and does not require us to specify alternative models. It can be used to get an indication of the plausibility of an existing sole model for some data, and thereby help decide whether we should search for a better alternative.

---

[8] The reasons may be unrelated to specific model predictions and data, and so transcend statistical considerations.

Model comparison

11.1 Bayesian model comparison

We used Bayesian model comparison in section 3.2 to interpret the positive/negative result of a medical test. In the present chapter we take this approach a step further and use it to compare parametrized models via the marginal likelihood (also called the evidence). I illustrate this using the coin tossing problem (binomial likelihood) and the line fitting problem (Gaussian likelihood). I will make some comparisons with frequentist hypothesis testing, and we will learn why maximum likelihood (and χ^2) may not be used for model selection. I conclude by highlighting some drawbacks of the Bayesian evidence, and will mention some other metrics that can be used to compare models.

Imagine we have a data set D and a model M that we think might describe the data. How good is the model? An interpretation of this question is: how probable is the model in light of the data, $P(M|D)$? Take the example of drawing playing cards from a deck. The chance of picking out the ace of spades (data D) from a normal deck of 52 cards (model M) is $P(D|M) = 1/52$. This information alone tells us nothing about whether the deck of cards is normal – it does not tell us $P(M|D)$ – because if M is true, each and every card has the same probability of being drawn, namely $1/52$. Our information doesn't change fundamentally even if we drew the same card k times (with replacement and shuffling), because every sequence of k cards has the same probability. Yet I am sure you would get suspicious if you drew the ace of spades two or three times in a row. Why? The reason is that you are implicitly thinking of another model – namely an abnormal deck with many aces of spades – under which the data you just got are more likely than under the normal deck model. Without considering the option of such an alternative model you cannot cast doubt on your multiple draws of the ace of spades, because it's as likely as any other sequence under model M. Without an alternative model, the data *must* have come from the one and only model you permit to exist, no matter how unlikely these data are.

We learned back in section 3.2 that in order to find $P(M|D)$ we need to know not only $P(D|M)$ but also $P(D|M')$ and $P(M)$ (see, for example, equation 3.3). This was illustrated in the breast cancer example, which showed that even though $P(D|M)$ can be very large – even equal to $1.0 - P(M|D)$ can still be very small. This is because $P(M|D)$ depends also on the prior probability of the alternative models for the data. If you draw the ace of spades several times and are suspicious, it is not *only* because of the data. It's also because your prior probability for the abnormal deck, $P(M')$, is not extremely small. If it were vanishingly small, or even zero, then no amount of data could convince you the deck was abnormal.

In the breast cancer example we could calculate $P(M|D)$ because we had only two models which were complementary, i.e. $P(M) + P(M') = 1$. In many situations we will have more than two models, and often they will not be an exhaustive set. In that case we cannot determine $P(M|D)$ (as was seen in section 3.2.1). But we saw that we could determine the ratio $P(M_1|D)/P(M_2|D)$ for any two models M_1 and M_2 to get the *posterior odds ratio*

$$R = \frac{P(D|M_1)P(M_1)}{P(D|M_2)P(M_2)}. \tag{11.1}$$

The term $P(D|M)$ is called the *evidence* for model M. It plays a key role in model comparison. If we can't decide a priori between the two models, then we set $P(M_1) = P(M_2)$. These cancel to leave the ratio of the evidences, which is called the *Bayes factor*

$$BF_{12} = \frac{P(D|M_1)}{P(D|M_2)} \tag{11.2}$$

of model 1 to 2. Note that $BF_{21} = BF_{12}^{-1}$. The usual way of doing Bayesian model comparison in practice is with Bayes factors, rather than determining the posterior model probabilities.

How much the odds ratio or Bayes factor must differ from unity before you call one of the models significantly better than the other is a personal choice. The statistics only does the calculations; the decisions you have to make yourself (but see section 3.4). Typically one would not make any claim about one model over another until the factor exceeds 10 or is less than 0.1. Anything within this range is normally considered as indistinguishable. Conservatively one only starts to make claims of real significance if the factor (or its inverse) is more extreme, say 50 or even more.[1]

There are two key aspects to Bayesian model assessment. First, it considers the probability of the data *conditional* on the model, $P(D|M)$. Since we've now obtained the data D, their probability per se is not relevant. The issue is rather which model is more likely to have produced them. The second key aspect is that it uses the *ratio* of probabilities (the posterior odds ratio or the Bayes factor). As we have seen in the deck of cards example, the individual, absolute values of the evidence are irrelevant. That's because the probability of any particular data set is usually very small, even under the true model (see section 5.2). This aspect makes Bayesian model assessment fundamentally different from frequentist hypothesis testing. As explained in section 10.1, because the frequentist approach only looks at the probability of one hypothesis, and because the probability of any particular data set is vanishingly small (zero, in fact, for real numbers), that approach must resort to the problematic concept of "more extreme data". The ratio of probabilities (or probability densities), in contrast, is well defined without having to resort to imaginary data.

---

[1] If you are not interested in the models themselves, but only in predictions from them, then you can avoid choosing a model by using instead a posterior-weighted combination of models.

11.1.1 The evidence as a marginal likelihood

The model M may contain one or more parameters θ. We have seen how, using the likelihood and prior, we can infer the posterior $P(\theta|D, M)$ for these parameters using Bayes equation,

$$P(\theta|D, M) = \frac{P(D|\theta, M)P(\theta|M)}{P(D|M)}. \tag{11.3}$$

The denominator, which we previously treated as just a normalization constant for the posterior, is the evidence. It follows from the rules of probability that

$$P(D|M) = \int \underbrace{P(D|\theta, M)}_{\text{likelihood}} \underbrace{P(\theta|M)}_{\text{prior}} d\theta. \tag{11.4}$$

The evidence is the integral of the likelihood over the prior, and for this reason is often called the *marginal likelihood*. It tells us how probable the data are under the model, independent of specific values of θ. Consider two models for a continuous-valued data set: one is a quadratic, the other is a sinusoid. The task of model comparison is to decide which model explains the data best, but without reference to a specific set of fitted parameters, because a specific set is not representative of the model as a whole. In particular, we should not first fit the models parameters by, say, maximum likelihood, and then compare the resulting fitted models. This is because no matter how complex or large the data set, I can always find a sinusoidal model that will fit every point exactly (I just need to make the frequency large enough). So it will always have the higher likelihood. More generally, by comparing maximum likelihood solutions we invariably favour the more flexible (more complex) model. We should instead find the model that gives the largest *average* likelihood, where the averaging is done over the prior. This is the evidence in equation 11.4. It can be thought of as an average of the predictive ability (likelihood) of the model over the plausible values of the model parameters (prior).

Note that when computing the evidence we must use a normalized likelihood, because its normalization constant is, in general, a function of the parameters θ. The prior must be normalized too, because the evidence is an integral over it. If it were not normalized then it would involve an unknown multiplicative constant a, i.e. $P(\theta|M) = aP^*(\theta|M)$, so the evidence would also involve this unknown constant. This would also not cancel in the Bayes factor, because the other model would have a different prior.

11.2 Example of an analytic evidence calculation: is a coin fair?

In section 5.1 we investigated whether a coin is fair ($p = 1/2$) or not ($p \neq 1/2$) by inferring the posterior PDF $P(p|n, r, \alpha, \beta)$, where n is the number of coin tosses, r the number of heads, and α and β are the parameters of the beta prior. A distribution peaked near $1/2$ suggests the coin is fair. In contrast, a PDF with essentially zero probability around this (e.g. the $r = 18$ panel of figure 5.2) would suggest the coin is not fair. Yet even armed with the posterior, the probability of p being *exactly* $1/2$ (or indeed any other number) is zero (see equation 10.1), so we cannot use the posterior PDF alone to say something like "the probability of this coin being fair is x". This is a bit of a false dichotomy though, as we are unlikely to be interested in whether a coin has exactly $p = 1/2$. Design biases aside, the reality of manufacturing means that no two coins will be *identical*. In practice we would probably be more interested in asking whether p lies within some narrow region, say 0.495–0.505, which for all practical purposes is essentially the same as $p = 0.5$. We could then integrate the normalized posterior PDF over this range to give x and quote this as the probability that the coin is more or less fair. The problem here is that we have made an additional assumption of what "essentially the same" is. Furthermore, for a given posterior, the narrower we choose to make the range, the smaller the probability.

A better way to answer this question is to use the evidence to compare two models. Here I will consider the following two models:

M_1: $p = \phi$, which corresponds to a prior $P(p|M_1) = \delta(p - \phi)$;
M_2: p is unknown, described by a beta distribution prior.
The likelihood is the binomial distribution

$$P_{\text{bin}}(r|p, n) = \frac{n!}{r!(n - r)!} p^r (1 - p)^{n-r}. \tag{11.5}$$

The prior for M_2 is

$$P(p|M_2) = \frac{1}{B(\alpha, \beta)} p^{\alpha-1}(1 - p)^{\beta-1} \quad \text{for} \quad \alpha > 0, \ \beta > 0, \tag{11.6}$$

where

$$B(\alpha, \beta) = \int_0^1 p^{\alpha-1}(1 - p)^{\beta-1} \, dp \tag{11.7}$$

is the beta function which, when α and β are integers, is

$$B(\alpha, \beta) = \frac{(\alpha - 1)!(\beta - 1)!}{(\alpha + \beta - 1)!}. \tag{11.8}$$

The evidence (equation 11.4) for either one of the models M is

$$P(D|M) = P(r|n, M) = \int_0^1 P_{\text{bin}}(r|p, n)P(p|M) \, dp. \tag{11.9}$$

For M_1 the prior is a delta function, so the evidence is simply

$$P(D|M_1) = \frac{n!}{r!(n - r)!} \phi^r (1 - \phi)^{n-r} = P_{\text{bin}}(r|\phi, n), \tag{11.10}$$

i.e. a binomial distribution with $p = \phi$. The evidence for M_2 is

$$\begin{aligned} P(D|M_2) &= \frac{n!}{r!(n - r)!} \frac{1}{B(\alpha, \beta)} \int_0^1 p^{r+\alpha-1}(1 - p)^{n-r+\beta-1} \, dp \\ &= \frac{n!}{r!(n - r)!} \frac{B(r + \alpha, n - r + \beta)}{B(\alpha, \beta)} \end{aligned} \tag{11.11}$$

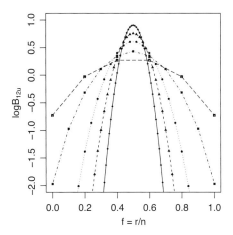

Fig. 11.1 The logarithm (base 10) of the Bayes factor for two models of the coin as a function of the ratio of heads r to total coin tosses n. M_1 assumes $p = 0.5$. M_{2u} assigns a uniform prior to p. The fives lines correspond to $n = 5, 10, 25, 50$, and 100 in order of increasingly narrow distribution.

which follows from the definition of the beta function (equation 11.7).[2] In both cases the evidence is the predictive distribution of the data – the number of heads, r – given the number of tosses and the model (which also encapsulates the corresponding prior). The Bayes factor for these two models is

$$B_{12} \equiv \frac{P(D|M_1)}{P(D|M_2)} = \frac{B(\alpha, \beta)}{B(r + \alpha, n - r + \beta)} \phi^r (1 - \phi)^{n-r}. \tag{11.12}$$

An interesting case is $\alpha = \beta = 1$, the uniform prior, which I'll call M_{2u}. In that case (using equation 11.8)

$$P(D|M_{2u}) = \frac{n!}{r!(n-r)!} \frac{r!(n-r)!}{(n+1)!} = \frac{1}{n+1}, \tag{11.13}$$

$$B_{12u} = (n+1)P_{\text{bin}}(r|\phi, n). \tag{11.14}$$

We may be surprised that the evidence for model M_{2u} is $(n+1)^{-1}$: it does not depend on r at all. It just decreases with an increasing number of coin tosses, regardless of their result. This actually makes sense. The more data we collect, the better we are able to determine the value of p. So the evidence for a model that has *no* preference for any p must decrease as we gather more data. And because it has no preference, it does not actually matter what those data are. This does not imply that the Bayes factor will always favour M_1 as we get more and more data. The evidence for M_1, and therefore the Bayes factor (B_{12u}), does depend on r.

Figure 11.1 plots B_{12u} as a function of the fraction of heads tossed $f = r/n$ for the

---

[2] This integral of the product of the beta and the binomial over all p gives a PDF in r, and is called the beta-binomial distribution.

case that M_1 represents a fair coin, $\phi = 1/2$. The plot shows this for various sample sizes n. The curves are of course symmetric about $f = 1/2$, because getting r heads has the same probability as getting $n - r$ heads when $p = 1/2$. For $n = 5$ (the broadest curve, with the six values $r = 0 \ldots 5$) the fair coin is only favoured when we get two ($f = 2/5$) or three ($f = 3/5$) heads, and then only marginally: the Bayes factor is just 1.9 in both cases. But even getting five or no heads does not weigh strongly against the coin being fair, as the Bayes factor is 0.19, i.e. M_{2u} is only favoured by a factor of about five. Increasing the amount of data (number of tosses) generally yields a more discriminative Bayes factor, which we see as a narrowing of the curve for increasing n. For example with $n = 50$, 30 heads ($f = 0.6$) gives a Bayes factor of 2.1, but 40 heads ($f = 0.8$) gives a Bayes factor of $1/2150$, which hugely favours M_{2u}. If we instead fix f to 0.4 and then increase the sample size n such that $r/n = 4/10$, $10/25$, $20/50$, and $40/100$, we get Bayes factors of 2.26, 2.53, 2.13, and 1.10 respectively, none of which are extreme enough to discriminate between the two models.[3] If we made $n = 100$ tosses we would need to get as many as 65 heads (or likewise as few as 35 heads) before the Bayes factor drops below 0.1 so that we can claim significant evidence in favour of M_{2u} over M_1. But even with exactly 50 heads in 100 tosses the Bayes factor is only 8.0, so 100 tosses is not enough to ever be confident that the coin is fair (with M_{2u} as the alternative). At least 156 tosses with exactly half heads is required before the Bayes factor exceeds 10.

Following this analysis you might be surprised at how extreme the data have to be in order to favour a model, and you may conclude from this that the evidence approach is conservative. But what this analysis really shows is that the M_{2u} prior is conservative: we are permitting all possible values of p to be equally likely a priori. That's quite an extreme state of ignorance (for a coin picked up at random), so you need a lot of data to overcome it. Values of r/n nearer to 1 or 0 give stronger evidence against M_1 and in favour of M_{2u} (and the Bayes factor varies more rapidly), because the further the data are from $r/n = 0.5$, the less plausible M_1 becomes.

The following code makes figure 11.1. If you want to investigate how the Bayes factor varies when we adopt a more conservative beta prior for model M_2, replace the line defining BF with equation 11.12 and insert your chosen values of α and β for the prior.

R file: `coin_evidence.R`

```
##### Calculate the evidence for the coin problem

phi <- 1/2
n    <- c(5,10,25,50,100)
pdf("coin_evidence.pdf", 4, 4)
par(mfrow=c(1,1), mgp=c(2.0,0.8,0), mar=c(3.5,3.5,1,1), oma=0.1*c(1,1,1,1))
plot(c(0,1), c(0,1), type="n", xlim=c(0,1), ylim=c(-2,1), xlab="f = r/n",
     ylab=expression(paste(log, B[12][u])))
for(i in 1:length(n)) {
  r   <- 0:n[i]
```

---

[3] The Bayes factor does not vary monotonically with sample size in this case: 4 out of 10 is (slightly) more discriminative than 40 out of 100 for these particular models (although if we compute further we find that 400 out of 1000 is hugely discriminative, with $B_{12u} = 4.6 \times 10^{-8}$). For more extreme results, e.g. $f = r/n = 0$, we do get a monotonic variation of the Bayes factor with n, as can be seen in figure 11.1.

```
f   <- r/n[i]
BF <- (n[i]+1)*dbinom(x=r, size=n[i], prob=phi)
points(f, log10(BF), cex=0.5, pch=13+i)
lines(f, log10(BF), lty=6-i)
}
dev.off()
```

11.3 Example of a numerical evidence calculation: is there evidence for a non-zero gradient?

In the previous section we were able to compute the evidences and therefore the Bayes factor analytically. Here we turn to a problem where we must instead use numerical (Monte Carlo) sampling.

Given a two-dimensional set of points $\{x_i, y_i\}$ with noise only on the y values (the standard deviation of which may or may not be known), we ask whether there is evidence for a linear correlation. We can frame this by asking "is there a significant probability for a non-zero gradient of a fitted line?". We could calculate the posterior probability density over the gradient, as we did in section 9.1. But the probability of any *exact* gradient is zero, and integrating over some narrow range of probability is also problematic (as discussed in section 11.2). A better approach is to ask "what is the Bayes factor $B_{12} = P(D|M_1)/P(D|M_2)$, where M_1 is a model with a zero gradient and M_2 is a model with an unknown gradient?". We would have to calculate Bayes factors if we were comparing models that did not share common parameters, such as a polynomial and a sinusoid. The case we are looking at here is actually one of *nested models*, because M_1 is a special case of M_2. In such cases the Bayes factor can be approximated using something called the Savage–Dickey density ratio (e.g. Trotta, 2007). But I will instead compute the marginal likelihood for both models as a means of illustrating the general approach.

As in section 9.1 the generative model is $f(x)$ and we assume an independent Gaussian likelihood for each data point y_i. Writing the prior as $P(\theta|M)$, the marginal likelihood (equation 11.4) is then

$$P(D|M) = \int \prod_i \frac{1}{\sigma\sqrt{2\pi}} \exp\left[-\frac{[y_i - f(x_i)]^2}{2\sigma^2}\right] P(\theta|M)\, d\theta \qquad (11.15)$$

where θ includes all the parameters in $f(x)$ as well as σ (which I assume to be the same for all data points). For the models and priors we consider, this integral cannot be performed analytically, so we use a Monte Carlo approximation. As explained in section 8.3 (in particular point 3 at the end of that section), the marginal likelihood can be estimated as

$$P(D|M) \simeq \frac{1}{N_s} \sum_{l=1}^{l=N_s} P(D|\theta_l, M) \qquad (11.16)$$

where the samples $\{\theta_l\}$ have been drawn from the prior. That is, the evidence is the average of the likelihood over a set of samples drawn from the prior.

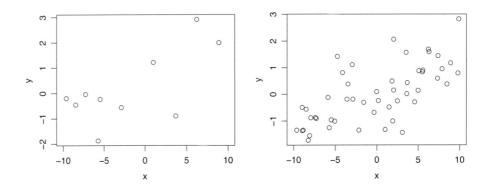

Fig. 11.2 Data drawn from a straight line model $y = b_0 + (\tan \alpha)x$ with $b_0 = 0$, $\tan \alpha = 0.1$, to which zero mean Gaussian noise with standard deviation $\sigma = 1.0$ have been added. Left: 10 data points. Right: 50 data points.

In section 9.1 we used an MCMC method to sample the posterior over the three pa-rameters of a straight line model: the intercept b_0; the gradient b_1 (or rather the angle $\alpha = \arctan(b_1)$); and the uncertainty σ on the y values. To calculate the evidence we must sample from the prior rather than from the posterior. In order to sample from a distribution it must be proper (normalizable).[4] This is the case for the priors we used on b_0 and α, because they were a Gaussian and a truncated uniform distribution respectively. But our prior for σ was improper; it was the Jeffreys prior with $P(\sigma) \propto 1/\sigma$. To make this a proper distribution I here set it to zero outside some finite range; see equation 5.21.

In contrast to the posterior, the prior is generally a simple function of the model parame-ters, and so is much easier to draw from. In the present case we have uniform distributions for α and $\log \sigma$, and Gaussian for b_0. Standard methods exist for sampling these; we do not need to use MCMC.

Let us use the evidence to compare two models:
M_2 is a general straight line, $y = b_0 + x \tan \alpha + \epsilon$, where b_0 and α are unknown parameters, and $\epsilon \sim \mathcal{N}(0, \sigma)$, which is the Gaussian random noise used to define the likelihood;
M_1 is the same as M_2, but with zero gradient, $\tan \alpha = 0$.

For the purpose of this demonstration I generate data at random. Ten x values are drawn from $\mathcal{U}(-10, 10)$, and the y value at each is computed from a straight line model with $b_0 = 0$ and $\tan \alpha = 0.1$, to which zero mean Gaussian noise with $\sigma = 1$ is added. M_2 is in principle a better description than M_1 for these data, but which model is favoured depends on the actual data; the noise is large so the data could, just by chance, favour M_1. The data are shown in the left panel of figure 11.2. Given only these data (but neither the model nor the noise level) I don't think it's at all obvious that a sloping line fits better than a horizontal one.[5]

---

[4] When we sampled the posterior in section 9.1 we only needed to compute the prior, not sample from it, so the fact that it was improper was not a problem. The likelihood ensured that the posterior was a proper distribution.
[5] If I was then told that the noise level was 5, which is larger than the standard deviation in the data, I would go for the horizontal line on the basis that it's simpler.

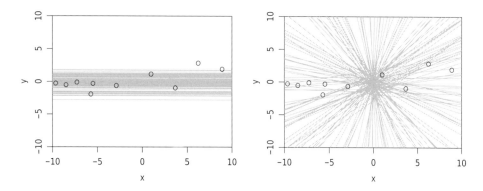

Fig. 11.3 Draws from the model priors shown as functions. Left: draws from M_1, which has $b_0 \sim \mathcal{N}(0,1)$ and $\alpha = 0$. Right: draws from M_2, which has $b_0 \sim \mathcal{N}(0,1)$ and $\alpha \sim \mathcal{U}(0, 2\pi)$. The same scales have been used on both axes and in both panels. The data (open circles) in both panels are those in the left panel of figure 11.2.

The R code to generate the data, sample the prior, calculate the evidence, and produce the plots is given in the following subsection. My parameter priors are

$$P(b_0) = \mathcal{N}(0,1) \tag{11.17}$$
$$P(\alpha) = \mathcal{U}(0, 2\pi) \tag{11.18}$$
$$P(\log \sigma) = \mathcal{U}(\log 0.5, \log 2). \tag{11.19}$$

By sampling the prior we are sampling over the prior set of possible functions, $b_0 + x \tan \alpha$, as well as the noise. It is instructive to plot what functions drawn from these priors look like. The right panel of figure 11.3 shows a random set of 100 functions drawn from the prior for M_2 (the values of $\log \sigma$ are not shown). I have plotted with equal axis scales so we can see the isotropy of the prior. Draws from the prior for M_1 are shown in the left panel.

Having drawn a large number (10^5) of parameter sets from the priors for the two models, we calculate the likelihood for each set using the data, and average these to estimate the evidence for that model (equation 11.16). The results are

$$\log P(D|M_1) = -8.33 \tag{11.20}$$
$$\log P(D|M_2) = -8.44 \tag{11.21}$$
$$\log B_{12} = 0.11 \tag{11.22}$$
$$B_{12} = 1.30. \tag{11.23}$$

This tells us that the data are 1.3 times more likely to come from M_1 than M_2. Assuming we have no prior reason to favour one model of the other, we adopt equal model priors $P(M_1) = P(M_2)$, in which case this result means we favour M_1 over M_2 by a factor of 1.3. This is extremely marginal: normally we want the Bayes factor to be larger than 10 (or less than 0.1) in order to claim significant support for one model over the other. Repeating this calculation with a different set of data can give quite different results however, because the true model has a gradient quite close to zero, and the noise is large: five other random sets of ten points give $B_{12} = 1.55, 8.29, 6.73, 0.88$, and 0.37. None of these would allow us to draw a significant conclusion, however.

This result is, of course, dependent on the choice of priors. In fact, the priors should be viewed as an intrinsic part of the model, because two very different choices creates two models with different capabilities. For example, by putting a very narrow prior on α around 0 for model M_2, we essentially end up with M_1. But these extremes aside, it is important to investigate how sensitive our results are to the choice of prior. Let's do this for the data in the left panel of figure 11.2. We first change the standard deviation of the prior on the intercept b_0 in the same way for both M_1 and M_2. The resulting Bayes factor is quite insensitive to this change. For example, decreasing the standard deviation by a factor of ten to 0.1 gives $B_{12} = 2.14$, and increasing it to 10 gives $B_{12} = 1.36$, both of which are insignificant. Other data sets could lead to different conclusions about the prior sensitivity, of course. Allowing σ to have much larger values, by increasing the upper limit on its prior, also has negligible effect on the Bayes factor. But forcing σ to be small does have a significant impact: if the lower and upper limits on $\log \sigma$ are set to 0.1 and 0.5 respectively, we get $B_{12} = 1.6 \times 10^{-6}$. This is because by forcing there to be little noise in the fit, a non-zero gradient can fit the data much better (if still poorly in an absolute sense).

With more data we should be able to distinguish better between competing models (unless neither model has anything to do with the data). Returning to the default values of the priors, I recalculate the evidence and Bayes factors but now using a particular set of 50 randomly drawn data points (shown in the right panel of figure 11.2). The results are

$$\log P(D|M_1) = -33.87 \tag{11.24}$$

$$\log P(D|M_2) = -29.37 \tag{11.25}$$

$$\log B_{12} = -4.50 \tag{11.26}$$

$$B_{12} = 3.15 \times 10^{-5}. \tag{11.27}$$

Now M_2 is favoured over M_1 – which is what we would expect – by a factor of about 31 000. Five other random sets of 50 points give $\log B_{12} = -1.66, -5.95, -3.56, -0.27$, and -2.11. The model I am generating the data from is very noisy, so the results are still highly variable. But we nonetheless see that the more data we have, the more stable the result is in the sense that we now tend to favour M_2 by a significant amount (except in one case here).[6] Remember that the Bayes factor (and Bayesian analysis in general) gives a result about the data we actually have, and not about some other data we might have got had we run the experiment again. The fact that there is variance in the data is accommodated by the likelihood. This variance (uncertainty) is incorporated into the width of the posterior, or into getting non-extreme Bayes factors. So while Bayes does not need to consider alternative data sets, it does not ignore the variance in the data.

In this example I have assumed that while the standard deviation of the noise σ is unknown, it is the same for all data points. We could relax this assumption and specify a separate noise parameter σ_i for each data point i. The approach remains the same: we specify a prior for each noise parameter (we could use the same prior for all) and marginal-

---

[6] I refrain from referring to M_2 as the "true" model because M_2 has an unspecified gradient, whereas the data are drawn from a model with $\tan \alpha = 0.1$.

ize over this larger set of parameters. This gives more parameters than data points, but we saw in section 9.1 that this presents no theoretical difficulties.

11.3.1 R code

The following R script produces everything discussed in the previous section. Most of it is concerned with producing the data and prior plots; the calculations themselves only take up a few lines. The code should be self-explanatory. You should experiment by changing the sampled data (change the seed), changing the amount of data, and changing their standard deviation. Try also changing the priors. Think carefully about what you are doing, and only change one thing at a time. Remember that the model uses only the measured, noisy data: the true model is actually irrelevant.

Be warned that this code can return log evidences of minus infinity! This occurs when all the likelihoods for a model are so small that they are numerically identical to zero, on account of the finite precision of a computer (see section 6.3.3). The evidence is then also zero. This will occur if the data are utterly implausible under one model or the other. Of course, if both models give numerically zero evidence (possible if the priors are chosen poorly for the overall properties of the data), then we still won't know which is better.

R file: `linearmodel_evidence.R`

```
##### Calculate the Bayesian evidence for two linear models of 2D data.
##### M1: alpha=0, M2: P(alpha) ~ 1.

library(gplots) # for plotCI

### Define likelihood

# Return log10(likelihood), a scalar.
# theta is the vector of model parameters, here c(b_0, alpha, log10(ysig)).
# data is the two-column matrix [x,y].
# dnorm(..., log=TRUE) returns log base e, so multiply by 1/ln(10) = 0.434
# to get log base 10
log.like <- function(theta, data) {
  # convert alpha to b_1 and log10(ysig) to ysig
  theta[2] <- tan(theta[2])
  theta[3] <- 10^theta[3]
  # likelihood
  modPred <- drop( theta[1:2] %*% t(cbind(1,data$x)) )
  # Dimensions in mixed vector/matrix products: [Ndat] = [P] %*% [P x Ndat]
  logLike <- (1/log(10))*sum( dnorm(modPred - data$y, mean=0,
                              sd=theta[3], log=TRUE) )
  return(logLike)
}

### Define true model and simulate experimental data from it

set.seed(75) # 75 gives data and plots in script
Ndat <- 10
x <- sort(runif(Ndat, -10, 10))
sigTrue <- 1
modMat <- c(0,0.1) # 1 x P vector: coefficients, b_p, of sum_{p=0} b_p*x^p
```

```
y <- cbind(1,x) %*% as.matrix(modMat) + rnorm(Ndat, 0, sigTrue) # noisy data
# Dimensions in matrix multiplication:
# [Ndat x 1] = [Ndat x P] %*% [P x 1] + [Ndat]
# cbind does logical thing combining scalar and vector; then vector addition
y <- drop(y) # converts into a vector
pdf("linearmodel_evidence_data_10.pdf", width=5, height=4)
par(mfrow=c(1,1), mar=c(3.5,3.0,0.5,0.5), oma=0.5*c(1,1,1,1),
    mgp=c(2.2,0.8,0), cex=1.2)
plot(x, y, xlim=c(-10,10))
#plotCI(x, y, xlim=c(-10, 10), uiw=sigTrue, gap=0) # data and true error bar
#abline(a=modMat[1], b=modMat[2], col="red") # true model
dev.off()

### Sample from prior

# Sample from prior.
# priorSamp is an array with dimensions (Nsamp, 3) containing the
# samples for b_0, alpha, log10(sigma)
set.seed(100)
Nsamp <- 1e5 # will need to be larger if the priors are broader
priorSamp <- cbind(rnorm(n=Nsamp, mean=0, sd=1),
                   runif(n=Nsamp, min=-pi/2, max=pi/2),
                   runif(n=Nsamp, min=log10(0.5), max=log10(2)))
sel <- sample.int(n=Nsamp, size=100) # 100 of the prior samples

# Plot data and overplot 100 prior models from M2
pdf("linearmodel_evidence_prior_models_M2.pdf", width=5, height=4)
par(mfrow=c(1,1), mar=c(3.5,3.0,0.5,0.5), oma=0.5*c(1,1,1,1),
    mgp=c(2.2,0.8,0), cex=1.2)
plot(x, y, type="n", xlim=c(-10,10), ylim=c(-10,10), xaxs="i", yaxs="i")
for(j in sel) {
  abline(a=priorSamp[j,1], b=tan(priorSamp[j,2]), col="grey")
}
points(x, y)
dev.off()

# Plot data and overplot prior models from M1
pdf("linearmodel_evidence_prior_models_M1.pdf", width=5, height=4)
par(mfrow=c(1,1), mar=c(3.5,3.0,0.5,0.5), oma=0.5*c(1,1,1,1),
    mgp=c(2.2,0.8,0), cex=1.2)
plot(x, y, type="n", xlim=c(-10,10), ylim=c(-10,10), xaxs="i", yaxs="i")
for(j in sel) {
  abline(a=priorSamp[j,1], b=0, col="grey")
}
points(x, y)
dev.off()

### Calculate likelihoods, evidences, and BF

data <- data.frame(cbind(x,y))

# Calculate likelihoods and evidence for model M2
logLikeM2 <- rep(NA, Nsamp)
for(j in 1:Nsamp) {
  logLikeM2[j] <- log.like(theta=priorSamp[j,], data)
}
```

```
logEvM2 <- log10(mean(10^logLikeM2))
# Calculate likelihoods and evidence for model M1
priorM1   <- cbind(priorSamp[,1], rep(0, Nsamp), priorSamp[,3])
logLikeM1 <- rep(NA, Nsamp)
for(j in 1:Nsamp) {
  logLikeM1[j] <- log.like(theta=priorM1[j,], data)
}
logEvM1 <- log10(mean(10^logLikeM1))
# Print results
cat("log10(Ev_1) = ", logEvM1, "\n")
cat("log10(Ev_2) = ", logEvM2, "\n")
cat("log10(BF_12) = ", logEvM1 - logEvM2, "\n")
cat("BF_12       = ", 10^(logEvM1 - logEvM2), "\n")
```

11.3.2 A frequentist hypothesis testing approach

We saw in section 10.4 how we can use the t test to perform a frequentist analysis of the significance of a non-zero gradient for the straight line. Applying this to the data generated in the previous section we get the following.

```
summary(lm(y ~ x, data=data))
  Coefficients:
            Estimate Std. Error t value Pr(>|t|)
  (Intercept) 0.48161    0.37076  1.299   0.2301
  x           0.14934    0.05713  2.614   0.0309 *
  ---
  Signif. codes:  0 '***' 0.001 '**' 0.01 '*' 0.05 '.' 0.1 ' ' 1

  Residual standard error: 1.117 on 8 degrees of freedom
  Multiple R-squared:   0.4607, Adjusted R-squared:  0.3932
  F-statistic: 6.833 on 1 and 8 DF,  p-value: 0.03094
```

The p value is 0.03, so we can reject the hypothesis that the gradient is zero at the 95% confidence level (but not at the 99% level), which suggests that we might need the gradient. This is quite an extreme conclusion compared to the Bayesian analysis that gave $B_{12} = 1.30$, i.e. we could not draw any conclusion. But remember that this t test reaches its conclusion without testing M_2.

The frequentist approach is quite easy, but this should not lure us into relying on it too much.[7] As was discussed in section 10.6 this t test does not tell us which model is better, because it does not explicitly test and compare the models. It only tells us how often we expect to get a gradient this far from zero among a large number of data sets drawn from the null hypothesis which, remember, involves fixing the intercept.

Of course, if we really believed our null hypothesis, then our best fit would be a horizontal line through the mean of the data, which is not equal to the intercept found by fitting a straight line. Thus we should really do a linear regression on the data after we have centred them, i.e. offset them to have zero mean in x and y. We then keep the intercept fixed to

---

[7] Remember that behind lm and the t test there is at least as much computer code as there is in the evidence calculation code (section 11.3.1). It is true, however, that the linear least squares fit and t test involve much less computation, because the evidence calculation requires a large number of likelihood computations.

zero and do a t test on the resulting gradient. If we do this on the data above, the result is as follows.

```
xc <- data$x - mean(data$x)
yc <- data$y - mean(data$y)
summary(lm(yc ~ xc - 1))
   Coefficients:
       Estimate Std. Error t value Pr(>|t|)
   xc  0.14934    0.05386   2.773   0.0217 *
   ---
   Signif. codes:  0 '***' 0.001 '**' 0.01 '*' 0.05 '.' 0.1 ' ' 1

   Residual standard error: 1.053 on 9 degrees of freedom
   Multiple R-squared:  0.4607,Adjusted R-squared:  0.4007
   F-statistic: 7.687 on 1 and 9 DF,  p-value: 0.02166
```

The p value is now smaller due to the extra degree of freedom. If we had not kept the gradient fixed in this example, the p value would have been the same as before, because translating the data in x and/or y does not change the best fitting straight line.

11.4 Comparing Gaussians (or other distributions)

We can use the approach of section 11.3 for essentially any model comparison problem. We saw in section 10.2 how we could use frequentist hypothesis testing to determine whether data drawn from a Gaussian distribution had a mean significantly different from a specified value μ. We did this both when the standard deviation was known (the z test) and unknown (t test). To do this with Bayesian model comparison we must define at least two models. The first, M_1, specifies that the data are drawn from a model with known mean μ. The second, M_2, specifies that the data are drawn from a model with unknown mean μ', which is assigned a prior distribution $P(\mu')$. By definition of the problem, the likelihood for one data point (in both cases) is a Gaussian with mean specified by the model, and standard deviation σ that is either known, or is unknown, in which case it must be assigned a prior $P(\sigma)$. The likelihood for N data points $D = \{x\}$ is

$$P(D|\theta, M) = \prod_i \frac{1}{\sigma\sqrt{2\pi}} \exp\left[-\frac{(x_i - m)^2}{2\sigma^2}\right] \tag{11.28}$$

where for M_1 we have $m = \mu$ and parameters $\theta = \sigma$, and for M_2 we have $m = \mu'$ and parameters $\theta = (\sigma, \mu')$. If instead σ were known then M_1 would have no free parameters and M_2 just one. In the general case of σ being unknown, the evidence for the two models is

$$P(D|M_1) = \int P(D|\sigma, M_1)P(\sigma) \, d\sigma \tag{11.29}$$

$$P(D|M_2) = \iint P(D|\sigma, \mu', M_2)P(\sigma)P(\mu') \, d\sigma \, d\mu'. \tag{11.30}$$

In general we will have to do these integrals numerically, as we did in section 11.3. If we adopt a Gaussian prior for μ' then the integral over μ' for M_2 will be analytic. If σ is known then this is equivalent to $P(\sigma)$ being a delta function, in which case we can just drop the integral over σ.

These expressions for the evidence are not specific to a Gaussian likelihood, so this method will work with other distributions too (just the meaning and the number of parameters may change). We can also generalize this approach in order to compare two samples of data, rather than one sample to a fixed mean. The procedure is always to define the model with its parameters and priors, and then to define the likelihood. Writing down the evidence, we will usually then have to solve the integrals numerically (by drawing from the priors, or by using one of the other methods described in section 11.6). A practical advantage of this over frequentist hypothesis testing is that it is a unified approach: we do not need to go searching for appropriate test statistics and their distributions.

11.5 How the evidence accounts for model complexity

The evidence tells us how well the data are predicted by the model. Suppose we have some data drawn from a process and we want to compare a simple model M_s with a more complex one M_c. The data might be the (x, y) data set like that considered above, in which case M_s might just be a horizontal line and M_c a fourth-order polynomial. Consider first models with fixed parameters. We can obviously imagine some specific data sets, in particular one drawn from the fourth-order polynomial, for which M_c will predict the data much better than M_s ever can, and so will have a much higher likelihood. But for many more data sets, M_c makes such a specific prediction that it will match the data poorly. M_s, in contrast, will explain few of the data sets really well, but it will never do very badly either. The reason for this is that all models (whether their parameters are fixed or not) have a fixed "budget" of predictive power to explain all data sets. This is indicated by the fact that the model evidence $P(D|M)$ is a *normalized* probability density function of the data D. Improved predictive power (a larger probability density) over some part of the data space must be compensated by a lower one elsewhere, in order for $P(D|M)$ to integrate to one. This is illustrated in figure 11.4.

Usually the parameters θ of a model are not fixed, and can be fitted to the data. In this case a more complex model is more flexible and so will generally fit a wider range of data sets better than a simpler model, in the sense of producing a larger likelihood. So when comparing our two models, perhaps we should first fit each model by finding the parameters that maximize its likelihood $P(D|\theta, M)$ and see which model gives the higher value of this. This is wrong, because it will invariably favour the more complex model simply because it is more flexible. Consider a set of ten data points in a two-dimensional plane. A fifth-order polynomial will fit these data better (give a larger maximum likelihood) than a quadratic. And a ninth-order polynomial will fit them perfectly, noise included! But

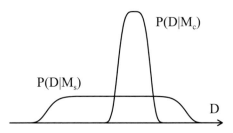

Fig. 11.4 Schematic illustration of the evidence for a simpler (M_s) and a more complex (M_c) model. The distributions show the probability of the data under the two models as a function of the measured data D.

this is not a "good" model in any useful sense of the term. Using the maximum likelihood to select a model would drive us to using ever increasingly complex models until we fit the data perfectly. (A model that fits the data "too well" is called an overfit.) Do not use maximum likelihood ratio tests for model selection!

In section 10.5 I explained how χ^2 is sometimes used to determine the goodness of fit of a given model, so we might think that the next step in that process is to find the model which has the smallest χ^2. But this is equally wrong, because χ^2 is just the negative log likelihood when we have a Gaussian error model; minimizing χ^2 is equivalent to maximizing the likelihood.

Maximum likelihood is the wrong thing to do for model comparison, because it ignores the different flexibilities of the models.

A closer analysis shows that the evidence takes into account the flexibility – or complexity – of the model, and so is an appropriate metric for model comparison. Taking the logarithm of equation 11.3 and rearranging we get

$$\log P(D|M) = \log P(D|\theta, M) + [\log P(\theta|M) - \log P(\theta|D, M)] \qquad (11.31)$$
$$\log(\text{evidence}) = \log(\text{likelihood}) + [\log(\text{prior}) - \log(\text{posterior})].$$

A complex model will generally have a posterior that is larger than the prior, and more so than a simple model. This is because (considering linear models) a more complex model has more parameters, and so has to spread its prior probability (which must integrate to unity) over a larger volume of parameter space than a model with fewer parameters. A typical value of its prior probability density will therefore be smaller. Furthermore, the more complex model can fit the data better, so its posterior probability density is typically larger. Both of these effects contribute to the term in square brackets in the above equation being more negative for a complex model than a simple model. This term acts as a penalty which reduces the evidence *for a given likelihood*. Thus for a more complex model to achieve a high evidence, it has to achieve a high enough likelihood in order to overcome this penalty. The evidence can therefore be seen as a combination of the fitting quality and a complexity penalty, and the best model will achieve the best trade-off between these two.

While the evidence involves a complexity penalty, it is not complexity per se that is penalized. What counts is how the plausibility of the model is changed in light of the data.

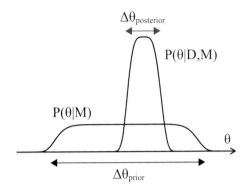

Fig. 11.5 Schematic illustration of the concept of the Occam factor. The prior $P(\theta|M)$ shrinks to the posterior $P(\theta|D,M)$ under the action of obtaining the data D.

This can be understood with the concept of the *Occam factor*. If the likelihood is dominated by a single peak at $\hat{\theta}$ over which the prior is more or less constant, then we can approximate the evidence as

$$P(D|M) = \int P(D|\theta, M) \times P(\theta|M)\ d\theta$$

$$\simeq P(D|\hat{\theta}, M)\Delta\theta_{\text{posterior}} \times \frac{1}{\Delta\theta_{\text{prior}}}$$

$$\simeq \underbrace{L(\hat{\theta})}_{\text{best fit likelihood}} \times \underbrace{\frac{\Delta\theta_{\text{posterior}}}{\Delta\theta_{\text{prior}}}}_{\text{Occam factor}} \qquad (11.32)$$

where $L(\hat{\theta}) = P(D|\hat{\theta}, M)$ is the likelihood at the best fit solution, $\Delta\theta_{\text{prior}}$ is the prior parameter range, and $\Delta\theta_{\text{posterior}}$ is the posterior parameter range (the width of the posterior). This is illustrated in figure 11.5. The Occam factor (which is always less than or equal to one) measures the amount by which the plausible parameter volume shrinks on account of the data. For given $L(\hat{\theta})$, a simple model will fit over a large part of the parameter space, so $\Delta\theta_{\text{posterior}} \simeq \Delta\theta_{\text{prior}}$ and the Occam factor is not significantly less than one. In contrast a more complex model, or one which has to be more finely tuned to fit the data, will have a larger shrinkage, so $\Delta\theta_{\text{posterior}} \ll \Delta\theta_{\text{prior}}$. Now the Occam factor is small and the evidence is reduced. Of course, if the fit is good enough then $L(\hat{\theta})$ will be large, perhaps large enough to dominate the Occam factor and to give the model a large evidence.

The Occam factor is related to the idea of *Occam's razor*, which essentially says that we should chose simpler models where possible. This does not mean that we should always favour the simpler model: some phenomena and data sets are complex, and these will need complex models to describe them well. Bayesian model comparison balances complexity and accuracy via the Occam factor. Depending on the prior adopted – and the prior plays a major role in determining how complex a model is – this approach certainly can identify a more complex model as being best.

The Bayesian evidence naturally accommodates model complexity when doing model comparison. There is no natural parallel to this in frequentist hypothesis testing, where complexity control is either ignored or is imposed in an ad hoc fashion. I stress again that one should not use the ratio of maximum likelihoods to choose between models, because this just favours the more complex model. It is equivalent to compressing the prior parameter range to an arbitrarily small range around the best fit parameters, thereby artificially elevating the Occam factor for that model.

The evidence also encapsulates the concept of bias-variance trade-off mentioned in section 4.8. The bias of a model is the difference between the expected (average) predictions and the true values. The variance is the variability of the predictions around the expected ones. A more complex model is one that can be made to fit the data more precisely, thereby achieving a low bias. But the fit is more sensitive to the exact data: a small change in the data produces a big change in the fitted model. It has a larger variance. A simpler model, in contrast, is less influenced by the data (e.g. no high order terms in the polynomial), so has lower variance. But it cannot be made to fit the data as precisely because it lacks those more flexible terms, resulting a larger bias. Ideally we want a model with low bias and low variance, but in general increasing the model complexity will decrease the bias at the cost of increased variance.

11.6 Other ways of computing the evidence

The direct way of computing the evidence is to sample the prior and to calculate the average likelihood at these samples, as we did in section 11.3. While priors are normally simple and easy to sample from (without MCMC), for complicated problems we may need a very large number of samples to get an accurate measure of the evidence. This will occur if the likelihood (seen as a function of the parameters) is much more peaked than the prior, in which case only a very small fraction of the samples will contribute significantly to the evidence. This problem gets exponentially worse (literally) as the number of parameters increases. We are therefore interested in numerically more stable measures.

There are a number of other ways of numerically approximating the evidence. One is based on the idea of importance sampling (section 8.4.2). Newton & Raftery (1994) show that given a set of samples $\{\theta_l\}$ drawn from the posterior (using MCMC), the evidence can be approximated as

$$P(D|\sigma, M) \simeq \left(\frac{1}{N_s} \sum_{l=1}^{N_s} P(D|\sigma, \theta_l, M)^{-1} \right)^{-1} \tag{11.33}$$

which is the harmonic mean of the likelihood at these samples. However, this method is widely criticized in the literature as inaccurate and unstable, and only converges to the true evidence with an impractically large number of samples in most cases. It is not recommended.

One of the attractions of the harmonic mean method is that is uses samples drawn from the posterior. A single sampling procedure can therefore be used both for estimating parameters and for approximating the evidence. Better methods than the harmonic mean exist which do this. One example is nested sampling (e.g. Skilling, 2004; Sivia & Skilling, 2006). Another is thermodynamic integration (e.g. Gregory, 2005; Lartillot & Philippe, 2006; Friel & Pettitt, 2008). This latter method requires running parallel MCMC chains, each of which involves the likelihood raised to a power between 0 and 1, a process known as parallel tempering. The integral of the mean log likelihood values over the chains gives the log evidence.

If the likelihood, when viewed as a function of the parameters $\boldsymbol{\theta}$, is dominated by a single narrow peak, then we may be able to approximate it as a multivariate Gaussian about its maximum $\hat{\boldsymbol{\theta}}$. In that case we can use the quadratic approximation of section 7.1 to write the likelihood as

$$P(D|\boldsymbol{\theta}, M) \simeq P(D|\hat{\boldsymbol{\theta}}, M) \exp\left(-\frac{1}{2}(\boldsymbol{\theta} - \hat{\boldsymbol{\theta}})^{\mathsf{T}} \Sigma_{\hat{\theta}}^{-1} (\boldsymbol{\theta} - \hat{\boldsymbol{\theta}})\right). \tag{11.34}$$

The maximum we can find by numerical optimization (see section 12.6). The covariance matrix $\Sigma_{\hat{\theta}}$ is equal to the negative inverse of the matrix of second derivatives of the log likelihood evaluated at $\hat{\boldsymbol{\theta}}$ (equation 7.15). The dependence on the data comes through this and $\hat{\boldsymbol{\theta}}$. Assuming the prior is reasonably constant across this narrow peak, we can approximate it as $P(\hat{\boldsymbol{\theta}}|M)$. The integration of the likelihood over this prior (equation 11.4) then just involves integrating the exponential term in equation 11.34 over $\boldsymbol{\theta}$. This is the Gaussian integral, the result of which is $(2\pi)^{J/2}|\Sigma_{\hat{\theta}}|$ where J is the dimensionality of the parameter space. The resulting approximation for the evidence is therefore

$$P(D|M) \simeq (2\pi)^{J/2}|\Sigma_{\hat{\theta}}|^{1/2} P(D|\hat{\boldsymbol{\theta}}, M) P(\hat{\boldsymbol{\theta}}|M). \tag{11.35}$$

This will only be a good approximation in the case of highly informative data, such that the approximations mentioned are valid.

Other methods for computing the evidence are discussed in Kass & Raftery (1995). In the next section I will summarize one other method, which I introduced in Bailer-Jones (2012).

11.6.1 The cross-validation likelihood

Definition

Let $D = (D_1, D_2, \ldots, D_N)$ denote a set of N measurements (they could be scalars or vectors). We group the N measurements into K disjoint partitions ($1 < K \leq N$). Denote the data in the kth partition as D_k and its complement as D_{-k}, i.e.

$$D = D_k \cup D_{-k}. \tag{11.36}$$

The principle of the cross-validation (CV) likelihood is to calculate the likelihood of D_k using D_{-k}, without having an additional dependence on a specific choice of model parameters θ. This likelihood is $P(D_k|D_{-k}, M)$. It tells us how well, under model M, some of the data are predicted by the other data. By marginalization

$$L_k \equiv P(D_k \mid D_{-k}, M) = \int P(D_k \mid D_{-k}, \theta, M) P(\theta \mid D_{-k}, M) \, d\theta$$

$$= \int \underbrace{P(D_k \mid \theta, M)}_{\text{likelihood}} \underbrace{P(\theta \mid D_{-k}, M)}_{\text{posterior}} \, d\theta \qquad (11.37)$$

where D_{-k} drops out of the first term because the model predictions are conditionally independent of these data when θ is specified. I call L_k the *partition likelihood*. It is the likelihood for data D_k marginalized over the posterior computed using all the other data D_{-k}. In practice we estimate the partition likelihood using a Monte Carlo approximation of this integral (cf. equation 8.9),

$$L_k \simeq \frac{1}{N_s} \sum_{l=1}^{N_s} P(D_k \mid \theta_l, M), \qquad (11.38)$$

where the N_s samples $\{\theta_l\}$ are drawn from the posterior $P(\theta \mid D_{-k}, M)$.

Combining the partition likelihoods over all K partitions in some way should give an overall measure of the fit of the model. As L_k is a product of the likelihoods for each data vector within a partition, it scales multiplicatively with the number of data vectors in partition k. This suggests that an appropriate combination is

$$L_{\text{CV}} = \prod_{k=1}^{K} L_k \qquad (11.39)$$

which I call the *K-fold CV likelihood*, for $1 < K \leq N$. For $K < N$ its value will depend on the choice of partitions. If $K = N$ there is one data vector per partition, a unique choice. I call this the *leave-one-out CV likelihood*.[8]

The posterior PDF in equation 11.37 is given by Bayes' theorem. As we are sampling, it is sufficient to use the unnormalized posterior, which is

$$P(\theta \mid D_{-k}, M) \propto P(D_{-k} \mid \theta, M) P(\theta \mid M). \qquad (11.40)$$

Relation to the evidence

Whereas the evidence involves integrating the likelihood (for D) over the prior (equation 11.4), the partition likelihood involves integrating the likelihood (for D_k) over the posterior (for D_{-k}) (equation 11.37). This is like using D_{-k} to build a new prior from "previous" data. We can use the product rule to write the partition likelihood as

$$L_k \equiv P(D_k \mid D_{-k}, M) = \frac{P(D \mid M)}{P(D_{-k} \mid M)}, \qquad (11.41)$$

i.e. the ratio of the evidence calculated over all the data to the evidence calculated on the subset of the data used in the posterior sampling.

---

[8] In principle we could also compute the K-fold CV likelihood with $K = 1$, in which case we use all of the data both to draw the posterior samples and to calculate the likelihood. This is not a useful measure of goodness-of-fit.

The K-fold CV likelihood is in fact equivalent to the evidence. Using equation 11.39 and 11.41 we can write

$$L_{CV} = \prod_k \frac{P(D|M)}{P(D_{-k}|M)} = \frac{P(D|M)^K}{\prod_k P(D_{-k}|M)}. \tag{11.42}$$

The denominator can be written as $P(D|M)^{K-1}$, as we can see if we consider the case $K = 3$ (dropping the M for brevity)

$$\prod_k P(D_{-k}) = P(D_1, D_2)P(D_2, D_3)P(D_1, D_3)$$

$$= P(D_1)^2 P(D_2)^2 P(D_3)^2$$
$$= P(D)^{K-1} \tag{11.43}$$

where we have used the fact that the data sets are disjoint and (assumed) independent. Thus it follows from equation 11.42 that

$$L_{CV} = \frac{P(D|M)^K}{P(D|M)^{K-1}} = P(D|M). \tag{11.44}$$

Why use the cross-validation likelihood?

Why would we use the cross-validation likelihood if it is equivalent to the evidence? The main reason is numerical accuracy. The evidence is computed numerically by sampling from the prior (equation 11.16). The prior is often a very broad function compared to the region over which the likelihood is non-negligible. Many of the likelihood evaluations contributing to the Monte Carlo sum are therefore negligible, so this sum could have a large numerical uncertainty. In particular, it is likely to be underestimated. The number of samples we require for a good estimation is unknown (it could be very large), and grows exponentially with the dimensionality of the parameter space. Samples from the posterior (typically obtained by MCMC), in contrast, are more likely to produce a large likelihood, because the posterior is the product of the likelihood with the prior. This is how the partition likelihood is calculated, so it is more likely to be numerically stable. The number of likelihood calculations we will need for the cross-validation likelihood is likely to be far smaller than the number required for the evidence to achieve a stable result.

11.7 Other measures for model comparison: AIC and BIC

The evidence has a simple, solid theoretical basis, but it can be slow to compute. Its dependence on the prior can also make it difficult to interpret when our prior information is hard to represent as a PDF. Other metrics have been developed which may be more convenient for model comparison. Two of these – AIC and BIC – use the maximum likelihood.

The *Akaike information criterion* (AIC) was derived by Akaike (1973, 1974) from concepts in information theory. We have seen that whereas a model with a larger likelihood will fit the data better, selecting models on this basis will just favour more complex models. Often (but not always) a more complex model is one with more parameters. So we could use the maximum likelihood penalised by the number of parameters as a measure of model quality. The AIC does precisely this and is defined as

$$\text{AIC} = -2\ln L_{\text{max}} + 2J \tag{11.45}$$

where J is the number of independently fitted parameters in the model. Like the evidence, the absolute value of AIC is of no relevance. Unlike the (logarithm of the) evidence, the more negative the AIC, the better the model. Note that the AIC is on the scale of twice the natural logarithm of the likelihood, whereas the evidence is linear in the likelihood. To use the AIC for comparing models we simply calculate it for each of the models of interest on the same data, and identify the best model as the one with the most negative AIC. Whether this is *significantly* better than the others is a choice. If we required a model to have ten times the likelihood of another model with the same number of parameters, for example, then we would require a difference in the AIC of at least $2\ln 10 = 4.6$. Note that one parameter has the same relevance as a factor e in likelihood. The number of parameters J in the definition of the AIC counts the presence of variance of the data (whether known or not) as one parameter. Thus if the model were a straight line in one dimension – which has two parameters – fit to noisy data, we would have $J = 3$. This idea of counting parameters is clearly rather simplistic, because a second order polynomial is much less flexible than a sinusoidal function, which also has three parameters.

A variant on AIC is the *Bayesian information criterion* which, as its name suggests, was derived from Bayesian considerations (Schwarz, 1978). It is defined as

$$\text{BIC} = -2\ln L_{\text{max}} + J\ln N \tag{11.46}$$

where N is the number of data points. The BIC generally imposes a larger penalty for the complexity of the model than does the AIC (because $\ln N > 2$ when $N > 8$).

To illustrate the AIC and BIC I calculate it for the two models used in the coin example in section 11.2. Model M_1 has the probability of getting a heads in a single toss fixed at $p = 1/2$. This model has no free parameters so $J = 0$. Model M_2 is the case that p is unknown so has $J = 1$. Actually, in section 11.2 model M_2 is the case that p is unknown *and* is described by a beta prior. But the AIC and BIC do not recognise prior distributions: they only take into account the number of free parameters in the model. Hence they treat all prior distributions over p as being the same model, including a prior that has a very narrow distribution about $p = 1/2$, even though conceptually this is much more like M_1. This shows the inflexibility of the AIC and BIC compared to calculating the evidence. It's the price we pay for their simple (and simplistic) measure of model complexity.

The likelihood for both models M_1 and M_2 is the binomial distribution, equation 11.5. Model M_1 has no free parameters, so its maximum likelihood L_{max} is the value of the binomial distribution at $p = 1/2$. The maximum likelihood for model M_2 can be found by differentiation (see section 4.4) to be at $p = r/n$. The number of data points N in the definitions of AIC and BIC is the number of coin tosses, n. The difference in the AIC and

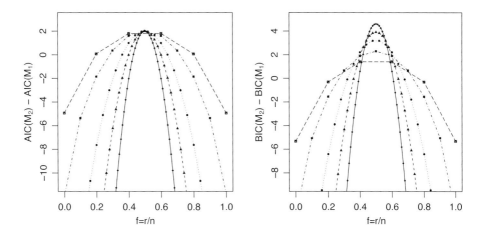

Fig. 11.6 Left: the difference between the AIC for M_2 (which has p as a free parameter) and M_1 ($p = 1/2$) in the coin problem from section 11.2. As in figure 11.1 the five lines are, in order of increasingly narrow distribution, for $n = 5, 10, 25, 50$, and 100. Note that the AIC – and thus the vertical axis here – is proportional to twice the natural logarithm of the likelihood, whereas the vertical axis in figure 11.1 is proportional to the base ten logarithm of the evidence (and therefore also the likelihood). Both plots cover a factor of order 1000 in likelihood (the range on the vertical axis here is about $2 \ln 1000 = 13.8$). Right: as the left panel, but for the BIC.

BIC for the two models is

$$
\begin{aligned}
\mathrm{AIC}(M_2) - \mathrm{AIC}(M_1) &= 2 \ln P_{\mathrm{bin}}(r \,|\, p = 1/2, n) - 2 \ln P_{\mathrm{bin}}(r \,|\, p = r/n, n) + 2 \\
\mathrm{BIC}(M_2) - \mathrm{BIC}(M_1) &= 2 \ln P_{\mathrm{bin}}(r \,|\, p = 1/2, n) - 2 \ln P_{\mathrm{bin}}(r \,|\, p = r/n, n) + \ln n.
\end{aligned}
\tag{11.47}
$$

We can now calculate these for different experimental setups and results – for different n and different numbers of heads r – and use the results to decide which model is favoured.

The results for the AIC are shown in the left panel of figure 11.6. We see a broadly similar behaviour as that obtained with Bayes factors (figure 11.1), although a closer look shows that the two metrics disagree about the relative performance of the two models as n changes. In this particular example, the variation of the BIC with the data (right panel of figure 11.6) is much more like the Bayes factor. But it must be stressed that this is not always the case. The code below produces the two plots.

R file: `coin_aic_bic.R`

```
##### Demonstration of AIC and BIC on coin modelling comparison problem

phi <- 1/2
n   <- c(5,10,25,50,100)

pdf("coin_aic.pdf", 4, 4)
par(mfrow=c(1,1), mgp=c(2.0,0.8,0), mar=c(3.5,3.5,1,1), oma=0.1*c(1,1,1,1))
```

```
plot(c(0,1), c(0,1), type="n", xlim=c(0,1), ylim=c(-11,3), xlab="f=r/n",
    ylab=expression(paste("AIC(",M[2],") - AIC(",M[1],")")))
for(i in 1:length(n)) {
  r  <- 0:n[i]
  f  <- r/n[i]
  aicM1 <- -2*dbinom(x=r, size=n[i], prob=1/2,    log=TRUE)
  aicM2 <- -2*dbinom(x=r, size=n[i], prob=r/n[i], log=TRUE) + 2
  points(f, aicM2-aicM1, cex=0.5, pch=13+i)
  lines(f,  aicM2-aicM1, lty=6-i)
}
dev.off()

pdf("coin_bic.pdf", 4, 4)
par(mfrow=c(1,1), mgp=c(2.0,0.8,0), mar=c(3.5,3.5,1,1), oma=0.1*c(1,1,1,1))
plot(c(0,1), c(0,1), type="n", xlim=c(0,1), ylim=c(-9,5), xlab="f=r/n",
    ylab=expression(paste("BIC(",M[2],") - BIC(",M[1],")")))
for(i in 1:length(n)) {
  r  <- 0:n[i]
  f  <- r/n[i]
  bicM1 <- -2*dbinom(x=r, size=n[i], prob=1/2,    log=TRUE)
  bicM2 <- -2*dbinom(x=r, size=n[i], prob=r/n[i], log=TRUE) + log(n[i])
  points(f, bicM2-bicM1, cex=0.5, pch=13+i)
  lines(f,  bicM2-bicM1, lty=6-i)
}
dev.off()
```

In the above example the maximum likelihood and therefore the AIC and BIC could be calculated analytically. But so could the evidence for the priors we chose, so the AIC and BIC offer no advantage in this case. In most real-world problems, all of these metrics would need to be computed numerically. Whereas computing the evidence normally involves drawing from the prior (to do a Monte Carlo integration), finding the maximum likelihood is an optimization process (section 12.6). The latter may be faster, and moreover does not require us to select a prior. Yet this is precisely the weak point of the AIC and BIC: they rely on just one value of the likelihood (the maximum) and use simplistic measures of model complexity based on the number of parameters and number of data points. Within a class of models (e.g. all polynomials) in which the models are nested this is not such a problem, because in comparing AIC or BIC we are only interested in the difference in the number of parameters between the models, and this will be clear. But overall, whereas the AIC and BIC are potentially more convenient than the evidence, they are not as general as a metric for model selection. Further discussion can be found in Kass & Raftery (1995).

11.8 The stopping problem

The stopping problem is a well-known issue that illustrates a fundamental problem with hypothesis testing using p values. It has been discussed by many authors in different contexts. This particular example is taken from Gregory (2005).

In an astronomical survey we observe $n = 102$ stars and determine that $r = 5$ of these are white dwarfs. Our hypothesis M is that 10% of stars in such surveys are white dwarfs. Do the data support this hypothesis?

11.8.1 The frequentist approach(es)

From a frequentist hypothesis testing point of view we want to do a two-sided test of our hypothesis that the true fraction of white dwarfs is $p = 0.1$. The test is two-sided because both high deviations and low deviations of the data from this value of p are evidence against M. We assume that all of the observations are independent.

We first give the data and this information to Alice. This is what she does.

Frequentist Alice. Having observed n stars, she says that the probability that r of these are white dwarfs is given by the binomial distribution

$$P(r|p,n) = \binom{n}{r} p^r (1-p)^{n-r}. \tag{11.48}$$

Thus (twice) the probability of observing five or fewer white dwarfs is

$$P_A = 2\sum_{r=0}^{r=5} P(r|p,n) = 0.102 \tag{11.49}$$

the factor of 2 arising because she is doing a two-sided test. This is Alice's p value. It is more than 0.05, so she cannot reject M at the 95% level (or any other higher level).

We then give the same information and data to Bob, but additionally tell him that before we started observing, we decided to keep observing stars until we had observed five white dwarfs, and then we stopped. This is what he does.

Frequentist Bob. Given this extra information, the random variable is now n, not r. The probability that there were $r - 1$ white dwarfs among the first $n - 1$ observations, and therefore also $(n - 1) - (r - 1)$ stars which were not white dwarfs, is given by a binomial distribution. The probability that the last observation was a white dwarf is p. Thus the probability of the data is

$$P(n|p,r) = \binom{n-1}{r-1} p^{(r-1)}(1-p)^{(n-1)-(r-1)} \times p$$

$$= \binom{n-1}{r-1} p^r (1-p)^{n-r}. \tag{11.50}$$

This, incidentally, is called the *negative binomial distribution*, and is a probability distribution in n, in contrast to the binomial distribution, which is a probability distribution in r. Thus (twice) the probability of observing 102 stars or more given that we observed five white dwarfs is

$$P_B = 2\sum_{n=102}^{n=\infty} P(n|p,r)$$

$$= 2\left(1 - \sum_{n=5}^{n=101} P(n|p,r)\right) = 0.044. \tag{11.51}$$

where again we have a factor of 2 due to the two-sided test. Bob's p value is less than 0.05, so he rejects M at the 95% confidence level (and any other lower level).

Who is right?

Frequentist hypothesis testing considers both analyses to be correct. The contradiction arises because in order to calculate a p value we must define a reference set of hypothetical data which were not, but which could have been, observed. This requires us to make assumptions about what could have been observed. Alice considers n to be fixed and r to be the random variable, so she takes as a reference set different unobserved values of r. Bob, in contrast, considers r to be fixed and n to be the random variable, so his reference set is all the possible values of n.

To interpret these results as a frequentist we would have to decide in advance (before we started the experiment) which approach we were going to take. Let's assume we decide to keep observing stars until we observe five white dwarfs. What do we do if we get bad weather, or a lack of funds prevents us from reaching this number? We could no longer do our original analysis, but we also know that a different analysis could give us the opposite result. We are in a dilemma and would have to throw away the data. But obviously the data don't suddenly become worthless: they are what they are. The problem is that this approach to the analysis is dependent on irrelevant information, namely which variable we choose to be random.

11.8.2 The Bayesian resolution

The frequentist approach runs into problems because it is asking the wrong question, namely the probability of getting the data, whereas what we should be doing is an inference of p, i.e. computing its PDF given the data. Let's now take this Bayesian approach in which our model M' is that p is unknown and is described by a prior $P(p|M')$. Both Alice and Bob want to determine

$$P(p|D, M') = \frac{1}{Z} P(D|p, M')P(p|M'). \tag{11.52}$$

Bayesian Alice. The likelihood is

$$P(D|p, M') = P(r|p, n) = \binom{n}{r} p^r (1-p)^{n-r} \tag{11.53}$$

so the posterior is

$$P(p|D, M') = \frac{p^r (1-p)^{n-r} P(p|M')}{\int p^r (1-p)^{n-r} P(p|M') \, dp} \tag{11.54}$$

where the binomial coefficient cancels out because it is independent of p.

Bayesian Bob. The likelihood is

$$P(D|p, M') = P(n|p, r) = \binom{n-1}{r-1} p^r (1-p)^{n-r}. \tag{11.55}$$

Taking the same prior as Alice, the posterior is

$$P(p\,|\,D, M') \;=\; \frac{p^r(1-p)^{n-r}P(p\,|\,M')}{\int p^r(1-p)^{n-r}P(p\,|\,M')\,dp} \tag{11.56}$$

and once again the binomial coefficient involving the factorials cancels out.

The Bayesian approach gives the same conclusion about the probability of p and therefore about how well the hypothesis is supported by the data. Alice and Bob now arrive at the same answer because both the binomial and negative binomial distributions have the same dependence on p. The two frequentist approaches differ from each other because they look at the distribution of the "random variable", either r or n, and this has a different dependence in the two cases. In a Bayesian approach the data is not a random variable that could have been something else. It is the data which are given, and the parameter which is unknown.

To test the original hypothesis we can look at the posterior PDF over p. This is shown in figure 11.7 using a uniform prior. The value of $p = 0.1$ is quite far down in the low tail of the distribution, thus casting doubt on the hypothesis.

An alternative approach is to compute the Bayes factor between a model with $p = 0.1$ and the model with a prior $P(p\,|\,M')$. The evidence is the integral of the product of the likelihood and prior. The binomial coefficient, which depends on r and n but not on p, factorizes out of the integral. When Alice forms her Bayes factor, the binomial coefficient cancels in the ratio. The same happens for Bob, leaving them both with the same dependence on p. So the two of them would compute identical Bayes factors also. This approach tells us which of the two models fits better, but not whether the original hypothesis is true or not. But even if we took one of the frequentist approaches we could also not say whether the hypothesis was true or not: we could only reject or fail to reject it at some confidence level (see sections 10.1 and 10.6).

The following R code calculates the frequentist and Bayesian solutions and produces the plot. The function dnbinom computes the density of the negative binomial distribution. To plot figure 11.7 I normalize the posterior numerically (so you can easily insert any prior), but as I use a uniform prior $P(p\,|\,M') = 1$, the normalization constant here is $B(r + 1, n - r + 1)$ (compare equation 11.7 with the denominator in equations 11.54 and 11.56).

R file: code/stopping_problem.R

```
##### The stopping problem

p <- 0.1
n <- 102 # no. trials
r <- 5   # no. successes

PA <- 2*sum(dbinom(x=0:r, size=n, prob=p))
# = 2*pbinom(q=r, size=n, prob=p)
PB <- 2*(1 - sum(dnbinom(x=0:(n-r-1), size=r, prob=0.1)))
# = 2*sum(dnbinom(x=(n-r):1e4, size=r, prob=0.1))
# where 1e4 is used as an approximation of infinity.
# Note the definitions adopted in dnbinom():
# x=n-r (no. failures) and size=r. Can check by comparing:
```

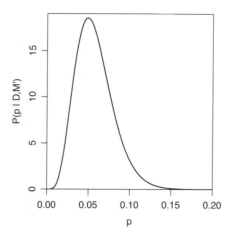

Fig. 11.7 Posterior PDF over p in the stopping problem for $n = 102$, $r = 5$ when using a uniform prior on p.

```
# choose(n-1, r-1)*p^r*(1-p)^(n-r)
# dnbinom(x=n-r, size=r, prob=p)

pdf("stopping_problem.pdf", 4, 4)
par(mfrow=c(1,1), mar=c(3.5,3.5,0.5,1), oma=0.5*c(1,1,1,1),
    mgp=c(2.2,0.8,0))
Nsamp   <- 1e4
deltap <- 1/Nsamp
pgrid  <- seq(from=1/(2*Nsamp), by=1/Nsamp, length.out=Nsamp)
pdense <- pgrid^r * (1-pgrid)^(n-r)    # with uniform prior
pdense <- pdense/(deltap*sum(pdense)) # normalize posterior
# could instead do analytically for a uniform prior: pdense/beta(r+1, n-r+1)
plot(pgrid, pdense, type="l", lwd=1.5, xaxs="i", yaxs="i", xlim=c(0,0.2),
     ylim=c(0,19), xlab="p", ylab="P(p | D,M')")
dev.off()
```

11.9 Issues with Bayesian model comparison

While Bayesian model comparison overcomes many of the problems of frequentist hypothesis testing, it is not without its own issues.

(1) The main issue is the dependence of the evidence on the prior. Conceptually priors are a good thing, as they are a way of formalizing the additional information beyond the data (such as plausibility constraints, invariances, etc.) that one invariably has. Moreover, we have seen in earlier chapters how priors are unavoidable, and can even be essential to get meaningful solutions at all. In parameter estimation, i.e. when computing the posterior PDF $P(\theta \mid D, M)$, we have seen how the dependence of the posteriors on the

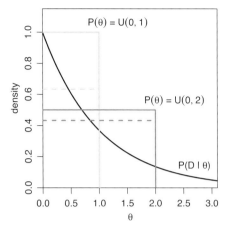

Fig. 11.8 Illustration of the dependence of the evidence on the prior. The black line shows the likelihood as a function of the parameter θ. The light grey and dark grey solid lines show uniform priors with ranges $(0, 1)$ and $(0, 2)$ respectively. The dashed lines show the two corresponding values of the evidences, the evidence being the integral of the likelihood over the prior.

prior becomes ever weaker as the data become more informative (e.g. figure 5.6). The evidence tends to have a stronger dependence on the prior. Consider a one parameter problem in which the likelihood $P(D|\theta, M)$ has an exponential dependence on the parameter, $e^{-\theta}$ (see figure 11.8). Suppose we want to adopt a uniform prior. As we must sample from the prior to compute the evidence, we need a proper prior, so we need to set limits on it. If we first choose $\mathcal{U}(0, 1)$ (the light grey line in the figure), then the evidence, which is the integral of the likelihood over this, is 0.63 (shown by the dashed line). But if we then decided to use a prior with twice the range, $\mathcal{U}(0, 2)$, the evidence becomes 0.42 (the dark grey lines in the figure). In general, changing the prior range changes the evidence. In principle this is what we want, because the prior is an integral part of the model: changing the prior changes the model. But it can be a problem in practice if we are not sure what prior range we should use. There is not much we can do about this other than do our best to specify plausible prior ranges based on the available data, and to investigate the sensitivity of the Bayes factors to changes in the priors. Clearly one should not use the narrowest priors imaginable for one model, and the broadest imaginable for another. If the likelihood is very peaked (informative data) and all plausible priors are essentially flat over this range, then the evidence is insensitive to the choice of prior. This is exploited by the quadratic approximation of the evidence (equation 11.35).

(2) For equal model priors the expression for the posterior probability of some model M_1 (see equation 3.8) is

$$P(M_1|D) = \frac{1}{1 + \sum_{k=2}^{K} BF_{k1}} \tag{11.57}$$

where BF_{k1} is the Bayes factor of model M_k with respect to model M_1. These are always positive, so the posterior probability of M_1 increases monotonically as the number of models included in the sum decreases. Thus if we (erroneously) do not include all plausible alternative models (those which don't have $BF_{k1} \simeq 0$) in our analysis, the sum is smaller than it should be, and the posterior probability of M_1 is artificially increased. This could happen if the models are complex, so we cannot be sure that we have an exhaustive set. It's not really a problem in practice, however, because we don't normally try to calculate model posterior probabilities. We normally just calculate posterior odds ratios or Bayes factors of a series of models relative to a baseline model. This allows us to identify the best of a set of given models, even when they are incomplete.

(3) Bayesian model comparison does what's written on the label: it compares models. It cannot be used to assess the "absolute" probability of an isolated model. From a Bayesian perspective this makes no sense, because if we only ever have one model, the data must have come from it. But in practice we often want to get some idea of whether just a single model is a plausible explanation of the data.

(4) Bayesian inference can be computationally intensive. While there are some analytic results (such as those in chapters 5 and 6, and reasonable approximations can sometimes be made to give analytic results), general problems require us to sample and integrate PDFs using numerical methods. The computational bottle-neck is often the calculation of the likelihood. This computational cost is the price we pay for the generality and power of the method.

(5) The entire Bayesian approach depends on the likelihood function. There may be problems where the likelihood is hard to define accurately, making any inferences based on it of limited value. In such cases we may be able to substitute the likelihood for a similarity measure and still obtain reliable results. One such approach is known as *Approximate Bayesian Computation* (ABC).

We will see some practical alternatives to using the evidence to control for model complexity, and therefore to select the most appropriate model, in chapter 12, using cross-validation and regularization together with basis functions.

## 12

Dealing with more complicated problems

In this final chapter we will learn ways to tackle more complicated problems. I first introduce the concepts of cross-validation and regularization as means for controlling the complexity of fitted models, taking again the example of curve fitting. Global polynomials have some unpleasant properties, so I will introduce more practical methods of curve fitting using local basis functions, in particular splines and regression kernels. I will finish up with an outline of two approaches that are useful when computing the entire posterior is difficult or unnecessary: numerical optimization for finding the maximum of the posterior and bootstrapping for estimating its variance.

12.1 Cross-validation

Suppose we have a data set $\{x, y\}$ for which there is unknown noise ϵ in the y values. We want to fit a polynomial function to this such that $y = f_J(x) + \epsilon$, where

$$f_J(x) = \sum_{j=0}^{J} x^j \beta_j \tag{12.1}$$

but we do not have a good idea of what order polynomial – value of J – to use. We cannot select the J that gives the best fit, because we could just keep increasing J until we fitted the data exactly. This is not what we want when we know there is noise in the data.

One approach is to calculate the Bayesian evidence for different order polynomials and select the one with the highest evidence. We did this in section 11.3 to decide whether a line with zero gradient (polynomial order zero) or finite gradient (order one) was better. However, using the evidence has some disadvantages.

- It is hard to decide on the priors for polynomial parameters, because our prior information probably doesn't come in the form of constraints on a polynomial. The impact of different values of the high order coefficients on the function is not very intuitive. (We will see in section 12.3 how we can overcome this to some extent.)
- The evidence is relatively slow to compute, because we often have to make millions or more likelihood calculations (unless we have a strong prior or can chose one that permits an analytic integration).
- If we cannot specify the noise model (for the uncertainties in y), then the likelihood is not defined so the evidence cannot be calculated.

Recall that we do not need to know the standard deviation of the noise: we saw in section 11.3 that once we have a noise model, e.g. a zero mean Gaussian with standard deviation σ, we can specify a prior on σ and marginalize over it. Often we can define a reasonable noise model, so the last point above is not usually a restriction in practice. Nonetheless, the following method will allow us to fit a suitable function, with complexity control, without an explicit noise model.

An alternative to using the evidence is cross-validation. Suppose we have N data points. We divide these at random into two sets, one called the *training set* D_{train} containing N_{train} data points, the other called the *test set* D_{test} containing N_{test} data points. We choose a polynomial order J, fit the line using the training set, then calculate the quality of the fit using the test set. If we're using least squares to fit the line (chapter 4), then the quality of the fit is measured by the sum of squared residuals on just the points in the test set

$$\text{SS}_{\text{test}}(J) = \sum_{i \in D_{\text{test}}} [y_i - f_J(x_i; D_{\text{train}})]^2 \qquad (12.2)$$

where what comes after the ";" in the function indicates here the data used to fit the model. We repeat this for a range of J, and then choose the model – value of J – that gives the smallest value of the test error. By evaluating the fit on a different set of data from what was used to make the fit, we avoid selecting a model that *overfits* the data. A very high order polynomial may fit the training data very well – better than a lower order polynomial – but if much of the variation it fits is due to noise rather than an intrinsic variation, then we will measure a large sum of squared residuals on the test set. This is because although the test set will show the same intrinsic variation, its noise will be different (because noise is random). By splitting the data into separate train and test sets we establish how well the model solution generalizes. This is a form of *regularization*, which we will discuss more in section 12.2.1.

A disadvantage of this approach is that by keeping back data from the fitting we will not get such a good fit. We mitigate this problem via a variation known as *leave-one-out cross-validation*. We now define N training sets such that the ith, call it D_{-i}, contains all data points *except* the single point (x_i, y_i). For a given model J we fit the curve separately for each training set, and for each of these calculate the residual on the single point that was not in the training set. The quality of fit is the sum of squared residuals for all these fits, i.e.

$$\text{SS}_{\text{CV}}(J) = \sum_{i=1}^{N} [y_i - f_J(x_i; D_{-i})]^2. \qquad (12.3)$$

To put the sum of squares on the scale of the data and so make it more interpretable we can form the root mean square (RMS) residual instead, $\sqrt{\text{SS}/N}$, which is a measure of the residual per point (cf. equation 4.25).

One can also make an intermediate approach called *K-fold cross-validation*, in which there are $K < N$ train/test sets of data with more than one data point in each test set. This is useful if N is too small for us to construct sufficiently large train and test sets, but too large for us to want to (or to need to) construct N models to do leave-one-out cross-validation.

Cross-validation effectively enables us to sample the variation in the underlying function *and* the noise using just the available data. The method does not distinguish between the two; nor does it need to. By evaluating the quality of the fit on the omitted data we are able to monitor this fit quality as a function of the complexity of the model. Models that are either too simple (low order) or too complex (high order) will both perform badly. This can be understood in terms of a trade-off between the bias and variance of the fit. Suppose we had N different sets of training data.

- If the model we are fitting has low order, then the N different fits on the N training sets will all be quite similar. Hence they will all produce similar sized residuals on the test set, so the variance of these residuals will be low. But their bias will be large, because the model is not able to fit the true complexity of the data.

- If the model has high order, then the N fits will now have much lower bias due to the increased flexibility of the model. But their variance will be comparatively large, because a small change in the training data will produce a significantly different fit (due to the sensitive high order terms), and thus quite different residuals on the test data.

We saw in section 4.8 that the expected squared residual is a combination of bias and variance. Thus the optimal fit – smallest residuals on the test data – will be achieved by a model lying between the two extremes.

The R code at the end of this section demonstrates cross-validation by fitting lines of order $J = 0$ to $J = 9$ to a set of $N = 25$ data points. These data come from a fourth-order polynomial to which I have added zero-mean Gaussian noise with standard deviation 250. The data are shown in the top-left panel of figure 12.1.

As we saw in section 4.5, the model can be written in matrix form as $\mathbf{f} = X\boldsymbol{\beta}$. In the code I use the function `poly` to define the polynomials of various orders. Given a scalar value x, `poly` (with the specification `raw=TRUE`) creates the vector of values $(x, x^2, \ldots, x^J)$, where J is specified by the parameter `degree`. Note that `poly` does not include the power of zero in its output. Given a vector of N elements, `poly` operates on each element separately to produce an $N \times J$ matrix X. My function `polyeval` does this and then multiplies the result by $\boldsymbol{\beta}$ to evaluate the model $X\boldsymbol{\beta}$ (the operator `%*%` does matrix/vector multiplication). `poly` is also used together with `lm` to fit the polynomials via least squares. Note how I set `subset=-i` in `lm` to exclude just point i from the fit for the leave-one-out cross-validation.

Each model is fit by minimizing the sum of squared residuals, for which an estimate of the noise is not required. The fits for each J are shown in the panels of figure 12.1 labelled $J = 0$ to $J = 9$. The bottom-right panel plots the RMS of the fits for each J, as defined earlier. For $J = 0$ we only have the offset term in the model, so the fit is just the mean of the training data. Interestingly in this case, $J = 0$ is marginally favoured over $J = 1$. We then see a significant drop in the RMS as J increases from $J = 1$ to $J = 4$, the true value. Fifth- and sixth-order polynomials give much the same performance, but as J increases beyond this, the solutions begin to get considerably worse due to the large variance of the fits – overfitting – which we can see in the panels at large x. It's clear too that these higher order polynomials will give very poor extrapolations of the data.

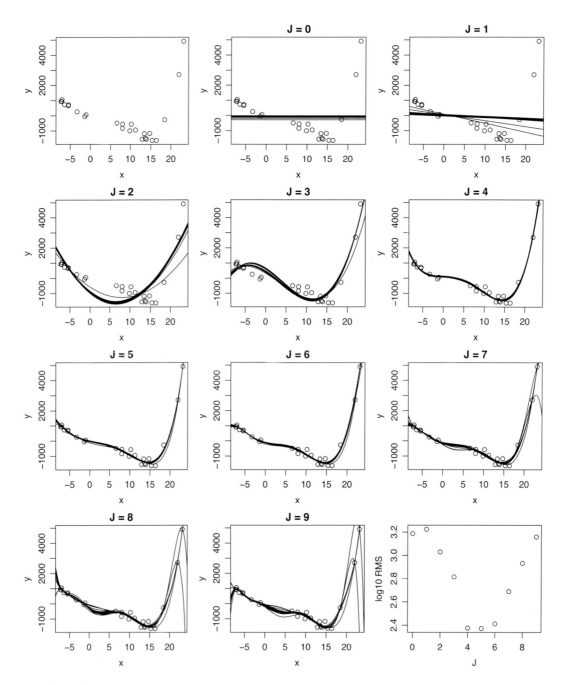

Fig. 12.1 Demonstration of leave-one-out cross-validation using polynomials. The top-left panel shows the data. The next ten panels labelled $J = 0$ to $J = 9$ show the $N = 25$ fits for each training data set for polynomials of order J. The bottom-right panel plots $\log \text{RMS}$ vs J, where $\text{RMS} = \sqrt{\text{SS}_{\text{CV}}/N}$ and SS_{CV} is defined in equation 12.3.

Cross-validation is frequently used with nonlinear models of high-dimensional data sets, such as neural networks, and many other machine learning methods.[1]

R file: CV_polynomials.R

```
##### Demonstration of cross-validation model selection using polynomials

# Function to evaluate polynomial with parameter beta at x.
# poly() as used here excludes x^0 - the constant, beta[1] - so I remove
# it from the matrix multiplication, then add it explicitly afterwards.
polyeval <- function(x, beta) {
  return(poly(x=x, degree=length(beta)-1, raw=TRUE) %*% beta[-1] + beta[1])
}

pdf("CV_polynomials.pdf", 10, 12)
par(mfrow=c(4,3), mar=c(3.5,3.5,1.5,0.5), oma=c(0.5,0.5,1,0.5),
    mgp=c(2.2,0.8,0), cex=1.0)
set.seed(63)

# Simulate data: 4th order polynomial
beta <- c(0, 5, 1, -2, 0.1)
Ndat  <- 25
sigma <- 250
x <-   runif(Ndat, min=-8, max=25)
y <-   polyeval(x, beta) + rnorm(Ndat, 0, sigma)
# xp, yp just for plotting
xp <- seq(from=-10, to=30, length.out=1e3)
yp <- polyeval(xp, beta)
plot(x,y)
#lines(xp, yp, col="red")

# Do CV. Plot all fits for each j in a separate panel
jmax <- 9 # evaluate all polynomials with order from 0 to jmax
# Do j=0 separately
rss0 <- 0
plot(x, y, main=c("J = 0"))
for(i in 1:Ndat) {
  pred <- mean(y[-i])
  rss0 <- rss0 + (pred-y[i])^2
  abline(h=pred)
}
# Do j=1:jmax
rss <- vector(mode="numeric", length=jmax)
for(j in 1:jmax) {
  rss[j] <- 0
  plot(x, y, main=paste("J =", j))
  for(i in 1:Ndat) {
    # poly() used in lm() does include x^0 term
    mod  <- lm(y ~ poly(x, j, raw=TRUE), subset=-i)
    pred <- predict(mod, newdata=data.frame(x=x[i]))
    rss[j] <- rss[j] + (pred - y[i])^2
    lines(xp, predict(mod, newdata=data.frame(x=xp)))
  }
}
```

---

[1] Indeed, classic feedforward neural networks are nothing more than rather flexible, nonlinear basis function models with parameters normally fit by optimization, with the complexity of the fit controlled by cross-validation.

```
}
plot(0:jmax, log10(sqrt(c(rss0, rss)/Ndat)), xlab="J", ylab="log10 RMS")
dev.off()
```

12.2 Regularization in regression

12.2.1 Regularization

A fundamental principle employed in science is the *principle of parsimony*. This states that we should try to explain data with the simplest model possible. One can think of arbitrarily complex and bizarre ways to explain data. Prior implausibility aside, experience shows that overly complex models make poor predictions. They fit the available data too precisely, so are unable to generalize to the broader situation. In other words, such models have high variance. In contrast, overly simple models will not explain the data well enough: they have a large bias. A model should be as simple as possible, but as complex as necessary.

Another name for this concept is *Occam's razor*: if each of a set of models explains the data equally well, choose the simplest, by which we mean the one with the fewest assumptions. This principle does not say we should always favour the simplest model. A complex problem may need a complex solution.

The cross-validation technique in the previous section controlled the complexity of the model (the order of the polynomial in the example used) by using the variance in the data, and seeing how well its solution generalized to data left out. We saw in section 11.5 how the Bayesian evidence automatically performs complexity control at the model level via the Occam factor, by trading-off model fit with model complexity. We can also apply this idea of regularization at the level of the individual model. We will see how to do this here and will then see that the procedure has a probabilistic interpretation.

Consider a data set $\{x_i, y_i\}$ (for $i = 1 \ldots N$) with unknown noise, to which we want to fit a polynomial of order J. We write this as $y_i = f(\mathbf{x}_i) + \epsilon$, where

$$f(\mathbf{x}_i) = \sum_{j=0}^{J} x_i^j \beta_j = \mathbf{x}_i^{\mathsf{T}} \boldsymbol{\beta} \tag{12.4}$$

with $\mathbf{x}_i^{\mathsf{T}} = (1, x_i, x_i^2, \ldots, x_i^J)$. Note the 1 in this vector, which multiplies β_0 to provide the constant term in the fit. The larger the absolute values of the polynomial coefficients in the fitted model, the more complex the function, because large coefficients produce more bendy curves (see figure 12.1). Thus a term such as

$$\sum_{j=0}^{J} \beta_j^2 \tag{12.5}$$

is a measure of the complexity of the model. As it stands, this expression is inconsistent, because each β_j is a coefficient of a different power of x and thus has different units. It makes no sense to sum terms with different units. We remedy this by *standardizing* the data. This means transforming each component of $\mathbf{x}$ to have zero mean and unit standard deviation over the N data points. Thus for $j = 1$ we do

$$x_i \rightarrow \frac{x_i - \bar{x}}{\sigma_x} \tag{12.6}$$

where $\bar{x}$ and σ_x are the mean and standard deviation (respectively) of $\{x_i\}$. We do the same for the other powers of x (components of $\mathbf{x}$). Once standardized, the components of $\mathbf{x}$ are unitless and each component of $\boldsymbol{\beta}$ has units y. If we also centre the $\{y\}$, $y_i \rightarrow y_i - \bar{y}$, then we don't need the constant term in the model, so we can set $\beta_0 = 0$ and drop the constant from the definition of $\mathbf{x}$.

A good fit demands that we minimize the residual sum of squares (RSS), but this will get smaller the larger J is. So instead of minimizing the RSS we minimize the objective function

$$\mathcal{E} = \sum_{i=1}^{N} [y_i - f_J(x_i)]^2 + \lambda \sum_{j=1}^{J} \beta_j^2 \tag{12.7}$$

where $\lambda \geq 0$. The first term is the RSS and the second term is our measure of model complexity. As we want to minimize $\mathcal{E}$, this corresponds to a complexity penalty. The size of the parameter λ determines the degree of the penalty: a larger λ means more penalty (more regularization), and so less complex solutions. As β_j^2 has units of y^2 (due to the standardization), λ is unitless. A relatively complex model will achieve a small value for the RSS by fitting the data closely, but it will have a large value of $\sum \beta_j^2$. The larger λ, the more this complexity term contributes to the objective function $\mathcal{E}$. By minimizing $\mathcal{E}$ with fixed λ, we achieve a set of coefficients $\{\beta_j\}$ that give us the optimal trade-off between fitting on the one hand and complexity on the other.

Fitting models with a penalty term is called *regularization*. In the context of least squares, this particular type of regularization is called *Tikhonov regularization*. Linear regression with this regularizer is called *ridge regression*, which we specify and solve in the next section.

12.2.2 Ridge regression

Section 4.5 showed the matrix formulation for ordinary least squares linear regression with J input variables. Section 4.6 showed how this can also be applied to polynomial regression in one-dimension with a Jth order polynomial.

Ridge regression can be seen as a generalization of these. Proceeding as in section 4.5, we first write the input data as an $N \times J$ design matrix X, in which each row is the vector of J features for input i. That is, the ith row is the vector of J input dimensions if we're doing linear multi-dimensional regression, or the vector $(x_i, x_i^2, \ldots, x_i^J)$ if we're doing polynomial regression. We standardize this, so each column of X has zero mean and unit standard deviation. $\mathbf{y}$ is the N-dimensional vector $(y_1, y_2, \ldots, y_N)$. This we centre, so it has zero mean. $\boldsymbol{\beta}$ is the J-dimensional vector of parameters, so the regularization term is $\boldsymbol{\beta}^\mathsf{T}\boldsymbol{\beta}$ (a scalar product).

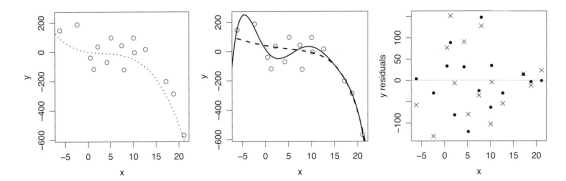

Fig. 12.2 Demonstration of ridge regression. Left: the data (open circles) have been computed from a third-order polynomial (dotted line) to which Gaussian noise have been added. Centre: ordinary least squares (solid line) and ridge regression (dashed line) fits to the data using sixth-order polynomials in both cases. Right: residuals for ordinary least squares (filled circles) and ridge regression (crosses).

Following from equation 12.7, we want to minimize

$$\mathcal{E} = (\mathbf{y} - X\beta)^{\mathsf{T}}(\mathbf{y} - X\beta) + \lambda\beta^{\mathsf{T}}\beta \tag{12.8}$$

with respect to β for a given λ. Differentiating and setting to zero we get

$$\mathbf{0} = -X^{\mathsf{T}}(\mathbf{y} - X\beta) + \lambda\beta$$
$$\mathbf{0} = -X^{\mathsf{T}}\mathbf{y} + X^{\mathsf{T}}X\beta + \lambda\beta$$
$$\beta = (X^{\mathsf{T}}X + \lambda I)^{-1}X^{\mathsf{T}}\mathbf{y} \tag{12.9}$$

where I is the identity matrix. Compare this with the ordinary least squares solution in equation 4.56. They are of course the same when $\lambda = 0$, as this corresponds to no regularization. Given that $|X^{\mathsf{T}}X + \lambda I| > |X^{\mathsf{T}}X|$ for $\lambda > 0$, we see that $|\beta|$ is smaller in ridge regression than in ordinary regression. Also, the term $(X^{\mathsf{T}}X + \lambda I)$ will always be invertible for $\lambda > 0$, so we will always get a solution, unlike with ordinary least squares. Thus ridge regression gives solutions to ill-posed problems, which is another purpose of regularization.

To use ridge regression effectively we have to decide on a sensible value of λ. Although λ is unitless, we can see from equation 12.7 that its value will not be invariant under an increase in the size of the data set, for example. But if we offset or rescale x or y, we don't need to change λ (on account of the standardization). Finding an appropriate value for λ can be done to some degree by trial and error. Once in the right vicinity we could use cross-validation (section 12.1) to find the most appropriate value among some set. Several of the R methods use *generalized cross-validation* to do this. This is an approximation to leave-one-out cross-validation which is generally faster to compute.[2]

[2] Generalized cross-validation essentially replaces the individual elements of the smoother matrix – defined later – with the average of the trace of this matrix.

order	β_{OLS}	β_{ridge}
1	−363	−45.7
2	1117	43.6
3	1538	−18.8
4	−9588	−36.9
5	11743	−54.4
6	−4654	−63.7

Table 12.1 Coefficients for the ordinary least squares and ridge regression solutions shown in figure 12.2.

Figure 12.2 compares the solutions obtained with ridge regression (using $\lambda = 1$) and with ordinary least squares (OLS) regression, using a sixth-order polynomial in both cases. The data are $N = 15$ points computed from a third-order (cubic) polynomial to which Gaussian noise have been added. The plot shows the original data (prior to standardization) and the fits have been transformed back to the non-standardized scales. The values of the resulting coefficients in the two cases are shown in Table 12.1 (these apply to the standardized data.) All of the coefficients from the ridge regression have much smaller absolute values than those from the ordinary regression. The sum of squares of the coefficients is 2.6×10^8 and 1.3×10^4 for ordinary least squares and ridge regression respectively. The RMS of the residuals for the fits is 63.7 and 80.0 respectively. So ridge regression does what it should: it produces a fit with a slightly larger RMS but with significantly smaller coefficients.

The R below code implements the above and produces figure 12.2. I encourage you to experiment with the code by changing the value of λ (on the line where betaRidge is computed), the order of the models (degree), and the amount and complexity of the data. In the code I again use the function poly and polyeval as discussed in the previous section. In the line

```
X <- scale(poly(x, degree=6, raw=TRUE))
```

poly is applied to all N (Ndat) values of x to produce an $N \times J$ matrix. This is then standardized using the function scale to produce the $N \times J$ design matrix X. The vector xp is a dense vector of size N_p that I use to compute the fitted functions in order to plot them. To compute this, xp (and its powers up to J) must be standardized using the means and standard deviations computed on the data; we must not compute new means and standard deviations! The numbers required for this can be extracted from the standardized design matrix X using the attr function. R is a bit counter-intuitive when it comes to mixed matrix/vector algebra. One might think that

```
XP - attr(X,"scaled:center")
```

would subtract the $1 \times J$ vector attr(X,"scaled:center") from each row of the $N_p \times J$

matrix XP. But R does column-wise operations, so we first need to transpose XP, then sub-tract the vector of means, divide by the vector of standard deviations, and finally transpose back again. This is done with the following code.

```
t( (t(XP) - attr(X,"scaled:center")) / attr(X,"scaled:scale") )
```

Note that the plots are of the raw (non-standardized) data, and the resulting curves are converted back from the centered coordinates so they can be plotted with these raw data.

R file: `ridge_regression.R`

```
##### Demonstration of ridge regression for 1D polynomials

# Function to evaluate polynomial with parameter beta at x.
# poly() as used here excludes x^0 - the constant, beta[1] - so I remove
# it from the matrix multiplication, then add it explicitly afterwards.
polyeval <- function(x, beta) {
  return(poly(x=x, degree=length(beta)-1, raw=TRUE) %*% beta[-1] + beta[1])
}

# Function to solve for ridge regression parameters
# given data vector y of length N, N*J data matrix X,
# and lambda (scalar).
param.ridge <- function(y, X, lambda) {
  return( solve(t(X)%*%X + diag(lambda, nrow=ncol(X))) %*% t(X) %*% y )
}

pdf("ridge_regression.pdf", 12, 4)
par(mfrow=c(1,3), mar=c(3.5,3.5,0.5,0.5), oma=c(0.5,0.1,0.5,0.5),
    mgp=c(2.2,0.8,0), cex=1.2)

set.seed(100)

# Simulate data: 3rd order (cubic) polynomial
beta <- c(0, -5, 1, -0.1)
Ndat  <- 15
sigma <- 100
x <-   runif(Ndat, min=-8, max=25)
y <-   polyeval(x, beta) + rnorm(Ndat, 0, sigma)
# xp, yp just for plotting
xp <- seq(from=-10, to=27, length.out=1e3)
yp <- polyeval(xp, beta)

# Plot data and true model
plot(x, y, ylim=c(-580, 240))
lines(xp, yp, lty=3, lwd=2)

# Build matrices, centre y, centre and scale each power of x. "degree" in
# poly defines the order of the polynomial we use in both solutions.
ys <- scale(y, scale=FALSE)
X  <- scale(poly(x, degree=6, raw=TRUE)) # Ndat * 6 matrix
XP <- poly(xp, degree=6, raw=TRUE)        # 1e3  * 6 matrix
XP <- t( (t(XP) - attr(X,"scaled:center")) / attr(X,"scaled:scale") )
# Solve for OLS and ridge regression parameters and calculate residuals
betaOLS  <- param.ridge(ys, X, lambda=0)
residOLS <- X %*% betaOLS - ys
```

```
betaRidge   <- param.ridge(ys, X, lambda=1)
residRidge <- X %*% betaRidge - ys

# Plot raw (non-standardized) data together with fitted curves.
# Latter is done by plotting de-centered model predictions at raw xp.
plot(x, y, ylim=c(-580, 240))
lines(xp, XP %*% betaOLS   + attr(ys, "scaled:center"), lwd=2)
lines(xp, XP %*% betaRidge + attr(ys, "scaled:center"), lwd=2.5, lty=2)

# Plot residuals
plot(x, y, type="n", ylim=range(c(residOLS, residRidge)),
     ylab="y residuals")
abline(h=0, col="grey")
points(x, residOLS,  pch=20)
points(x, residRidge, pch=4)

dev.off()

# Print coefficients, sum of squares of coefficients, and RMS of residuals
format(data.frame(betaOLS, betaRidge), digits=3)
cat("beta^2 = ", sum(betaOLS^2), sum(betaRidge^2), "\n")
cat("RMS =    ", sqrt(mean(residOLS^2)), sqrt(mean(residRidge^2)), "\n")
```

12.2.3 Probabilistic interpretation of regularization

We saw in section 4.4.2 how ordinary least squares was equivalent to maximum likelihood for a Gaussian likelihood. Following the same approach, we can interpret the regularizer in ridge regression as a prior on the model parameters. Taking the logarithm of Bayes' theorem gives

$$\ln P(\boldsymbol{\beta}|\mathbf{y}) = \ln P(\mathbf{y}|\boldsymbol{\beta}) + \ln P(\boldsymbol{\beta}) + \text{constant} \tag{12.10}$$

where for brevity I have dropped the conditioning on both the model M and the input data X. The constant (the evidence) is independent of $\boldsymbol{\beta}$. Suppose the likelihood $P(\mathbf{y}|\boldsymbol{\beta})$ is Gaussian with covariance matrix Σ (equation 4.58) and we adopt a prior

$$P(\boldsymbol{\beta}) = \exp\left(-\frac{1}{2}\lambda\boldsymbol{\beta}^\mathsf{T}\boldsymbol{\beta}\right). \tag{12.11}$$

Equation 12.10 then becomes

$$\ln P(\boldsymbol{\beta}|\mathbf{y}) = -\frac{1}{2}(\mathbf{y} - X\boldsymbol{\beta})^\mathsf{T}\Sigma^{-1}(\mathbf{y} - X\boldsymbol{\beta}) - \frac{1}{2}\lambda\boldsymbol{\beta}^\mathsf{T}\boldsymbol{\beta} + \text{constant} \tag{12.12}$$

where the constant still only includes terms independent of $\boldsymbol{\beta}$. Comparing this to equation 12.8 (with $\Sigma = I$), we see that the objective function used in ridge regression is equal to -2 times the logarithm of the posterior PDF over the model parameters (to within an additive constant). Thus by minimizing the objective function we maximize the posterior. This is true even when the data have (common) error bars σ, as the scaling provided by a diagonal covariance matrix ($\Sigma = \sigma^2 I$) simply changes the relative sizes of the posterior and prior terms, a freedom which can be absorbed into the regularization parameter λ. For the more general case of different sized error bars and/or covariance, we can modify the objective function in ridge regression to be the negative of equation 12.12.

Thus we can interpret the regularizer as a prior on the solution that would otherwise be obtained with maximum likelihood. We have seen the effect of priors in several places in this book. Maximum likelihood estimates sometimes suffer from high variance on account of overfitting data, so the prior can play an important role by reducing this variance.

12.3 Regression with basis functions

A polynomial – equation 12.1 – is an example of a *basis function*. Another well-known one is a Fourier series, useful for representing periodic functions, which is

$$f(x) = a_0 + \sum_{j=1}^{J} a_j \cos\left(\frac{2\pi j x}{L}\right) + b_j \sin\left(\frac{2\pi j x}{L}\right) \tag{12.13}$$

for some constant L. The coefficients of this basis function are $\{a_j, b_j\}$. A basis function representation is anything that represents a function as a linear combination of functions, so in general we can write it as

$$f(x) = \sum_{j=1}^{J} \beta_j h_j(x). \tag{12.14}$$

12.3.1 Splines

Polynomials are not very nice functions to work with. Low order polynomials are not flexible enough to fit most data, yet once we use orders beyond a few, polynomials produce wild variations in a desperate attempt to fit the given data points. This is exacerbated by the fact that the polynomials we have been using are global: they extend across the whole data range. Often we would like to have a curve that gives increased curvature (flexibility) just in certain parts of the data space, without producing a big wobble somewhere else. This can be achieved using localized basis functions. A common example is splines.

The idea of a spline is to split the data into regions and to fit each separately. We consider them in one dimension. We define K points – called knots – along the x-axis at $x = \{v_k\}$. These knots divide the data space into $K + 1$ regions, whereby the first region spans the range $-\infty$ to v_1 and the last region spans the range v_K to $+\infty$. We now define $K+1$ basis functions, each of which describes the function over just one region.

A simple spline is a piecewise constant spline, in which we fit a constant separately to each region. The next simplest polynomial is a piecewise linear spline, in which we fit a straight line to each region, and so on for higher orders. Examples of such splines are shown in figure 12.3. Such piecewise functions (of any order) have the problem that they are not continuous at the knots, because each region is fit separately. One usually wouldn't even call such functions "splines".

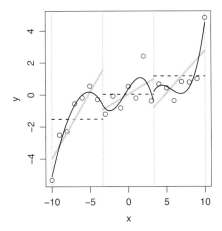

Fig. 12.3 Piecewise constant (dashed black line), linear (grey line), and cubic (solid black curve) functions fitted to the data (open circles) independently in each region between the knots (indicated by the vertical dotted lines). This region-independent fitting causes the functions to be discontinuous at the knots (even for the cubic function at the second knot: we just got lucky that it is nearly continuous in this example).

We can force the function to be continuous at the knots by defining a different set of basis functions. This will be easier if I first define the following notation. For any scalar a let

$$(a)_+ = \begin{cases} a & \text{if} \quad a > 0 \\ 0 & \text{otherwise.} \end{cases} \tag{12.15}$$

One possible basis function set to produce a *continuous* piecewise linear function is

$$\begin{aligned} h_1(x) &= 1 \\ h_2(x) &= x \\ h_{k+2}(x) &= (x - v_k)_+ \quad \text{where} \quad k = 1 \dots K. \end{aligned} \tag{12.16}$$

This defines $K + 2$ basis functions, essentially one for each of the $K + 1$ regions plus a constant offset. In fact the last basis function spans the last region, the one before last the last two regions, etc., but this amounts to (and achieves) the same thing. The function itself is given by equation 12.14. $K + 1$ independent linear fits would require $2(K + 1)$ parameters. But with K knots we have K continuity constraints and so K fewer free parameters. Thus we have $2(K + 1) - K = K + 2$ parameters, which is the number of coefficients of the basis function in equation 12.16 ($J = K + 2$). We can then proceed to infer these parameters using, for example, least squares.

The grey line in figure 12.4 shows an example of such a continuous piecewise linear spline with four knots. We can see how the continuity conditions at the knots modify the fit from comparison to figure 12.3: the boundary conditions force the fit in the central region

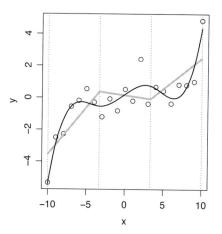

Fig. 12.4 Continuous linear spline (grey line) and continuous cubic spline (black line), each with four knots at the positions shown by the vertical dotted lines, fitted to 21 data points (open circles).

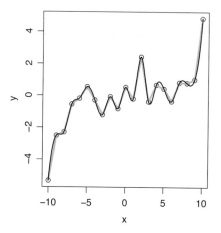

Fig. 12.5 As figure 12.4 but now with a knot at every data point, so that the function interpolates the points exactly.

to have a negative gradient, for example. The grey line in figure 12.5 shows the case when there is a knot at each data point. The function of course fits the data exactly.

Although such functions are continuous at the knots, they are not smooth, because their derivatives are not continuous. We can enforce this by using higher order piecewise functions. A quadratic spline has continuous gradients (first derivatives) but still has discontinuous second derivatives at the knots. Such fits look odd. A cubic spline – using cubic polynomials – is continuous in the zeroth, first, and second derivatives, and is the lowest order spline that looks smooth to the human brain (aliens may think otherwise). The basis functions are

$$
\begin{aligned}
h_1(x) &= 1 \\
h_2(x) &= x \\
h_3(x) &= x^2 \\
h_4(x) &= x^3 \\
h_{k+4}(x) &= (x - v_k)_+^3 \quad \text{where} \quad k = 1 \ldots K.
\end{aligned}
\tag{12.17}
$$

An arbitrary cubic equation would require four parameters in each region, thus requiring a total of $4(K + 1)$ parameters. The cubic spline has three continuity constrains for each knot, and thus $3K$ fewer parameters, leaving a total of $4(K+1) - 3K = K+4$ parameters, which is equal to the number of basis functions ($J = K + 4$). The black line figure 12.4 shows an example of this with four knots.

The more knots we choose, the better the curve will fit the data. If we set the number of knots equal to the number of data points, then the spline will fit the data exactly, as can be seen in figure 12.5.

12.3.2 Smoothing splines

A practical issue with splines is deciding how many knots to use. The more we use the better the fit, to the point of an exact fit when $K = N$, the number of data points. As the data are presumably noisy, this will be an overfit. A solution to this is to use *smoothing splines*. We put a knot at every data point ($K = N$) but then penalize complex solutions through the addition of a regularizer (section 12.2.1). A measure of the complexity of a function is its degree of curvature, which in turn is measured by the magnitude of the second derivative. Thus a suitable regularizer is

$$
\int \left(\frac{d^2 f}{dx^2} \right)^2 dx.
\tag{12.18}
$$

A smoothing spline uses a cubic spline basis (equation 12.17) to minimize

$$
\mathcal{E} = \sum_i [y_i - f(x_i)]^2 + \lambda \int \left(\frac{d^2 f}{dt^2} \right)^2 dt.
\tag{12.19}
$$

where t is a dummy variable for x. Compare this with the ridge regression objective function (equation 12.8). Instead of penalizing the magnitude of the polynomial coefficients equally for all components of the basis function, a smoothing spline penalizes the overall curvature of the function. It is instructive to write this objective function in matrix format. For N data points the basis function expansion (equation 12.14) can be written $\mathbf{f} = H\boldsymbol{\beta}$ where H is the $N \times J$ matrix of basis function evaluations. Equation 12.19 can then be written as

$$
\mathcal{E} = (\mathbf{y} - H\boldsymbol{\beta})^{\mathsf{T}}(\mathbf{y} - H\boldsymbol{\beta}) + \lambda \boldsymbol{\beta}^{\mathsf{T}} \Omega \boldsymbol{\beta}
\tag{12.20}
$$

where Ω is a $J \times J$ matrix with elements

$$
\Omega_{j,j'} = \int \frac{d^2 h_j}{dt^2} \frac{d^2 h_{j'}}{dt^2} dt.
\tag{12.21}
$$

The regularizer may be written like this because the function is a linear combination of the basis functions.[3] Differentiating the objective function with respect to β and setting to zero we get the solution for β

$$0 = -H^{\mathsf{T}}(\mathbf{y} - H\beta) + \lambda\Omega\beta$$
$$0 = -H^{\mathsf{T}}\mathbf{y} + (H^{\mathsf{T}}H + \lambda\Omega)\beta$$
$$\beta = (H^{\mathsf{T}}H + \lambda\Omega)^{-1}H^{\mathsf{T}}\mathbf{y}. \tag{12.22}$$

This has the same form as the solution for ridge regression (equation 12.9), except that the data are now expressed through basis functions, and the regularizer is not the identity matrix but the matrix of integrals of the products of the second derivatives of the basis functions.

The values of the model function at the N inputs are

$$\mathbf{f}(\mathbf{x}) = H\beta \tag{12.23}$$

which can be written

$$\mathbf{f}(\mathbf{x}) = S\mathbf{y} \quad \text{where}$$
$$S = H(H^{\mathsf{T}}H + \lambda\Omega)^{-1}H^{\mathsf{T}}. \tag{12.24}$$

The $N \times N$ matrix S is called the *smoother matrix*. It depends only on the fixed input data $\{x\}$ and the regularization constant λ, but not on $\mathbf{y}$. Equation 12.24 tells us that the fitted curve is therefore linear in $\mathbf{y}$.

We saw in section 4.5 that the model predictions arising from ordinary least squares could be written in a similar form to this, namely $\mathbf{f}(\mathbf{x}) = X\beta$, and its corresponding smoother matrix is $S_{\mathrm{OLS}} = X(X^{\mathsf{T}}X)^{-1}X^{\mathsf{T}}$ (see equation 4.57). Assuming $(X^{\mathsf{T}}X)$ is not a singular matrix, this has a unique solution with $N - J$ degrees of freedom, where J is the number of parameters in the model. It turns out that the degrees of freedom is also given by the trace of the matrix, $\mathrm{trace}(S_{\mathrm{OLS}})$. So for any model fit that can be written as $S\mathbf{y}$ where S does not depend on $\mathbf{y}$, we can define the *effective degrees of freedom* as $\mathrm{trace}(S)$. For splines, S depends only on $\{x\}$ and λ, so for given input data, specifying the effective degrees of freedom is equivalent to specifying λ. This can be a more intuitive way of setting the degree of regularization.

Figure 12.6 shows smoothing splines with different degrees of smoothing.

12.3.3 R code

You can use the following code to experiment with the splines discussed in the previous section. Run as is, it will create the plots shown in figures 12.3 to 12.6. The data are x and y. Curves are plotted by evaluating them on a grid of x points defined as the vector xp.

To fit the discontinuous splines (figure 12.3), I explicitly fit functions separately to the regions between the knots. I define these regions using two lists: xSeg to partition the data x and xpSeg to partition the plotting points xp. xSeg[[s]] indexes the points in x

---

[3] This also takes care of the potentially different units of the different basis functions h_j, because β_j has units y/h_j. The regularizer λ has units x^3.

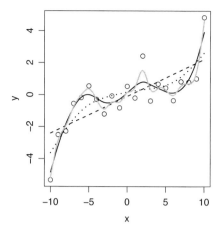

Fig. 12.6 A smoothing spline fitted to 21 data points (open circles) using four different values for the effective degrees of freedom (amounts of smoothing): 2 (dashed black line), 4 (dotted black line), 8 (solid black line), 15 (grey line). A spline with 21 effective degrees of freedom would fit the points exactly.

corresponding to region s (1, 2, or 3), and likewise for xpSeg[[s]]. The data are fitted using lm on the subset of the x data in that region. xp[xpSeg[[s]]] gives the x values of the points in the region at which we wish to evaluate the function, which is done with predict. The plot is then made with lines.

To fit and plot the continuous splines (figure 12.4) I again use predict and lm, but now with the spline basis functions defined by bs. Combined with lm this automatically divides the data up into the regions specified by the knots, fits them with the boundary conditions appropriate to the order (degree), and predicts only at the xp values within the region.

For the exact splines (figure 12.5) I use the function approxfun, which does exact linear (or constant) interpolation, and splinefun, which does exact cubic spline interpolation.

For the smoothing splines (figure 12.6) I use smooth.spline, which allows us to specify the amount of smoothing using the degrees of freedom. If you want to specify the number of knots instead, the syntax (for four knots) is as follows.

```
smooth.spline(x, y, spar=0, cv=FALSE, all.knots=FALSE, nknots=4)
```

R file: splines.R

```
##### Demonstration of 1D splines

library(splines) # for bs

# Simulate data
set.seed(101)
x <- -10:10
y <- sin(2*pi*x/10) + 0.005*x^3 + 1*rnorm(length(x), 0, 1)
xp <- seq(from=min(x), to=max(x), length.out=1e3) # for plotting splines
```

```
# Discontinuous splines of orders 0,1,3 with definable knots (here 4)
# I use lm() to explicitly fit within each region.
pdf("splines_discontinuous.pdf", 4, 4)
par(mfrow=c(1,1), mar=c(3.5,3.5,0.5,1), oma=c(0.5,0.5,0.5,0.5),
    mgp=c(2.2,0.8,0), cex=1.0)
plot(x,y)
knots=c(-3.33, 3.33)
abline(v=c(knots, range(x)), lty=3)
xSeg <- list(which(x<=knots[1]), which(x>knots[1] & x<knots[2]),
             which(x>=knots[2]))
xpSeg <- list(which(xp<=knots[1]), which(xp>knots[1] & xp<knots[2]),
              which(xp>=knots[2]))
for(s in 1:3) { # loop over the three subregions
  lines(xp[xpSeg[[s]]], predict(lm(y ~ 1, subset=xSeg[[s]]),
                                newdata=data.frame(x=xp[xpSeg[[s]]])),
        lwd=1.5, lty=2)
  lines(xp[xpSeg[[s]]], predict(lm(y ~ x, subset=xSeg[[s]]),
                                newdata=data.frame(x=xp[xpSeg[[s]]])),
        lwd=3, col="grey70")
  lines(xp[xpSeg[[s]]], predict(lm(y ~ x + I(x^2) + I(x^3),
                                   subset=xSeg[[s]]),
                                newdata=data.frame(x=xp[xpSeg[[s]]])),
        lwd=1.5, col="black")
}
dev.off()

# Continuous splines of orders 1,3 with definable knots (here 4)
# bs() automatically uses extreme data points as additional knots.
pdf("splines_continuous.pdf", 4, 4)
par(mfrow=c(1,1), mar=c(3.5,3.5,0.5,1), oma=c(0.5,0.5,0.5,0.5),
    mgp=c(2.2,0.8,0), cex=1.0)
plot(x,y)
knots=c(-3.33, 3.33)
abline(v=c(knots, range(x)), lty=3)
lines(xp, predict(lm(y ~ bs(x, knots=knots, degree=1)),
                  newdata=data.frame(x=xp)), lwd=3, col="grey70")
lines(xp, predict(lm(y ~ bs(x, knots=knots, degree=3)),
                  newdata=data.frame(x=xp)), lwd=1.5, col="black")
dev.off()

# Exact splines (i.e knot at each point) of orders 1,3
pdf("splines_exact.pdf", 4, 4)
par(mfrow=c(1,1), mar=c(3.5,3.5,0.5,1), oma=c(0.5,0.5,0.5,0.5),
    mgp=c(2.2,0.8,0), cex=1.0)
plot(x,y)
# Plots exact linear spline then exact cubic spline
lines(xp, approxfun(x, y, method="linear")(xp), lwd=3, col="grey70")
lines(xp, splinefun(x, y)(xp), lwd=1.5, col="black")
dev.off()

# Smoothing splines with various degrees of freedom
pdf("splines_smoothing.pdf", 4, 4)
par(mfrow=c(1,1), mar=c(3.5,3.5,0.5,1), oma=c(0.5,0.5,0.5,0.5),
    mgp=c(2.2,0.8,0), cex=1.0)
plot(x,y)
```

```
lines(predict(smooth.spline(x, y, df= 2), xp), lwd=1.5, lty=2, col="black")
lines(predict(smooth.spline(x, y, df= 4), xp), lwd=2,   lty=3, col="black")
lines(predict(smooth.spline(x, y, df= 8), xp), lwd=1.5, lty=1, col="black")
lines(predict(smooth.spline(x, y, df=15), xp), lwd=2,   lty=1, col="grey70")
dev.off()
```

12.4 Regression kernels

The smoothing spline achieves flexibility by defining a knot at every data point and then fitting a cubic function to each with two sets of constraints: (1) continuity of the zeroth, first, and second derivatives; (2) a regularization term. There are other approaches to achieve a smooth fit, and one is to use kernels. I introduced kernels in section 7.2.2 as a way of doing density estimation. A kernel is a weighted function of data in which the weights depend on the distance of the data points from the point of evaluation. Let's see how we can use kernels for doing local regression.

Consider a set of N data points $\{x_i, y_i\}$ for which we want to get an estimate of the regression function $f(x)$. A simple kernel estimate of the function at any x is just the mean of all those points within a distance λ, i.e.

$$f(x) = \frac{1}{N_\lambda} \sum_{i \in D_\lambda(x)} y_i \tag{12.25}$$

where $D_\lambda(x)$ is that set of points, and N_λ is its size. While this gives an estimate anywhere, we do not infer a global parametric function for the data, and also not an additive model like a basis function expansion. Rather we evaluate the function at specific points. To plot our estimate of the function, we simply evaluate it on a sufficiently dense and broad grid of points. For this reason, this approach to regression is often called a *non-parametric model*. We can think of moving a box over the data and taking the average in it as we go. This will smooth out the variations in the data. We can also use a variable size box by using a nearest neighbour kernel instead, as defined in section 7.2.2.

We are not limited to taking the mean of the points within the kernel. We could take the median instead, in which case we end up with a *median filter*. We could also fit a straight line, or quadratic, or higher order polynomial to the points within the kernel, all of which are examples of *local polynomial regression*. In practice we would not usually go beyond a quadratic function, and linear often suffices. Note that even if we fit a linear function locally, the resulting function is not globally linear.

The problem with these kinds of functions/filters is that they are often not very smooth, because as we move the filter, points jump in and out of the kernel, changing the mean, median, or polynomial fit, sometimes by large amounts. A solution to this is to assign weights to data points that decrease smoothly with increasing distance from the centre of the kernel. Thus we can generalize the above considerations to estimate the function as

$$f(x) = \frac{\sum_{i=1}^{N} K(x, x_i) \, y_i}{\sum_{i=1}^{N} K(x, x_i)} \tag{12.26}$$

which is the *Nadaraya–Watson kernel estimator*. Defining

$$u = \frac{x - x_i}{\lambda} \tag{12.27}$$

we normally write the kernel as $K(u)$, where λ is its bandwidth. One choice for the kernel is a Gaussian

$$K(u) = \frac{1}{\sqrt{2\pi}} \exp\left(-\frac{1}{2}u^2\right). \tag{12.28}$$

We don't actually need the normalization constant for regression because it will cancel out in equation 12.26, but I will write all kernels here as normalized. The Gaussian kernel has the disadvantage of infinite support (it is non-zero everywhere): distant points can influence the function, which is often undesirable. For this reason we often use truncated kernels. A popular choice is the Epanechnikov kernel (see figure 12.7)

$$K(u) = \begin{cases} \frac{3}{4}(1 - u^2) & \text{if} \quad |u| \leq 1 \\ 0 & \text{otherwise.} \end{cases} \tag{12.29}$$

The larger the bandwidth the greater the number of points in the kernel, so the more the estimator will smooth out the data (smaller variance, but larger bias). A small kernel, in contrast, will follow the variations in the data more closely (larger variance, but smaller bias). We saw in section 7.2.2 one method for determining the bandwidth from the data using the L^2 risk function. We could also use a kernel with a variable-sized bandwidth, which is what we get with the k nearest neighbours kernel. This can be useful if the density of the data along with x-axis is very non-uniform. With a fixed bandwidth (and kernel with finite support), λ must be at least as large as half the separation between the two most-separated neighbouring points in order to get an estimate everywhere. Yet this may be so large that it would over smooth the data where it is dense (leading to a large bias).

The following R code experiments with kernels for one-dimensional regression, by applying three different kernel methods to smooth the data points (open circles) shown in figure 12.8. The code first uses ksmooth (from the stats library) to do zeroth-order ($f(x) = b_0$) kernel regression with a constant Nadaraya–Watson kernel (kernel="box"). We see from the figure that the resulting curve is discontinuous. If you experiment with the size of the bandwidth you will find that these discontinuities are determined by the density of the data, and do not really vanish even if you increase the bandwidth. If you change the kernel to do Gaussian (kernel="normal") rather than constant weighting, you get a much smoother curve.

The code then uses locpoly (from the KernSmooth library) to do first-order ($f(x) = b_0 + b_1 x$) kernel regression, again with constant weighting. Increasing the order of this polynomial (with degree) does not have much impact on the shape of the curve.

The third method is loess (from the stats library). This smooths the data by fitting a first-order polynomial to the 20 nearest neighbours (set by span). The loess function

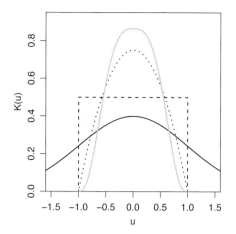

Four different normalized kernels: uniform (dashed black line), Gaussian (solid black line), Epanechnikov (dotted black line), and tricubic (grey line). Apart from the Gaussian, these are non-zero only for $|u| < 1$.

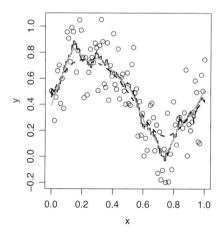

Three different kernel smoothers fitted to some data points (open circles). Solid black line: constant kernel with a bandwith of 0.1 (using `ksmooth`). Dashed black line: local linear function regression with Gaussian kernel weighting with a bandwith of 0.1 (using `locpoly`). Grey line: local linear function regression with constant kernel weighting using nearest 20 neighbours (using `loess`).

weights the points using a tricubic kernel

$$K(u) = \begin{cases} \dfrac{70}{81}(1 - |u|^3)^3 & \text{if} \quad |u| \le 1 \\ 0 & \text{otherwise.} \end{cases} \tag{12.30}$$

This is also shown in figure 12.7. It has a slightly nicer behaviour near to $|u| = 1$ than the Epanechnikov kernel.

I encourage you to experiment with changing the parameters in this code, in particular the degree of polynomials (by setting degree to 0, 1, or 2) and the bandwidth of the kernels (by setting bandwidth). Check the help pages of the methods to see how the bandwidths are defined. For all three cases the smoothing function is evaluated on a dense grid of points (much denser than the data). I then use lines to connect these in the plot to give the appearance of a continuous line. For ksmooth and loess I define this grid as xp, with 1000 equally spaced points. For locpoly the grid is defined internally to the function; the default is 400 equally spaced points.

R file: regression_kernels.R

```
##### Demonstration of regression kernels

library(KernSmooth) # for locpoly

pdf("regression_kernels.pdf", 4, 4)
par(mfrow=c(1,1), mar=c(3.5,3.5,0.5,1), oma=c(0.5,0.5,0.5,0.5),
    mgp=c(2.2,0.8,0), cex=1.0)

# Simulate data
x <- seq(from=0, to=1, length.out=100)
f <- function(x){0.5 + 0.4*sin(2*pi*x)}
set.seed(10)
y <- f(x) + rnorm(n=x, mean=0, sd=0.2)
plot(x, y)

# Constant regression using a constant kernel of specified bandwidth.
xp <- seq(from=min(x), to=max(x), length.out=1e3)
lines(xp, ksmooth(x=x, y=y, kernel="box", bandwidth=0.1, x.points=xp)$y,
      lwd=1.5)

# Local polynomial (oder=degree) regression using a Gaussian kernel
# with specified bandwidth and polynomial degree (here constant)
mod <- locpoly(x=x, y=y, degree=1, bandwidth=0.1, gridsize=1e3)
lines(mod$x, mod$y, lwd=2, lty=2)

# Local polynomial (oder=degree) regression using fraction span of data
# around each prediction point, i.e. span*length(x) nearest neighbours.
# It won't work if span is too small.
xp <- seq(from=min(x), to=max(x), length.out=1e3)
yp <- predict(loess(y ~ x, span=0.2, degree=1), newdata=data.frame(x=xp))
lines(xp, yp, lwd=2, col="grey60")

dev.off()
```

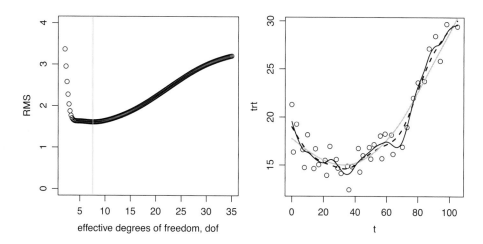

Fig. 12.9 Applying a smoothing spline to the rat diet data. The left panel shows how the root mean square (RMS) of the residuals varies as a function of the degrees of freedom in the smoothing spline, computed using cross-validation. The minimum, marked with a vertical dashed line, is at 7.6. The black dashed line in the right panel is the corresponding smoothing spline. The grey line and solid black line are smoothers with half and double this number of degrees of freedom respectively.

12.5 A non-parametric smoothing problem

I'll wrap up this analysis of models for curve fitting by applying both smoothing splines and a kernel regression smoother to the same data and comparing the results. For this we will use the "rat diet" (!) data set embedded in the `fields` package in R. The scientific content doesn't concern us here. We are just interested in getting a smooth, plausible fit of the variable `rat.diet$trt` to the variable `rat.diet$t` for the $N = 39$ data points.

I start with the smoothing spline (section 12.3.2), whereby I now use the `sreg` function in the package `fields`. I control the amount of smoothing through the effective degrees of freedom (dof). I vary this from 2 to 35 in steps of 0.2. For each value of dof, I use leave-one-out cross-validation to calculate the RMS of the fits. That is, for each dof I fit N cubic splines, each using a different set of $N - 1$ data points, and I calculate the residual on the one point left out for that N. These are combined to form the RMS. The results of this are shown in the left panel of figure 12.9. The smallest RMS of 1.61 is achieved with $dof = 7.6$. The data are shown in the right panel, overplotted with this best fitting smoother, as well as that smoother with dof equal to half and twice the optimal value. Fewer degrees of freedom corresponds to a smoother line. The splines are plotted by computing them on a dense grid of 1000 points. The R code to do this investigation and to make this plot is below.

R file: `ratdiet_splines.R`

```
##### Application of smoothing splines to the rat.diet data

library(fields) # for sreg and rat.diet
Ndat <- nrow(rat.diet)

# Calculate RMS using LOO-CV
dofValues <- seq(2,35,0.2)
Ndof <- length(dofValues)
rss   <- rep(x=0, times=Ndof)
for (k in 1:Ndof) {
  for(i in 1:Ndat) {
    mod  <- sreg(rat.diet$t[-i], rat.diet$trt[-i], df=dofValues[k])
    pred <- predict(mod, rat.diet$t[i])
    rss[k] <- rss[k] + (pred - rat.diet$trt[i])^2
  }
}
rms <- sqrt(rss/Ndat)

pdf("ratdiet_splines.pdf", 8, 4)
par(mfrow=c(1,2), mar=c(3.5,3.5,0.5,0.5), oma=c(0.5,0.5,0.5,0.5),
    mgp=c(2.2,0.8,0), cex=1.0)

# Plot RMS vs dof
plot(dofValues, rms, xlab="effective degrees of freedom, dof", ylab="RMS",
     ylim=c(0,4))
bd <- dofValues[which.min(rms)]
abline(v=bd, col="grey")
text(bd, 5, bd, pos=4)
cat(bd, rms[which.min(rms)])

# Plot data with three different spline smoothers
plot(rat.diet$t, rat.diet$trt, xlab="t", ylab="trt")
xp <- seq(from=min(rat.diet$t), to=max(rat.diet$t), length.out=1e3)
lines(xp, predict(sreg(rat.diet$t, rat.diet$trt, df=bd),   xp), lwd=2,
      lty=2, col="black")
lines(xp, predict(sreg(rat.diet$t, rat.diet$trt, df=bd/2), xp), lwd=2,
      lty=1, col="grey70")
lines(xp, predict(sreg(rat.diet$t, rat.diet$trt, df=2*bd), xp), lwd=1.5,
      lty=1, col="black")

dev.off()
```

I now apply a kernel smoother (section 12.4) to the same data, using the nearest neighbour kernel and doing local linear function regression. I again use the function loess. In its tricubic kernel (equation 12.30), λ is the distance to the furthest neighbour. The amount of smoothing is controlled via the number of nearest neighbours k. Again I use leave-one-out cross-validation to calculate the RMS of the fits, and plot this now against the number of neighbours. The number of neighbours is specified using span, as the fraction of the data available (i.e. k/Ndat). The result is shown in the left panel of figure 12.10. We see that the minimum is much flatter than the corresponding plot from the smoothing splines (figure 12.9). The optimal value is at $k = 16$ neighbours and achieves an RMS of 1.48, slightly smaller than with the smoothing splines. The data are shown in the right panel, overplotted with this best fitting smoother, as well as the smoother with half and twice as many neigh-

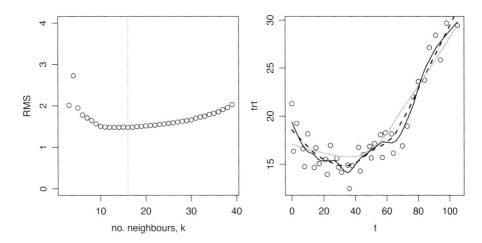

Fig. 12.10 Applying the kernel regression method to the rat diet data. The left panel shows how the root mean square of the residuals varies as a function of the number of neighbours used in the kernel. The vertical range is the same as in figure 12.9 to ease comparison. The point at $k = 2$ is off the scale with an RMS of 18.4 (it would be the left-most point). The minimum, marked with a vertical dashed line, is at 16 neighbours. The black dashed line in the right panel is the corresponding regression curve. The solid black line and grey line are smoothers with half (8) and double (32) this number of neighbours respectively.

bours. More neighbours corresponds to a smoother line. As before, the smoother is plotted by computing it on a dense grid of 1000 points. This is all done by the R code below.

R file: `ratdiet_kernelregression.R`

```
##### Application of kernel regression to the rat.diet data

library(fields) # for rat.diet
attach(rat.diet)
Ndat <- length(t)

# Calculate RMS using LOO-CV
# loess does not work with just one nearest neighbour, so I just do 2:Ndat.
# It also does not permit extrapolation, so I don't include the two test
# sets which would each be just the two extreme points. One could set
# surface="direct" as a loess option, but this gives worse fits for low k.
rss <- rep(NA, Ndat)
kRange <- 2:Ndat
for(k in kRange) {
  rss[k] <- 0
  for(i in 2:(Ndat-1)){ # RSS will be sum of squares of Ndat-2 residuals
    pred   <- predict(loess(trt ~ t, span=k/Ndat, degree=1, subset=-i),
                      newdata=data.frame(t=t[i]))
    rss[k] <- rss[k] + (pred - trt[i])^2
  }
}
```

```
}
rms <- sqrt((rss[kRange])/(Ndat-2))

pdf("ratdiet_kernelregression.pdf", 8, 4)
par(mfrow=c(1,2), mar=c(3.5,3.5,0.5,0.5), oma=c(0.5,0.5,0.5,0.5),
    mgp=c(2.2,0.8,0), cex=1.0)

# Plot RMS vs number of neighbours
plot(kRange, rms, xlab="no. neighbours, k", ylab="RMS", ylim=c(0,4))
bd <- kRange[which.min(rms)]
abline(v=bd, col="grey")
text(bd, 5, bd, pos=4)
cat(bd, rms[which.min(rms)])

# Plot data with three different kernel smoothers
plot(t, trt)
xp <- seq(from=min(t), to=max(t), length.out=1e3)
yp <- predict(loess(trt ~ t, span=16/Ndat, degree=1),
              newdata=data.frame(t=xp))
lines(xp, yp, lwd=2,   lty=2, col="black")
yp <- predict(loess(trt ~ t, span=32/Ndat, degree=1),
              newdata=data.frame(t=xp))
lines(xp, yp, lwd=2,   lty=1, col="grey70")
yp <- predict(loess(trt ~ t, span=8/Ndat, degree=1),
              newdata=data.frame(t=xp))
lines(xp, yp, lwd=1.5, lty=1, col="black")

dev.off()
detach(rat.diet)
```

12.6 Numerical optimization (mode finding)

When faced with a parameter inference problem, we ideally want to find the full posterior PDF over the model parameters. But for high-dimensional problems this may be too difficult or too time consuming. For other problems the posterior may be so sharply peaked that finding the full posterior is unnecessary: the maximum plus some estimate of its width is a sufficiently good approximation (as obtained by the quadratic approximation in section 7.1, for example). This can occur when the data are highly informative.

In such situations we will want to find the maximum of the posterior (or likelihood), which we discussed in section 4.4. In principle this is found by differentiation, but for most problems there will not be an analytic solution. We must instead proceed iteratively. The principle is straightforward. We define a starting point, take a sensible step in the parameter space, re-evaluate the function, and either terminate because we're sufficiently close to a maximum, or we iterate. Like Monte Carlo (chapter 8) this is a step-wise approach, but now we are no longer interested in a representative sampling of the distribution; we just want to get to a maximum (within some tolerance) as fast as possible. It is important to realise that generic posteriors are multimodal; an optimization method will find a local maximum, which is not necessarily the global maximum. In general it is impossible to know, without evaluating the posterior everywhere, whether the maximum found is the global one or just a local one. Nonetheless, if the maximum found is "high enough" it may be an adequate solution. Posteriors can have complex shapes, especially in high dimensions, so a local or global maximum could lie far from the mean or median.

There is a vast literature on numerical optimization methods so I will just give a brief and selective introduction here.

Let $f(\theta)$ be the function we wish to maximize with respect to θ. If θ is a scalar, then straightforward methods like the golden section search can be used to find the maximum once it is known to lie within some interval. Here I consider the more general (and difficult) problem in which θ is an N-dimensional vector (so I write it in boldface to emphasize this) in which case it is no longer possible to bracket the maximum with a finite number of points.

A method that just uses evaluations of the function is the *simplex method*, also known as the Nelder-Mead or amoeba method. The simplex in N dimensions is an object with $N + 1$ vertices. In two dimensions this is a triangle, in three a tetrahedron, and so on. We first initialize the problem by selecting $N + 1$ values of θ and computing $f(\theta)$ at these. We identify the "worst" of these (the one with the lowest value of f) and compute a new point by reflecting this point through the centroid of the other N points. If the value of f at the new point is higher than at the other points, then it replaces the worst point. If not, there is a more involved scheme for deciding how update the points. This procedure is iterated, with the result that the simplex steps slowly towards the maximum. Various criteria exist to determine exactly how the simplex behaves and how the points are updated, with the goal of achieving faster convergence.

Methods which take advantage of knowledge of the gradient ∇f may be more efficient. The simplest of these is *gradient ascent*. If the current position is $f(\theta_i)$, then we take a small step in the direction of positive gradient to give the next position

$$\theta_{i+1} = \theta_i + \alpha\,\nabla f(\theta_i) \tag{12.31}$$

where α is a small positive quantity. (The positive sign in front of α should be switched to a negative sign if we wish to minimize f.) Note that α is not dimensionless: an optimal size depends both on the scale of the parameters and the gradient. It could be adapted at each iteration. A larger α will make larger steps and potentially achieve faster convergence, but if the gradient is changing rapidly this could lead to stepping over the maximum and then continuing the search in the wrong part of the parameter space. While simple, gradient ascent can be notoriously slow to converge if the search finds itself on a nearly flat ridge. Like all gradient-based methods it can also get stuck at local maxima. Various ways exist to circumvent this, such as adding to the update a proportion of the gradient from the previous iteration, $\nabla f(\theta_{i-1})$, or restarting the algorithm at a different point. I will use gradient descent in section 12.7.

Gradient methods can be accelerated further by using the second derivatives of f, if they exist. This is done by *Newton's method*. If Ω_i is the Hessian matrix – the matrix of all second partial derivatives – evaluated at θ_i, then the update rule is

$$\theta_{i+1} = \theta_i - \alpha\,\Omega_i^{-1}\nabla f(\theta_i) \tag{12.32}$$

again for a small positive α. (Note the negative sign, which is present for both maximization and minimization.) There are variations on this approach, such as the Gauss–Newton or Levenberg–Marquardt algorithms.

All numerical optimization methods face the problem of when to declare convergence. The *exact* maximum is unattainable, and anyway not necessary. But we need to get close enough. Common convergence criteria are that the absolute or relative value of the change in f compared to the previous iteration is below some pre-defined value.

There exist more advanced optimization methods, such as BFGS, genetic algorithms, simulated annealing, and particle swarm. The methods vary according to what information is available (e.g. gradients), whether constraints are involved, how the data are represented, etc. Even if gradients are not available in an analytic form it may be possible to calculate them numerically. The primary goal of these methods is always the same, however: how to find the highest (local) optimum with the fewest function evaluations.

12.7 Bootstrap resampling

The very fact that we use a numerical method to find the maximum of the posterior often implies that we don't have a simple expression for the width of the posterior, and so don't have a measure of the precision of estimate of the maximum.

A conceptually simple way to determine the accuracy of an estimator is to recalculate it using different subsets of the data, and then to calculate the standard deviation or a confidence interval over the set of estimators. Such an approach is purely empirical, in the sense that it does not assume a model for the distribution of the data. It is therefore rather general. Suppose we have a set of N measurements $\{x_i\}$. We draw N points at random from this set *with replacement*. This sample of data is called the *bootstrap sample*. We then calculate our estimator from this sample; call it $\hat{x}$. It could be something straightforward, like the mean or mode, or it could be a more complicated function. We then repeat this K times, to give K bootstrap samples and a corresponding set of K estimators $\{\hat{x}_k\}$. The standard deviation of this set is a measure of the uncertainty of our estimator, with the estimator itself being obtained from the original set of data.

In this process our data $\{x_i\}$ are essentially acting as a discrete distribution that approximates the true, unknown distribution of x (assuming that the N measurements are independent). This is why we make the draws with replacement: we don't change a distribution just because we've drawn from it. Any technique that involves estimating a quantity by sampling with replacement is called *bootstrapping*, although it is usually used to obtain an empirical estimate of the uncertainty in an estimator.

One application could be to find the uncertainty on the result of numerically optimizing a posterior. We repeat the optimization K times, each time using a different bootstrap sample of the data to estimate the posterior. The standard deviation of these maxima is a bootstrap estimate.

To illustrate this, let us return to the problem introduced in section 3.5, of estimating the distance to a star from its parallax. Suppose that instead of a single parallax measurement we now have N independent parallax measurements of the same star, the set $D = \{\varpi_i\}$. Let us again adopt a Gaussian noise model with a common standard deviation σ_ϖ for all measurements. The likelihood of each measurement is given by equation 3.20, and the likelihood of the complete data set is the product of such terms. I adopt the prior in equation 3.31. The unnormalized posterior is

$$P^*(r\,|\,D) = r^2 \exp(-r/L) \prod_{i=1}^{N} \exp\left[-\frac{1}{2\sigma_\varpi^2} \left(\varpi_i - \frac{1}{r} \right)^2 \right] \qquad (12.33)$$

$$\ln P^*(r\,|\,D) = 2\ln r - \frac{r}{L} - \frac{1}{2\sigma_\varpi^2} \sum_{i=1}^{N} \left(\varpi_i - \frac{1}{r} \right)^2 \qquad (12.34)$$

where the neglected normalization constant is a function of D, σ_ϖ, and L, but not of r. We will take the maximum of the posterior as our estimator of r (for which the normalization constant is not required). It's easier to find this by maximizing $\ln P^*$. Differentiating equation 12.34 gives

$$\frac{d\ln P^*}{dr} = \frac{2}{r} - \frac{1}{L} - \frac{N}{\sigma_\varpi^2 r^2} \left(\langle \varpi \rangle - \frac{1}{r} \right) \qquad (12.35)$$

where $\langle \varpi \rangle = (1/N) \sum_i \varpi_i$ is the mean of the parallax measurements. We could solve this algebraically by setting the gradient to zero, because multiplying by r^3 gives us a cubic equation.[4] But for illustration purposes I will use the numerical gradient ascent method mentioned in the previous section.

The R code at the end of this section performs the bootstrapping for a set of $N = 10$ measurements. As before, parallaxes are in arcseconds and distances are in parsecs. The script first defines simple functions for returning the prior, likelihood, unnormalized posterior, and gradient of the logarithm of the unnormalized posterior. The function post.grid evaluates the posterior on a regular grid, scaling it so the maximum is unity. This is only used for plotting. draw.sample draws a bootstrap sample of size N from the data. The gradient.ascent function is generic for any one-dimensional function. A suitable value of the α parameter depends on both the scale of the gradient and the size of sensible steps to make in r. In this simple case it is best found by trial and error. It turns out that a value of order of 10^3 gives good convergence. Convergence is defined as having been achieved when the fractional change in gradient from the previous iteration is less than rel.tol. Its value is therefore scale independent. The maximum number of iterations is set to be so large that the convergence criterion is almost always reached first. Having defined these functions, the script goes on to define some data, set the parameters, and then to find numerically the maximum of the posterior for each of a large number K (Kboot in the code) of bootstrap samples. I set the initial value of the parameter in the optimization procedure to the inverse of the median of the bootstrapped data sample, which is a reasonable guess at the star's distance.

The left panel of figure 12.11 shows the unnormalized posterior (equation 12.33) using all of the data as well as posteriors for seven example bootstrap samples. For each of the

---

[4] Equation 12.35 is the same as equation 19 in Bailer-Jones (2015) – which is for the single measurement case – when replacing ϖ with $\langle \varpi \rangle$ and σ_ϖ^2 with σ_ϖ^2/N in the latter equation.

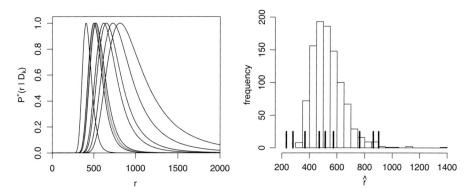

Fig. 12.11 Use of bootstrapping to estimate the standard deviation of the maximum of a distribution, in this case $P^*(r|D)$, the (unnormalized) posterior PDF of equation 12.33. Left: the thick grey line is the posterior calculated using all the data D. The black lines are posteriors calculated using seven example bootstrap samples of the data D_k ($k = 1\ldots7$). All distributions are scaled to have their maximum at one. Right: histogram of the maxima of the posterior for $K = 1000$ different bootstrap samples. For comparison the data D (plotted as $1/\varpi_i$) are shown with thick black vertical lines (the tenth point at $1/\varpi_i = 2780$ is not shown).

$k = 1\ldots K$ bootstrap samples, the code locates the maximum of the posterior $\hat{r}_k$ by gradient ascent. The initial and final parameter values as well as the number of iterations done are written to the screen, so you can see whether convergence was reached in each case. The set of all K such maxima is $\{\hat{r}_k\}$ (here $K = 1000$). A histogram of $\{\hat{r}_k\}$ is shown in the right panel of figure 12.11. The standard deviation σ of this set is the bootstrap estimate of the standard deviation in our distance estimate. We find $\sigma = 113.4$.

Our distance estimate is the mode[5] of the posterior for the original data set, which is 511.3. The quantity σ is a measure of how accurately we can determine r from the data. It is *not* a measure of the "standard error in the mean" of $\{\hat{r}_k\}$ (see section 2.4). Increasing K will increase the precision with which we can estimate σ. That is, the variance in σ will reduce as $1/K$ in accordance with the central limit theorem (section 2.3), something we can verify by bootstrapping too. But increasing K will not continuously reduce the value of σ itself, because the scale of σ is determined by the ten data points we have. The bootstrap resampling quantifies how much the posterior mode varies on account of us having a finite amount of noisy data.

The quantity σ is also not a measure of the standard deviation of the posterior computed on the complete data set, even though in this particular case they have a similar size.[6] That they are not conceptually the same should be clear from the fact that our σ is related to

---

[5] This was found by gradient ascent and is the same as the analytic solution to equation 12.35 to the first decimal place.

[6] It is estimated numerically at the end of the R code below to be 115.6. This requires that we integrate the posterior. I do this by computing it on a dense grid, and then use equations 5.7 and 5.8 to calculate the variance.

the mode. We could instead have chosen to make bootstrap estimates of the variance of the mean, or of the 95% quantile, or of something else.

It is instructive to repeat this experiment with a different sized data set. Selecting just the first three data points of data and re-running the code, gradient ascent estimates the mode of the posterior to be 1567. This is far from the previous value due to the small amount of noisy data. The standard deviation of the bootstrap sample is now $\sigma = 276$. For comparison, the standard deviation of the posterior (computed from these three data points) is 370. As bootstrapping is done with replacement, the number of unique bootstrap data samples is given by equation 1.78 with $n = r = N$. For $N = 3$ the number of unique bootstrap samples is just ten. As the numerical optimization is deterministic and the starting point is always the same for a given bootstrap sample, there are likewise now just ten unique values in $\{\hat{r}_k\}$ (which you can verify with unique(rMode)). It doesn't make much sense to use bootstrap on such small samples. For $N = 10$ the number of unique samples rises to 92 378.

When we have a functional form for the posterior PDF – as in this example – it would be preferable to compute its standard deviation or a confidence interval (section 5.5) and use this a measure of the uncertainty our distance estimate, rather than using the bootstrap. I only use the bootstrap here to illustrate it on a problem for which we can verify the solution by a more direct method. Bootstrap is most useful when we only have a set of data, and do not have the functional form of the distribution the samples were drawn from. For example, if we only had the ten parallax measurements, and didn't know the likelihood or prior, we could estimate their mean and then estimate the uncertainty in this using the standard deviation of a set of bootstrap samples. This is done by the following, whereby I simply use the inverse parallax as the distance estimator.

```
rMean <- numeric(Kboot)
for(k in 1:Kboot) {
  rMean[k] <- 1/mean(draw.sample(data))
}
cat(1/mean(data), "+/-", sd(rMean), "\n")
```

This gives 492 ± 97. The mode and its bootstrapped standard deviation computed above was 511 ± 113.

R file: bootstrap.R

```
##### Demonstration of bootstrapping and gradient ascent

### Define functions

# Prior (normalized)
d.prior3 <- function(r, rlen) ifelse(r>0, (1/(2*rlen^3))*r^2*exp(-r/rlen), 0)

# Likelihood of one data point
d.like <- function(w, r, wsd) dnorm(x=w, mean=1/r, sd=wsd)

# Posterior (unnormalized) of set of data
d.post <- function(data, r, wsd, rlen) {
  prod(d.like(w=data, r, wsd))*d.prior3(r, rlen)
}
```

```
# Gradient of natural logarithm of d.post w.r.t r
grad.log.d.post <- function(r, d.post, data, wsd, rlen) {
  N <- length(data)
  dlnPdr <- 2/r - 1/rlen - (mean(data) - 1/r)*N/(wsd^2*r^2)
  return(dlnPdr)
  #return(dlnPdr * d.post(data, r, wsd, rlen))
}

# Calculate posterior on dense grid, scaled so mode=1.
# Return dataframe of r, posterior
post.grid <- function(data, wsd, rlen) {
  r <- seq(from=0, to=2*rlen, length.out=1e4)
  post <- numeric(length(r))
  for(i in 1:length(r)) {
    post[i] <- d.post(data=data, r=r[i], wsd, rlen)
  }
  return(data.frame(r=r, post=post/max(post)))
}

# Draw bootstrap sample
draw.sample <- function(data) sample(x=data, size=length(data),
                                     replace=TRUE)

# Gradient ascent (generic)
# Returns two element list:
# - par is optimized value of parameter
# - lastIt is iteration number reached
# gradFunc is a function which returns the gradient of the function to
# maximize. Its first argument must be the (scalar) value of the parameter.
# param is the intitial value of the parameter.
# '...' is used to pass anything else gradFunc needs (data etc).
# Search until relative change in gradient in less than rel.tol
# or maxIts iterations is reached.
grad.ascent <- function(gradFunc, param, ...) {
  rel.tol <- 1e-5
  alpha   <- 1e3
  maxIts  <- 1e4
  for(i in 1:maxIts) {
    paramNext <- param + alpha*gradFunc(param, ...)
    #cat(i, param, paramNext, "\n")
    if(abs(paramNext/param - 1) < rel.tol) {
      break
    } else {
      param <- paramNext
    }
  }
  return(list(par=paramNext, lastIt=i))
}

### Apply method

# Data (parallaxes in arcseconds; so distances are in parsecs)
wsd   <- 1.0e-3
data <- 1e-3*c(0.36, 1.11, 1.16, 1.31, 1.74, 1.94, 2.12, 2.72, 3.56, 4.30)
#data <- rnorm(n=10, mean=1/500, sd=wsd) # Above data were drawn like this
```

```r
# Set up parameters
rlen  <- 1000 # prior length scale
Kboot <- 1000 # no. bootstrap samples

# Plot some example posteriors
pdf("bootstrap_posteriors.pdf", width=5, height=4)
par(mfrow=c(1,1), mar=c(3.5,4.2,1,1.25), oma=0.1*c(1,1,1,1),
    mgp=c(2.3,0.9,0), cex=1.15)
z <- post.grid(data, wsd, rlen)
plot(z$r, z$post, type="l", lwd=3, col="grey70",
     xlab="r", ylab=expression(paste(P^symbol("*"), "(r | ", D[k], ")")),
     ylim=c(0,1.05), xaxs="i", yaxs="i")
set.seed(555)
for(j in 1:7) {
  dataSample <- draw.sample(data)
  z <- post.grid(dataSample, wsd, rlen)
  lines(z$r, z$post, type="l", lwd=1.2)
}
dev.off()

# Draw bootstrap samples and for each calculate the maximum of the
# log posterior using gradient ascent
rMode <- numeric(Kboot)
for(k in 1:Kboot) {
  dataSample <- draw.sample(data)
  rInit <- 1/median(dataSample) # initial distance for optimization
  opt <- grad.ascent(gradFunc=grad.log.d.post, param=rInit, d.post,
                     dataSample, wsd, rlen)
  cat(rInit, opt$par, opt$lastIt, "\n") # print initial, final, no. its
  rMode[k] <- opt$par
}
cat("sd of bootstrap samples =", sd(rMode), "\n")
pdf("bootstrap_histogram.pdf", 5,4)
par(mfrow=c(1,1), mar=c(3.5,3.5,1,1), oma=0.1*c(1,1,1,1), mgp=c(2.2,0.8,0),
    cex=1.15)
hist(rMode, breaks=25, xlim=c(200, 1400), main="",
     xlab=expression(hat(r)), ylab="frequency")
segments(x0=1/data, y0=0, x1=1/data, y1=25, lwd=3)
dev.off()

# Estimate distance to the star using the original data set
opt <- grad.ascent(gradFunc=grad.log.d.post, param=rInit, d.post, data,
                   wsd, rlen)
cat("Numerical estimate of mode =", opt$par, "\n")

# For comparison, compute sd of posterior from all data (need to normalize
# it first). r must extend to where post is very small
r <- seq(from=0, to=2*rlen, by=0.1)
post <- numeric(length(r))
for(i in 1:length(r)) {
  post[i] <- d.post(data=data, r=r[i], wsd, rlen)
}
rMean <- sum(post*r)/sum(post)
plot(r, post, type="l") # ensure we have sampled the full posterior
cat("sd of posterior =", sqrt(sum(post*(r-rMean)^2)/sum(post)), "\n")
```

References

Akaike H., 1973, Information theory and an extension of the maximum likelihood principle, in Petrov B.N. & Csáki F., Second International Symposium on Information Theory, Tsahkadsor, Armenia, USSR, 1971, Budapest: Akadémiai Kiadó, pp. 267–281

Akaike H., 1974, A new look at the statistical model identification, *IEEE Transactions on Automatic Control* **19**, 716–723

Bailer-Jones C.A.L., 2012, A Bayesian method for the analysis of deterministic and stochastic time series, *Astronomy & Astrophysics* **546**, A89

Bailer-Jones C.A.L., 2015, Estimating distances from parallaxes, *Publications of the Astronomical Society of the Pacific* **127**, 994–1009

Cox R.T., 1946, Probability, frequency and reasonable expectation, *American Journal of Physics* **14**, 1–13

Deming W.E., 1943, *Statistical Adjustment of Data*, Wiley

Foreman-Mackey D., Hogg D.W., Lang D., Goodman J., 2013, emcee: The MCMC Hammer, *Publications of the Astronomical Society of the Pacific* **125**, 306–312

Friel N., Pettitt A.N., 2008, Marginal Likelihood estimation via power posteriors, *Journal of the Royal Statistical Society B* **70**, 589

Gelman A., Rubin D.B., 1992, Inference from iterative simulation using multiple sequences, *Statistical Science* **7**, 457–511

Gigerenzer G., 2002, *Reckoning with Risk*, Penguin

Goodman J., Weare J., 2010, Ensemble samplers with affine invariance, *Communications in Applied Mathematics and Computational Science* **5**, 65

Gregory P., 2005, *Bayesian Logical Data Analysis for the Physical Sciences*, Cambridge University Press

Jaynes E.T. 1973, The well-posed problem, *Foundations of Physics* **3**, 477–493

Jeffreys H., 1961, *Theory of Probability*, Cambridge University Press, 3rd edition

Kass R., Raftery A., 1995, Bayes factors, *Journal of the American Statistical Association* **90**, 773

Kass R.E., Wasserman L., 1996, The selection of prior distributions by formal rules, *Journal of the American Statistical Association* **91**, 1343–1369

Kass R.E., Carlin B.P., Gelman A., Neal R.M, 1998, Markov Chain Monte Carlo in practice: A roundtable discussion, *The American Statistician* **52**, 93–100

Kolmogorov A.N., 1933, *Grundbegriffe der Wahrscheinlichkeitsrechnung*, 1933, Springer (English translation: *Foundations of the Theory of Probability*, 2013, Martino Fine Books)

Kruschke J.K., 2015, *Doing Bayesian Data Analysis*, Elsevier, 2nd edition

Lartillot N., Philippe H., 2006, Computing Bayes factors using thermodynamic integration, *Systematic Biology* **55**, 195

MacKay D.J.C., 2003, *Information Theory, Inference and Learning Algorithms*, Cambridge University Press

McElreath R., 2016, *Statistical Rethinking: a Bayesian Course with Examples in R and Stan*, CRC Press

Newton M.A., Raftery A.E., 1994, Approximate Bayesian inference with the weighted likelihood bootstrap, *Journal of the Royal Statistical Society B* **56**, 3–48

Robert C.P., Chopin N., Rousseau J., 2009, Harold Jeffreys's Theory of Probability revisited, *Statistical Science* **24**, 141–172

Schwarz G., 1978, Estimating the dimension of a model, *The Annals of Statistics* **6**, 461–464

Sivia D.S., Skilling J., 2006, *Data Analysis: a Bayesian Tutorial*, Oxford University Press, 2nd edition

Skilling J., 2004, in Fischer R., Preuss R., von Toussaint U. (eds), AIP Conf. Proc. Vol. 735, Bayesian Inference and Maximum Entropy Methods in Science and Engineering, p. 395

Trotta R., 2007, Applications of Bayesian model selection to cosmological parameters, *Monthly Notices of the Royal Astronomical Society* **378**, 72

Venables W.N., Ripley B.D., 2002, *Modern Applied Statistics with S*, Springer, 4th edition

Wald A., 1943, A method of estimating plane vulnerability based on damage of survivors, Statistical Research Group, Columbia University, CRC 432

Index

Only the first or most substantial use of R functions are indexed (under "R functions"). Where they occur in named files which spread over more than one page, the reference is to the first page.